FOURTH
EDITION

MODERN TECHNICAL WRITING

Theodore A. Sherman
University of Idaho

Simon S. Johnson
Oregon State University

Prentice-Hall, Inc., Englewood Cliffs, New Jersey 07632

Library of Congress Cataloging in Publication Data

SHERMAN, THEODORE A.
  Modern technical writing.

  Bibliography
  Includes index.
  1. Technical writing. I. Johnson, Simon S.
  II. Title.
  T11.S52 1983   808'.0666   82-21576
  ISBN 0-13-598748-2

Editorial/production supervision and
  interior design: Chrys Chrzanowski
Cover design: Wanda Lubelska
Manufacturing buyer: Harry P. Baisley

© 1983, 1975, 1966, 1955 by Prentice-Hall, Inc.,
Englewood Cliffs, New Jersey 07632

Printed in the United States of America

10 9 8 7 6 5 4

ISBN 0-13-598748-2

PRENTICE-HALL INTERNATIONAL, INC., *London*
PRENTICE-HALL OF AUSTRALIA PTY. LIMITED, *Sydney*
EDITORA PRENTICE-HALL DO BRASIL, LTDA., *Rio de Janeiro*
PRENTICE-HALL CANADA INC., *Toronto*
PRENTICE-HALL OF INDIA PRIVATE LIMITED, *New Delhi*
PRENTICE-HALL OF JAPAN, INC., *Tokyo*
PRENTICE-HALL OF SOUTHEAST ASIA PTE. LTD., *Singapore*
WHITEHALL BOOKS LIMITED, *Wellington, New Zealand*

Contents

APPENDICES

Preface

The fourth edition of *Modern Technical Writing* consists of four parts: *Technical Writing in General; Reports, Proposals, and Oral Presentation; Business Correspondence;* and a *Handbook of Fundamentals*. This is the same overall plan that was used in earlier editions except that the material has been updated or rewritten to keep abreast of current practices of the better writers in government, education, science, and industry.

The logic of this arrangement is that some of the various skills needed for technical writing can best be acquired if first studied separately and then put to use in the writing of reports and other longer pieces. The parts are written, however, in a manner that permits readers or instructors to take them up in a different order if they wish.

Experience has shown that *Modern Technical Writing* has four kinds of readers. The largest group are students in business, technical, or scientific fields. They are people who do not plan to make a living from writing, but they recognize the important part clear communication will play in their professional lives. A second group are students who plan to make a career in technical writing itself. These students realize that the need for communication in all professions has created a growing field with great opportunities for anyone willing to pursue them. A third group of readers are men and women already entered in careers in business, industry, or government who wish to improve their communications skills through course work, seminars, or through reading on their own. And finally, an increasing number of individuals in countries outside the United States who are engaged in international commerce have found *Modern Technical Writing* useful in learning the best methods of communicating the technical information of their fields to audiences in many languages. We have kept in mind each of these groups during the writing of this edition, and also, of course, we have kept in mind the many teachers who select this text for use in their classrooms.

Part I opens with a brief introductory chapter that discusses the importance of writing in the world today and how readers might best profit from reading this text. Then, new to this edition, is a chapter about the Process of Writing. This discussion, which shows how each of the other elements in the book fits into the overall writing process, was added because many readers come to technical writing with little experience in writing. The discussion also allows us to stress elements such as reader adaptation, rewriting, and editing—subjects that had been suffused throughout the text before but never discussed for themselves. The book then takes up Style and Organization, covering them in that order because skill or lack of skill in style affects all writing, while much technical writing is so short as to offer few problems of organization. These chapters are followed by Mechanics, which covers matters of form that are peculiar to technical writing or appear in it frequently.

The chapter on Special Areas, which follows, performs a dual function. It provides writing assignments that may be used while the study of style, organization, and mechanics is still under way, and it explains ways of handling certain writing tasks that may arise during the writing of longer forms such as reports and proposals. We have retained the treatment of article writing in this chapter because even though readers may not be ready to write a technical article at this stage, they can learn much about technical writing by studying the articles they read in their technical fields.

The emphasis in Part II remains much as it was, despite some reorganization. The chapter on Procedure in Writing a Report now follows the discussion of the importance and qualities of reports, a change suggested by users of the book. The current placement allows readers to expand their understanding of reports by seeing how one is constructed. Users of the book who follow the chapters in order may now begin working on a larger report earlier than in the past. The new organization also makes clearer the relationship between Special Elements in Reports and Formal and Semiformal Reports. The discussion in the former concerns all reports, and the treatment of the latter points out those elements that make a report important enough to be called formal.

The chapter, Reports—Their Variation in Form and Purpose, includes many new examples and a new discussion of feasibility reports and environmental impact statements. We added the discussion not because we believe any special classification creates a "new" type, but because of increasing interest in the areas that might be included under these terms. We also added a short research proposal to the chapter on Proposals to reflect the great importance of that area of technical writing.

Oral Presentation of Technical Information, new to the previous edition, remains largely unchanged. Although a study of this chapter is no substitute for training in public speaking, we believe that its recommendations can nevertheless be of substantial assistance to those who use this book on the numerous occasions when they will present their ideas in person before a small group or a large audience.

Changes have been made throughout Part III, *Business Correspondence,* to reflect the evolving practice of business men and women in all parts of the world. Some of the changes grow out of the increasing informality found even in established industries; other changes grow out of the increasing role of women in business; still others grow out of recognition in business of the value of time-saving practices and devices. In the text, we point out where recent practices differ from established habits so that readers may see how their experiences fit into current business procedures.

In difficult economic times, job hunting becomes a greater worry to students and older workers alike. This problem is reflected in Letters Concerning Employment and Resumés, which includes a new treatment of letters of application, an expanded treatment of resumé writing, and a discussion of types of resumés.

The general subject of business correspondence remains important because, though strictly speaking not all letters are technical writing, those who do technical writing will find it necessary to write more and more letters as their careers continue. Consequently, they should be made aware of the difference between the skills called for in letter writing and those that they need in the other writing that their work involves.

The *Handbook of Fundamentals* has been updated. It is intended primarily for purposes of reference when a writer is doubtful about some matter that concerns correctness and effectiveness of writing in general. Its attitude is liberal; but in fairness to those who use it, some of the traditional taboos are pointed out because they may still be taken seriously by people whom those who use this book will address.

In all parts of the book new ideas have been added and changes have been made wherever close scrutiny revealed an opportunity to make the book more effective. The exercises are of the same kind used before, but consist of new material except on occasions where something in the previous editions brought out a point so well that no satisfactory replacement could be obtained or devised. Also, many of the subjects for writing assignments are new except where they are broad enough to allow students considerable freedom in utilizing material from their own special interests.

The specimens of various kinds of technical writing are largely new, and in every case have been written recently enough to be in line with current practice. Particular pains were taken in obtaining reports and proposals to use as specimens. All of those included were written for actual use in government, science, or industry, and are shown in facsimile to let students see an actual cross section of material produced by professional workers in fields they themselves may enter. No specimen is the product of professional technical writers and artists. On the contrary, they are the work of professional people who must write as part of their jobs but whose main occupation is not writing.

These specimens were obtained from many sources, and those that are presented were chosen because they seemed to represent best what is being produced in the field. They are of various types and concern various fields. Geographically, they come from places as far apart as Alaska and New York.

They comprise only a small percentage of those received. Examples not included were also valuable in that they too were examined with care so that what the book says about reports and proposals would correspond with current practices. Also, they were a source of subject matter in devising exercises and assignments that would be most like what the student will be doing on the job.

Throughout the book—in the text proper and in assignments, exercises, and specimens—material will be found that concerns subjects which have emerged as vital in recent years, not only to students, but to people throughout the world. Ecology, pollution control, and energy sources are typical examples of these interests.

The number of persons who have been of service while revision was under way has been so great that listing all of them would be impractical, but a particular debt of gratitude is owed each of the following: Allen S. Janssen, Dean (Emeritus) of the College of Engineering, University of Idaho; John Banks, President of the National Steel and Shipbuilding Company of San Diego; Ronald Hurlbutt, Traffic and Engineering Department Manager, Oakland, California; William Smith, City Engineer, Moscow, Idaho; Diane Kireilis and Elbert G. Melcher, Rocky Mountain Center on Environment; Robert Larson, Ken Seaman, and Harry Sloffkiss, National Assessment of Educational Progress; Edmund E. Tylutki, Associate Professor of Biology, University of Idaho; Kenneth D. Briggs, Pacific Missile Test Center; Kiku Hayashi, The Boeing Company; W. L. Garner, Sandia Laboratories; and A. Morrie Craig, Oregon State University School of Veterinary Medicine.

In addition, the corporations and governmental agencies listed below were extremely cooperative: The Army Corps of Engineers; The Asphalt Institute, College Park, Maryland; Boeing Company; Brooklyn Union Gas; Chevron Chemical Company; Douglas United Nuclear Industries, Inc.; DeKalb Agrisearch, Inc.; Earl V. Miller Engineers, Phoenix, Arizona; General Electric Company; International Snowmobile Industry Association; Montgomery County, Maryland, Division of Transportation Engineering; National Assessment of Educational Progress; National Steel and Shipbuilding Company; the City of Oakland, California, Traffic Engineering Department; The Potlatch Corporation; Rocky Mountain Center on Environment; 3M Company; T.R.W. Corporation; U.S. Gypsum; United States Environmental Protection Agency; United States Forest Service; University of Washington College of Engineering; and Washington State University College of Engineering Extension Service.

The faculties of the University of Idaho and of Oregon State University have given encouragement and cooperation at all times.

Theodore A. Sherman
Simon S. Johnson

I

TECHNICAL WRITING IN GENERAL

1 | Introduction

In order to profit by a study of technical writing you will find it necessary to recognize one fact at the outset: Writing will be a part of your work that can do most to advance your career. If you now have a job, you may already appreciate the importance of writing in your daily activities. Skill in writing must be regarded as a professional tool, ranked equally with your other professional skills and knowledge. If you think of writing as something to do on the job rather than only in the classroom—a skill to convey information for practical use by an employer or client rather than just to demonstrate academic proficiency—you will have motivation for improvement. And if you are motivated to make a genuine effort to improve your writing, you have every reason to expect that the time you devote will contribute to a successful career.

No one who has been exposed to the comments of employers can doubt that the preceding statements are justified. In articles, public addresses, and personal remarks, high-ranking executives in industry and government have long been stressing the value of good writing. For example, in a personal letter one industrial executive stated, "To rise to an executive position one must be able to present ideas effectively, first on paper and then orally. If the ideas are not presented well on paper, the chances are that the writer will never get a chance to present them orally." Another executive said that engineers who could not write good reports might as well reconcile themselves to little better than drafting jobs for the rest of their lives. And the chief engineer of a nationwide company commented that if its field engineers could write more clearly, the company's long-distance telephone bills would be reduced by one-fourth.

Why do people in positions of authority consider skill in writing so important? The reason is clear to anyone familiar with the demands made on a worker with technical training. A few years ago, a Westinghouse study showed that engineers spend approximately one day in five writing, and a more recent study found that engineers in positions of authority spend about one third of their time producing reports and another third reading reports produced by

3

others. In agriculture, forestry, bacteriology, mining, metallurgy, public health, or any other technical field, the facts are the same: Writing is essential in almost every important project. Before work is started, the possibilities of a project must be stated and evaluated, and plans must be made in writing. When the work is under way, memoranda and letters must be exchanged as problems arise, and written reports must be made so that those in authority can learn how the work is progressing. When the project has been completed, the results must be recorded. Often, indeed, the only tangible result of a long and expensive project will be a written report; and if that report is ineffective, the benefits of all the time and energy expended may be jeopardized.

Since your writing is important to your employers, it is also important to you. Certainly it is one part of your work that is sure to reach the attention of your superiors on every level. If you write badly, not only will your weakness in writing be recorded against you, but your ability in other areas may be obscured. On the other hand, if you write effectively, you will not only receive credit for being a good writer, but will also find it easier to gain recognition for your other abilities. Many a person first catches the attention of those upon whom promotion depends by means of well-written reports. Unfortunately, many persons entering a technical profession find progress difficult because they had not realized the necessity of learning to use language correctly and effectively.

There is, however, a more basic reason for learning to write well. It was described by a famous scientist, author of more than 170 scientific books and papers: "The mere process of writing is one of the most powerful tools we have for clarifying our own thinking," he told listeners. "I seldom get to the level of a publishable manuscript without a great deal of self-torture and at least three drafts. My desk is littered with rejected attempts as I proceed. But there is a reward. I am never as clear about a matter as when I have just finished writing about it. The writing process itself produces that clarity. Indeed, I often write memoranda to myself solely for the purpose of clearing up my own thinking."

Clarity in thinking and writing has always been important, of course, but the importance has increased in recent years as the United States has moved toward an economy based on information. In fact, a recent Department of Commerce study indicated that more than half of the gross national product of the United States is now derived from the gathering, storage, retrieval, and dissemination of information.

One result of this change in the American economy has been the rise of technical writing as a career. While many readers of this book will be interested in improving their writing for use in scientific or technical fields, others will be planning to go into the field of technical writing itself. And those who have not yet decided upon a career may well think about this growing field. Job opportunities are excellent for those able to communicate technical subjects to various audiences, especially in electronics, energy, and computers. And in addition to openings in industry, there are jobs in radio, television, and news-papers as well as in the more than 1000 American magazines with a technical basis.

A technical writer's job varies considerably from company to company and field to field. For instance, in one company the beginning technical writer may spend much of the day at a desk writing, rewriting, or editing. But this is the exception, for most technical writers spend time discussing projects with colleagues and superiors, gathering information from engineers and technicians, designing layouts and scheduling printing, supervising illustrations and tables, finally writing and editing. Because it is a growing field, almost every technical writing job tends to become what an individual makes of it. Variety and responsibility are perhaps the most attractive characteristics of the profession in this age of specialization. Most technical writers are involved with writing projects from the inception when they are little more than ideas all the way through to the finished product as pamphlets, brochures, manuals, or reports. A technical writer, therefore, derives a great deal of personal satisfaction from his or her work.

Two other kinds of satisfaction should also be mentioned. The first was described by a publications manager at Intel who started his career as an engineer. He said that as an engineer he worked on specific projects for extended periods while the state of the art expanded to the point where he was practically illiterate in major parts of his own field. Now as a technical writer, he said, "I'm always on the cutting edge of technology. Because I deal with reports covering all aspects of the field, I am more knowledgeable about a subject than many of the engineers working in the area." Job satisfaction also comes in the form of salary. Many companies, especially in the field of electronics, pay entry-level technical writers on the same scale as entry-level engineers.

Putting technical writing's importance aside, there is the question of its relationship to writing in general. Certainly the qualities cultivated in general writing are also important in technical writing. As a technical writer you will continue to work toward correctness in grammar, punctuation, and spelling; and whatever skills you have acquired in making your sentences and paragraphs effective will improve your technical writing. Sound organization, desirable in writing of any sort, is a matter of special concern; and the same might be said of clearness, simplicity, and directness of style. All in all, skillful, effective technical writing is built upon the foundation of skill in writing in general.

Still, general writing ability does not in itself guarantee satisfactory technical writing. The latter is governed by a sense of values that other writing does not usually emphasize so strongly; it often calls for the use of special forms; it requires familiarity with some special techniques.

Dealing as it so often does with the sciences, theoretical or applied, technical writing sets an unusually high value on objectivity, meticulous accuracy, and restraint. It is directed to the reader's mind and makes little effort to appeal to emotions. Its purposes are utilitarian, and it is usually intended for readers who already have, to some degree, a special interest in the subject matter. Consequently, it does not try to be colorful and entertaining to the point of becoming flashy and superficial.

Some of the forms in which technical writing appears are the bulletins and pamphlets issued by experiment stations and government bureaus, and the technical or semitechnical papers and articles read at meetings or published in technical periodicals. Many of these forms do not have to be studied as separate types, but there is one form of technical writing—the report—that is particularly important and calls for special study. Though not every technical writer is called upon to produce all of the forms listed, almost all must write reports. Because of this fact and because reports vary so widely, they are given special emphasis in this book. Special attention is also given to proposals, which industry and government also regard as immensely important.

As you work in any of the forms of technical writing, you will at various times need to give instruction, write a definition, explain a process, describe an object, or analyze a condition or problem. The problems that arise on each of these occasions receive individual attention in this book.

You will also find it necessary to know how to use numbered lists, tables, and figures. You will need unusual familiarity with the conventions that govern abbreviations and the use of figures or words for numbers, especially in cases when the conventions of technical writing differ from those that govern writing in general. You will need to acquire skill in the use of topical headings so that the reader can easily grasp the organization of your work and find required information. And further, since the soundness of technical writing must bear up under critical appraisal, you will need to master the process of documentation.

The treatment of technical writing in this book is divided into three parts. The first is devoted to matters that concern all forms of technical writing. The process every writer goes through in the production of a manuscript is described in Chapter 2, to show how each step fits into a finished composition. Style is discussed next because style affects the quality of whatever is written, even a memorandum of two or three sentences. Next, suggestions are made about organization. Succeeding chapters cover mechanical form, special areas, and the use of tables and figures.

The second part of the book is devoted to reports and proposals, which deserve special emphasis for reasons already mentioned.

The third section deals with business correspondence. Though letters do not always deal with technical subjects, they assuredly comprise a substantial portion of the writing demanded of almost anyone in a technical profession, and therefore merit attention.

The three main parts of the book are followed by a Handbook of Fundamentals so that if your general ability in writing is weak, you will have access to the information you are most likely to need in overcoming that weakness.

All in all, the material in this book should provide the help you will need in acquiring a reasonable degree of skill in technical writing. To be sure, neither a book nor a course can develop your full writing potential. Only long experience and extensive practice can do that. But study of the subject can at least do the following for you:

1. Make you aware of the standards your employers expect you to meet—standards of accuracy and compliance with instructions as well as standards of writing;
2. Familiarize you with practical, tested writing procedures;
3. Give you a clearer picture of the manner in which writing fits into the total pattern of professional activity;
4. Reduce the likelihood that at the beginning of your professional work you will form undesirable habits that may harden until they become difficult to break;
5. Leave you a better writer than you were before; but even more important,
6. Lay a foundation for self-improvement, so that your skill in technical writing, like your other professional skills, will steadily increase as your career continues.

Another comment is in order. In the present day, characterized as it is by rapid change, questions are constantly raised about whether various time-honored courses justify retention in one or another college curriculum. In fact, authorities tell us that information has a half-life of less than five years in most technical fields. That is, in five years, half of what you learn today in a technical class will be out of date, irrelevant, or wrong. Yet no one has said that *writing* will cease to be an essential skill; and though language is not static, you may take comfort in the fact that skill in writing acquired today will not become outdated.

Thus the ability to write effectively is well worth the time you will spend in acquiring it. Long after most of the things you are learning have ceased to be useful, you will still be using whatever ability you have acquired in using the English language. It would be an overstatement, perhaps, to say that a career in a technical profession is impossible if you cannot write effectively. It is no overstatement, however, to say that weakness in writing is a handicap that will weaken your qualifications for many desirable positions, and that skill in writing is an asset that can make your professional advancement faster and easier.

Finally, the skills you acquire in writing can go beyond helping you in your job or your employer in commercial pursuits. While personal ambition is not the only motive that influences conduct, a desire for professional advancement is not incompatible with a desire to contribute to the general welfare. The problems plaguing the modern world are far too complicated to be solved by crusading zeal alone. Every conceivable profession will be involved in their solution. As a professional-level employee, you will have access to the latest developments and knowledge in your technical field. It will be these developments and knowledge that will solve the problems of a modern world. The skills in technical writing, then, are skills in communication, and before society can effect changes, it must receive communication of the knowledge necessary to make the changes. You, as an individual, can contribute more toward solving problems and will have more influence on others if you achieve a recognized professional standing. Thus, in urging you to achieve skill in writing in order to advance your career, we are urging you also to consider the possibility of more effective efforts toward the improvement of the human condition.

2 | The Writing Process

Many people look on writing ability as a talent given to a few, like a talent for music or painting. When these people read well-written prose, they may notice how the right words seem to fall into place, how the thought flows naturally from sentence to sentence, how paragraphs build upon each other in such a way that no other arrangement seems possible. When these people try to write, however, words don't come at all, thought dries up, and the idea of constructing an entire paragraph causes tears of frustration to flow far easier than sentences. You may have had some of these feelings yourself when faced with the task of writing a report or even a memorandum, and you may have envied the apparent facility others seem to have with words, unaware that they too have struggled. For writing is a process, and struggle is part of the process that all writers experience. Ernest Hemingway told George Plimpton that he rewrote the last page of *A Farewell to Arms* thirty-nine times, and when asked what the problem was, replied, "Getting the words right."

It is true that verbal ability differs among individuals, just as does musical ability or the ability to draw. Some individuals grow up in an atmosphere of literacy. They have read regularly from childhood on, have discussed ideas and heard ideas being discussed, and have written whenever they had the chance. Naturally, it is easier for them to use words than for someone who manages to reach college or even a job without ever having written a sustained composition. But even if you are in the latter category, the situation is far from hopeless. Unlike ability in music, for instance, which may require practice with an instrument, verbal ability requires the use of an instrument you already know: words. You have been talking all your life and are quite at ease in most situations when you speak. You can utilize this ability to make writing easier, as will be explained later. For both types of writers—those who find writing relatively painless, and those who would rather have a tooth pulled without anesthetic— ways are available to make writing less formidable.

Writing is a process, and the process can be learned. The process involves four major stages: (1) prewriting, (2) writing, (3) rewriting, and (4) editing. Most good writers spend an appropriate amount of effort in each stage, which makes subsequent stages easier. Attempts to skip a stage usually create more problems rather than saving time. The rest of this chapter is devoted to a discussion of these stages.

PREWRITING

The term "prewriting" covers all those activities good writers perform before they actually start to write. Sometimes, when they have only a short piece to write, they perform them all without thinking about them consciously. But when faced with a large writing task, they know that time spent here saves effort later.

Reader Adaptation

Perhaps the most important task undertaken before writing, and one often overlooked by beginning writers, is the determination of the intended reader. Accomplished writers understand that until they identify the reader and know the reader's purpose in reading, they cannot begin to write. This fact is so important it cannot be overemphasized: All written work must be adapted to the needs, desires, and background of the reader or readers.

If you are a student or are starting on your first job after graduation, most of your formal writing has probably been in school. The writing situation in a classroom is one, however, that rarely exists on the job. In the classroom your reader is usually the teacher, who probably knows more about the subject than you do. Your purpose in writing is to prove that you have mastered an adequate amount of material. Since you are not attempting to convey new information but merely to show your mastery of the subject, you try to do so by writing everything you can find about it as coherently as possible.

On the job, however, your reader may know a good deal about your subject or may know very little, but most readers will not know what you know. If they did, there would be little point in your writing and no point in their reading. Therefore, to help determine what the reader knows, the first question you must ask yourself before writing is:

Who is my primary reader?

You will need to determine such factors as the reader's education and experience, knowledge about the material and attitude toward it, and position and relationship to you. In addition, because in technical writing most papers go beyond first readers to second or third readers, and then into files to be pulled out months or years later for yet other readers, the writer must also ask:

Who are my secondary readers?

For example, many complex jobs require progress reports that are handed to immediate supervisors. The writer knows who these readers are and exactly how much they know. Usually, the writer has no trouble telling these readers what they need. However, progress reports are frequently used in constructing final reports, or are sent up the administrative ladder to higher levels. The writer who does not take into account these secondary readers—who may be far from secondary in importance—may fail to include adequate information or perhaps fail to stress points that the primary reader would understand but secondary readers would overlook.

You may conclude from this that the safest course would be to fall back on the habits learned in school and include everything you know about the subject, but such a strategy obscures important information with excessive detail, and annoys readers who are pressed for time.

Whether they have in mind either primary or secondary readers, successful technical writers usually classify those who will read their work as either "experts" or "lay readers."

Expert. The expert is a reader who is well-versed in the subject matter of the paper, and therefore wants you to present facts and get to *your* point as soon as possible. Experts are capable of making up their own minds about the subject, so if they want your opinions, they want them separate from your observations. There are three kinds of experts that you will be writing for:

1. *Administrator/politician.* These readers make policy decisions, and therefore are interested in the effects of actions you describe. They may or may not be well-educated. Often they will not be familiar with the most recent developments in your field but they will be well-versed in the implications of those developments. Depending on the situation, they may or may not be familiar with the technical terminology of your field. They tend to be interested in summarized information and in conclusions, rather than in how something works.
2. *Engineer/researcher.* These readers make practical decisions. They are almost always well-trained, familiar with recent developments, and aware of technical terminology. When reading reports directly related to their fields, they tend to be interested in how things work and what has been accomplished.
3. *Technicians.* These readers make things happen, and need to know what to do. They may or may not be well-educated or trained, but they are interested in the why of a project as well as the what.

Lay Reader. Lay readers are the hardest to write for because their purpose in reading and knowledge of the subject may be the most difficult to determine, but as the nuclear power industry discovered, you ignore the lay reader at your own peril. In a democracy, it is the lay reader who makes the final decisions about possible action. Lay readers are not stupid. They are merely people who have not specialized in your field. They may range from people with little formal education to experts in other fields. As writers we should remember that all of us are lay readers in fields other than our own and should write with that fact in mind.

Scope and Purpose

Once you know who your primary and secondary readers are, you must ask yourself two more questions about the scope of your project:

How much do the readers know?
and
How much do the readers want to know?

The latter question is closely related to another question that must be answered before other writing questions can be addressed:

What is the purpose of this paper?

You should determine what the reader should know and be able to do after reading and what attitude the reader should have. That is, will the reader use the information for decision-making or in actually performing the tasks described. Obviously, a description of a bicycle intended to enable someone to repair it would include different information, presented in a different manner, from a description of a bicycle intended to facilitate a decision about the safety hazards of certain construction options, such as the length of the handlebar stem or the use of center-pull brakes.

We will address these questions again and again throughout this book, for the answers to them affect every aspect of the writing process. They determine how much and what kind of information you include, which words you should define and which you choose to avoid entirely. They determine the length of sentences and of paragraphs, and even the kinds of sentences you use. They determine the organization of paragraphs and of the complete paper itself. In fact, the answers to these questions are so important that we cannot stress this fact strongly enough:

> Until you know who your readers will be, the reason or purpose that causes them to read, and the result you hope to obtain, you cannot successfully write a technical paper.

Scheduling

Another important prewriting task is scheduling. You may have a date when the completed product is required. In such a case, you need to work back from that date to determine how much time you have for each required task. For instance, if the manuscript will be reproduced commercially, the printer has to have time to reproduce it. Whether or not you hire a typist, time has to be allotted for typing. Although this kind of scheduling may seem obvious, all too often beginning writers forget to allow adequate time for each task, thus adding haste to other problems of writing; or worse, they miss the due date because even though they finished the manuscript on time, they did not allow time for others to perform their tasks. A schedule need not be elaborate if only one person is involved, but on occasion, it might cover an entire wall. A simple schedule for a ten-week project might look like:

Gathering information	———three weeks
Organizing material	———one week
Writing	———two weeks
Rewriting & editing	———one week
Drawing charts & graphs	———one week
Typing & proofreading	———one week
Reproduction & binding	———one week
Delivery	———due date

A more elaborate schedule might include dates for individual components of the report; dates for printers, layout, and design; or dates for parts of the job that need to be finished before writing can commence. The purpose is the same: to allow all people involved enough time to finish each task with as little pressure as possible and to allow all parts of the project to end on time. Of course, once the schedule is made, if the deadlines are not met, the advantages are lost.

Gathering Information

By establishing the above steps, you can often save yourself time in the next step: gathering information. Knowing the scope of your paper, for instance, will indicate how much detail and what kind to gather. Knowing your reader may well give you indications on sources for information. For instance, if an immediate supervisor wants to know the status of a project, project workers might be adequate sources of information, whereas for an administrator, you might have to go to printed sources for background. In all cases, if you see this step as part of the writing process rather than as a separate task, your search will be better directed. For a discussion of sources of information for technical reports, refer to the section "Gathering Information" in the chapter on "Procedure in Writing A Report."

Note-taking is part of the information-gathering process. With the advent of the copying machine, note-taking has changed considerably. Now, rather than jotting ideas from a source, most researchers simply photocopy the page or pages they feel they might need. There is still one old danger, however:

Nothing is more frustrating when you sit down to write than discovering that you do not remember where you found an important piece of information. Always, when taking notes or photocopying, write down the complete bibliographic data. See the section on "Documentation" for instructions on how to do this.

Organizing

The subject of organization is so central to the writing process that we have devoted an entire chapter to a discussion of how and why to organize. We mention it here because it is part of the prewriting process. Time spent organizing your information can save you struggle in the writing stages. But organization is not frozen when you start to write. Always keep looking for better ways as you work with the material.

Letting the Subconscious Work

If you allow yourself adequate time for each stage of the writing process, you can benefit from one of the best aids you have: your subconscious. Once you have covered the steps above, let your mind work on its own. Recent research indicates that logical, analytical, sequential thinking may take place in the left brain, as does most verbal activity, while the right brain may process information holistically, intuitively, imaginatively. The sudden insights of Archimedes, Newton, or Einstein are thought to be examples of the synthesizing powers of the right brain. You can gain access to these powers through many paths.

Often, people with a problem decide to "sleep on it." They think about it just before dropping off to sleep, and when they wake in the morning, they have arrived at a solution. Another method is talking. Talk to friends or colleagues about your project. Often in the discussion, because you are not concerned with just the problem, ways of presenting the information "jump out" at you. Many a writer, in discussing a current project, discovers a solution to a problem and rushes off to the typewriter to record it. A third method is "brainstorming." In brainstorming, the writer, alone or with others, jots down ideas one after another as fast as possible without being critical at all, looking them over later to discover which are useful.

Of course, do not confine your use of the subconscious to the prewriting stage. You should attempt to take advantage of this aid at all stages.

WRITING

The second stage in the process is the actual writing of the first draft. Since *Modern Technical Writing* is concerned particularly with this and subsequent stages, the discussion here will be briefer than its importance would otherwise warrant, because so many chapters are devoted to aspects of this stage.

If your preparation has been adequate in the prewriting stage, the actual writing may come easier for you, but most of us have been so conditioned by criticism from teachers and editors that when faced with a blank sheet of paper, we freeze. A good way to avoid such a writer's block is to accept the fact that your first draft is just that—a first draft. Do not be critical at this stage. Definitely do not be hypercritical. No one else will see the first draft, so don't worry about spelling, sentence structure, grammar. It is more important that you get something on paper than that you get it perfect. In fact, it is not even important that you have all the information. *Time* magazine has incorporated into its standard work process an approach for avoiding this writer's block: When writers do not have information, they fill in the words *TO KOME, KOMING,* or *TK* (as in "$TK millions will be needed for the cleanup"), and after the first draft is finished, they seek the missing details. The words are spelled with a *K* to reduce the danger of their being overlooked and finding their way into the printed magazine. It often turns out that the missing information preventing you from writing is not needed at all.

Some authorities suggest that you write your first draft in a single sitting from start to finish. They believe that such an approach helps provide a unified tone to a work. This is good advice for beginning writers faced with a short piece and large amount of time. Few of us have such opportunities, but if you do, take advantage of them.

When faced with a larger project, you may start at the beginning and work your way through. Or you may start at the end and write that first. Some writers suggest doing the abstract first on the grounds that a draft of the abstract will guide them through the rest of the project. Most professional writers, however, start at a place in the work they know will be easiest to write. Then they go to the next easiest. In this way, they build momentum, start accumulating a body of work, and give their subconscious longer to work on the harder parts.

In any writing project, you are apt to hit a spot you cannot work through. Many writers struggle for hours or days, like Joseph Grand in Camus' *La Peste,* who never gets beyond his first sentence. When you find yourself stuck, skip over that area and go to a section that you know you can do. Then come back later to work on the difficult section. You may discover that once you have written succeeding parts, the difficult section is easier.

In technical writing on the job, you have an advantage that other writers do not have. Much of the material you work with has been worked with before, in manuals, sales brochures, and the like. You can use these sources to help you find ways to express your ideas. But do not assume that the expression of an idea in such a source is as a matter of course suitable, for *your* piece may have different readers and a different purpose.

REWRITING

Many writers claim that writing is really rewriting, the next stage in the process. In fact, few substantial works ever reach completion in one draft. After you have finished your first draft, let it "cool." When you come back to it after a lapse of time, you are more apt to see what the words on paper actually say rather than what you thought you wrote.

Many beginning writers do not understand the purpose of rewriting. They think that if they correct spelling errors and awkward constructions, they have done the job. But more than that, at this stage you need to ask yourself if you have found the best possible presentation for your purpose and readers. Would another organization be more effective? Can information or sections be omitted? Is information missing? At this stage you will check for accuracy of fact, correctness of grammar, completeness, and logic. Here too you check for style, as described in the next chapter.

Professional writers use several methods to aid in rewriting. Some rewrite with a pencil, working between the lines and in the margins. Others use the "cut and paste" method, whereby they actually cut up the manuscript and reorganize sentences, paragraphs, and sections by physically moving them from place to place, and if they don't like the new order, to a third or fourth position. And some writers simply retype, having learned to use the typewriter so well that it is faster to run a page through the typewriter than to use another method.

The new word processing machines eventually will make retypists of all writers. These machines show an entire page at a time on a display screen. The writer electronically moves the words around and never touches a piece of paper. But until you have access to such a machine, try the methods described above to find which seems most comfortable.

For a beginning writer, the best aid in rewriting is the ear. As mentioned before, you have been speaking all your life and are probably more at ease with the spoken word than the written. Thus your ear will catch problems faster than your eye. To take advantage of this fact, read the manuscript aloud. If it sounds awkward or incorrect to your ear, it probably is; change it. Written work is not the same as speech, of course, but if you find that you cannot read a passage aloud without difficulty, you probably should not write it either.

When you have been through the entire section you are working on several times, retype it. Mistakes that are hidden among the scribbles, cross-overs, and cut pieces often come to light on a clean copy. When you get the clean copy, go over it with the same diligence you went over the first draft. Only when you find few possibilities for improvement should you consider typing a final copy.

Many engineers and professional people, when told what rewriting entails, complain that they can't do all of it. As one sanitation engineer complained, "The company doesn't give me enough time for one draft, much less two or three." It is true that many companies are shortsighted about writing. They often fail to realize that major exposure to customers is through written reports, letters, brochures, and the like. But at the same time, many professional people forget that the major record of their performance is the reports they write. You can not afford to take less than enough time to do a good job. A compensating thought is that with practice, writing does become easier.

EDITING

One of the final jobs before reproducing the copy is editing. Although the term editing often refers to work done by an editor before work done by someone else is printed, we are concerned here with the job you or a colleague does on your own work before making a final copy. At the editing stage, you look for matters of format, company style, spelling, and grammar, as well as all the elements mentioned in the previous sections of this chapter. This is the time to bring your critical abilities to bear on the paper. Ask yourself if the paper has the proper focus to answer the questions posed in the prewriting stage. If the answer is yes, then your function is to see that no errors find their way into print. If the answer is no, the paper should go through another revision.

Editing is often done by a person other than the writer. In this case, the editor should remember that language often allows several effective ways to express the same idea, and it is not the editor's job to recast the manuscript, but to suggest improvements that are consonant with the author's approach and intent. If a problem is just one of personal style, the author's wishes should prevail. In that narrow area between style and correct grammar, the editor should insist only upon the latter. At the same time, a new reader can often spot problems that the original writer missed because of familiarity with the material. Both the writer and the editor should remain open to suggestions.

At some point in the process, a final copy is ready for delivery or repro-duction. Even though, like most people, you may be sick of your project at this stage, you cannot let down your guard. This is the last chance to correct those typographical errors or omissions that always seem to find their way into print. In order to get a new look at the copy, some proofreaders read pages out of order. Often a new person reading the copy will find errors that all others working on a project missed. A careful rereading here often pays for itself by saving you, your company, or your profession from embarrassment.

Peer Review

Similar in many ways to editing, peer review is a stage many scientific and technical papers undergo before publication. In this stage, an expert in the field, usually not connected with the project, is asked to read and comment on a manuscript. This section is included here because it often becomes part of the writing process, especially when suggestions for changes are made, as they often are.

The Technical Writing Division of Sandia Laboratories has published an approach to peer review that describes the process in a way that is helpful to both reviewer and writer.

> A manuscript should be read at least twice. The first reading should be done with an open mind since the reviewer needs to respond to the work in its entirety. The second and subsequent readings require the reviewer to write out reactions and suggest specific improvements. Vague statements such as "this could be better" are not very helpful. After you have read through the report for the first time, put it down and reflect: Were the author's objectives achieved? Were there any glaring omissions? Would the target readers feel educated or insulted? Would the work reported be a credit to the author and organization? Would the reader react negatively to an injudicious statement?

These are questions writers should ask themselves, of course, but involvement with the material often prevents them from providing as objective an answer as may be necessary. The Sandia description stresses one point that every editor, reviewer, and writer should remember: "Critiquing involves a psychology that can be appreciated by imagining yourself in the position of being critiqued, and by remembering how you respond when writing of yours is reviewed."

Team Writing

Team writing presents special editing problems—and in industry an extensive piece is usually produced by a team. Sometimes material may be picked up from previous publications, sometimes two or more writers work together on each section, sometimes writers work on separate sections and the person with the editorial function puts the sections together. The major problem is consistency. When two or more persons are writing, they are bound to present information differently. The reader, however, reads a single work, and will be confused if similar information is not presented consistently. (This is a practical reason why writers of technical material should refrain from using a distinctive, personal style). The editor must see that the work reads as if a single person wrote it. The writers must understand and be tolerant of the changes made for this purpose.

Usually, one person has final authority for the project, so disagreement does not take excessive time or create hard feelings. Such a procedure works best if the authority is rotated through the staff so that each person has an opportunity to oversee a project. As a writer on a team, keep a copy of your work so you can compare your efforts with the final copy. In this way, you can improve your writing skills for the next project.

Now that you understand the process of writing, you can approach your own writing projects with greater confidence. The remainder of the text deals with the parts of the process and with the products you hope to write.

ASSIGNMENTS

The following assignments provide an opportunity to review and practice the material covered in the chapter. Readers may choose to complete all or only a portion of the assignments at this time. In some cases, the assignments can be completed now, and repeated again using different material after covering later chapters. For instance, the peer review and rewriting assignments could be repeated profitably following almost any of the writing assignments in the text, and the *Reader Analysis Worksheet* is intended for use with any writing assignment. Throughout the text, assignments and exercises are provided as a guide for possible work, but readers should feel free to improvise or change the suggestions to meet their own special needs. For example, if you would rather not use a description of your writing experience for a peer review (Assignment 2), another sample of your writing would serve as well. The purpose of the assignment, however, should be followed to ensure that the material in the chapter is actually covered.

Assignment 1

Write a brief paper describing the training you have had in writing, including courses in school and experience on the job. Describe the kinds of papers you have written and the audiences you have tried to reach. Now that you understand better the complete writing process, discuss the areas you have covered well in the past and the areas you have slighted. Conclude your paper with a set of specific goals to improve your writing. The purpose of this assignment is to draw your attention to your writing problems, so that you may better focus your studies.

Assignment 2

Exchange the paper written in Assignment 1 with a friend or colleague for a peer review. The reviewer should follow the suggestions in the chapter on editing and reviewing, but for this assignment should focus primarily on the concreteness of the paper. Questions the reviewer might ask include:

1. Are the examples specific enough?
2. Are the goals specific?
3. Are the goals achievable or are they generalized expressions of hope (e.g., I hope to improve)?
4. Have areas been omitted from consideration?

The reviewer should make specific suggestions for improvement, including suggestions of other areas that may need to be included.

Assignment 3

Study the suggestions made by the reviewer in Assignment 2 and rewrite the paper describing your writing experience. If you feel that the reviewer has misunderstood a point you were attempting to make, restructure the paper to make the point easier to grasp.

Assignment 4

Work out a simplified schedule for a writing project. You need not have a subject to write about for this assignment, but you will need to know the date when the project must be completed. Work back from the deadline, giving yourself dates for completion of each stage of the project. If you do not have a subject, include in your schedule adequate time to find one. This assignment should alert you to the time available for each stage in order to finish your project.

Assignment 5

Copy the following *Reader Analysis Worksheet* on a separate sheet of paper and fill it out, using as much space as necessary for completeness. Since this assignment provides practice in analyzing the report writing situation, you may describe the readers for a report, letter, or memorandum you have written in the past, or you may describe the readers for a report project you have to complete in the future.

Reader Analysis Worksheet

Subject of report:

Title: (You may use a working title rather than the final title.)

Due Date:

Primary reader: (Give name if you know it, otherwise give position or other specific identifying features.)

    Education/experience: (List general educational level as well as education or experience directly concerned with the report subject.)

    Relation to writer: (Is the reader higher or lower in the organization than the writer, or in a separate organization?)

    Other relevant information: (Is the reader obliged to read the entire report or only part of it, or must the reader be persuaded to read by the presentation?)

Secondary readers:

    Education/experience:

    Relation to writer:

    Other relevant information:

Scope:

    Knowledge of report subject:

        Primary reader: (Describe in as much detail as you can.)

        Secondary readers:

    How much do the readers want or need to know?

        Primary reader:

        Secondary reader:

    Attitude about the subject: (Do the readers approve of the suggestions to be made or will they resist them?)

Purpose: (Specify the main purpose of the report)

Why should the readers study this subject?

What should the readers know after reading?

What should the readers do after reading?

What attitude should the readers have?

Is the final attitude a change?

List any minor purposes:

Assignment 6

Find a short article in a scientific or technical journal and analyze the intended audience. Then rewrite the material for a different audience. For instance, you might take a brief article from *Science, Astronautics and Aeronautics,* or *American Journal of Physics* and rewrite the material for a general college class or for a grade school class. Be sure to convey the information accurately, even if you omit detail. This assignment should give you practice in using different approaches for different audiences.

3 | Effective Style

Many an inexperienced writer gives little thought to style, being concerned only with getting ideas down on paper without making errors. If you adopt this attitude—if you think of style as merely an ornament—you are neglecting one of the major opportunities to make your writing more effective. Of all the weaknesses that annoy those who read technical writing, ineffective style is one of the most damaging.

The importance of style in making communication effective becomes clear if you realize that everything you say *has a style*. Whether or not you are aware of the effect of your style, your reader responds to the way the information is presented as well as to the information itself. Indeed, sometimes the reader will respond more to the manner of presentation than to the information. Thus, we take up the subject of style before dealing with other elements of writing.

What do we mean by style? It is style that makes the difference between the enthusiasm of a patent medicine advertisement and the restraint of an article in a medical journal, or between the fervor of an attorney pleading a case and the analytical impersonality of a judge in the charge to the jury. In short, style concerns not what a writer says, but how it is said.

Style is like clothing, in that suitability varies with the occasion, thus absolute rules cannot be given. Tennis whites, for instance, are perfectly acceptable on the courts, not appropriate on a ski slope in January, and usually not acceptable at an interview for a position in a corporation. Likewise, the way you talk at a party with friends might be perfectly acceptable, yet talking in the same manner during a court hearing, even if the content of your discussion were relevant, might not produce the result you wanted. The same is true in writing. Your presentation should vary, depending on the purpose of the paper and the reader you hope to reach. What might be appropriate for the engineering staff might be completely ineffective, and therefore inappropriate, for technicians or executives, or for an audience of lay readers.

In technical writing you will not need to develop a distinctive personal style. Rather, you can concentrate on such qualities as clearness, directness, conciseness, and readability. Technical style should ordinarily be neither so informal as to sound unprofessional nor so excessively formal as to sound stilted. It should lean toward restraint, for though a reasonable degree of enthusiasm is sometimes an asset, an excess suggests bias. It should not be charged with emotional intensity, for it usually deals with questions that are answered by the mind rather than by emotions.

Yet even in the limited context of technical writing the subject of style is complex. A long list of desirable qualities compete for attention. This list includes not only those mentioned—clearness, directness, conciseness, and readability—but also simplicity, precision, and specific, concrete, understandable diction. Some of these qualities overlap. Some of them are the result of others. (For example, readability and clearness are the end results of attention to directness and simplicity.) Also, unfortunately, an excessive effort to attain one quality may cause the loss of other qualities equally desirable. (For example, up to a certain point conciseness improves readability, but carrying conciseness too far will make your writing less readable.)

This chapter will concentrate attention on what you can actually *do* in order to make your style more effective. At a working level, you have two major questions to think about as you try to improve your style: *How can I be sure that my diction (choice of words) will make my style more effective?* and *How can the construction of sentences make my style more effective?*

The sections that follow are based on these questions, but the subject of conciseness involves both diction and sentence structure to such an extent that it is discussed separately.

EFFECT OF DICTION ON STYLE

If diction is to do its part in making your style effective, you will need to bear in mind the following points until you can apply the right principles automatically:

Use concrete, specific language.
Avoid needlessly technical language and jargon.
Use an appropriate level of formality.
Avoid stiff, pompous language.

Use of Concrete, Specific Language

In any writing, technical or otherwise, concrete and specific language makes a writer's style more vigorous and colorful. In technical writing it is especially desirable because it not only increases liveliness but also gives information more effectively and reduces the likelihood of vagueness or misunderstanding.

Concrete and *specific* are not identical in meaning, as you can see when you consider their respective antonyms, *abstract* and *general*. Concrete terms are those that refer to things that can be perceived by the five senses; specific terms are those referring to definite or defined things or situations. In any practical effort to improve your style you should try, at the same time, to find words that are both concrete and specific.

The improvement that can result from such an effort can be seen by comparing the following specimens.

Abstract and general: We plan to present a proposal covering all important aspects of the problem.

Concrete and specific: On October 10 we shall present a proposal covering cost, materials, methodology, and personnel.

Abstract and general: The projected lifespan of the building is adequate for the cost.

Concrete and specific: The projected 25-year usefulness of the warehouse will allow monthly mortgage payments and still provide a 4 to 7 percent margin of profit.

Abstract and general: Moderate winds, when accompanied by new snow and near-thawing temperatures, often create avalanche conditions.

Concrete and specific: Winds between 15 and 30 mph, when accompanied by snow and temperatures between 10° F and 30° F, often create unstable slabs in avalanche-starting zones.

The last pair of examples illustrates again the situation discussed under Reader Adaptation in the preceding chapter. The second of the two examples is definitely more concrete and specific than the first; however, the first would be perfectly acceptable where a high level of technical information was not necessary, while the second would be better for a technical audience. Thus you must remember, when applying any rule, to keep your audience in mind. Technical writers must constantly be alert to this kind of choice. They base their decisions, even in choosing among concrete terms, on the purpose of the writing and the needs of their readers. Note the contrast in these additional examples, where the same term can be specific for one audience and yet general for another, depending on the kind of information the readers require.

GENERAL	SPECIFIC
vehicle	fork lift, dolly, truck, automobile
truck	pickup, tractor-trailer, railway truck, stake truck
pickup	$1/2$-ton, $3/4$-ton, longbed, six-passenger
receptacle	jar, beaker, trash can, flask
flask	Erlenmeyer, pocket, volumetric, founding

As we will point out later, use of the concrete, specific diction recommended above is valuable not only for its own sake, but also because it contributes to directness.

Vagueness is a serious defect in technical work. Often, to be sure, vagueness is not just a matter of style. You will sometimes be tempted to write vaguely in a conscious or unconscious effort to conceal your lack of specific information. Yet there will probably be times when you have enough information to avoid vagueness but may neglect to use that information. You may write, "*Small* samples are taken at *frequent* intervals," when you know the facts well enough to say, "Samples of *100 cc* are taken at *half-hour* intervals"; or you may write, "A *fairly large* piece of steel is placed in the machine," when you might have written, "A *12-inch* piece of steel." Or you might use, "The pouring of concrete is *seriously* behind schedule" when you might have used "*three weeks* behind schedule."

The last example brings up another consideration. The engineer who supplied it commented, "I want my people to tell me the objective facts; I'll draw my own conclusions about whether three weeks is or is not a serious delay." His attitude should not be taken, however, to mean that *interpretation* of specific facts is never desirable. For some readers, interpretation is needed. An audience might be told that a certain number of pounds of soil per acre had been lost because of erosion, and be unable to judge whether the amount was alarming. Consequently, a writer might do well to add that this loss amounted to half an inch, and that the loss was serious because the topsoil was only ten inches deep. The significant point is that in offering this interpretation a writer would be providing *additional* objective facts rather than omitting them in favor of vague expressions of unsubstantiated interpretation.

Avoidance of Needlessly Technical Language and Jargon

In your efforts to be concrete and specific, beware of going to the opposite extreme. Technical language has come into existence because it is often needed for clear, effective communication. Some subjects would be almost impossible to discuss without using technical terminology, for a single technical term will frequently convey an idea that would otherwise call for a long phrase or clause.

Nevertheless, technical language should not be used when it can be avoided. The purpose of writing is not just to get a message down on paper, but to get it into the mind of the reader; and you may be sure your reader will not always possess your own specialized vocabulary. One of the most common complaints against technical writing is that those who produce it apparently believe that everyone who reads it is a specialist in the subject concerned.

Moreover, the best authorities on writing agree that even when you are writing solely for technically trained readers, it is best to use technical language only when you need it. It is not good style to use *ozostomia* when you might say *bad breath,* or to refer to *imbricate scales* when you might call them *overlapping*. *Scutellate* could be replaced by *flash* on almost any conceivable occasion, and *cachexia* means no more than *wasting away*. And the person who wrote *reducing the mean diameter of the spray spectrum* might better have used *making the spray finer*—and avoided the embarrassment of having his style ridiculed in a technical magazine.

The warning against overly technical language applies not only to words that are included in standard English but also to the technical jargon that specialists in any line of activity use when they talk informally among themselves in a shop or laboratory. Such terms as *nukes* and *specs* are examples. For a general audience it might be better to use *nuclear weapons* and *specifications*. Similarly, *sink street* and *lead sled,* common jargon among sailplane devotees, are less suitable than *area of descending air currents* and *heavy sailplane*. To be sure, such language might lend flavor to a popular article and may be natural and unobjectionable in personal conversation or informal memoranda to specific readers. It is not suitable, however, when standard English is called for. You may be sure that the conversational language scientists use in the laboratory seldom appears in their technical articles or formal reports.

Jargon is not limited to new words in informal, conversational usage. It also includes the use of standard English terms in ways that seem intended to make a simple statement sound special. An example of this kind of jargon is the word *impact* used as a verb. There is nothing wrong with *impact* in such a sentence as, "Auto exhaust hydrocarbons have an adverse impact on air quality," but the engineer who wrote, "Smoke from the Cyclone Peak burn impacted the air quality over Central City," polluted the English language while objecting to pollution of the air. *Affected* would be less ostentatious, while *reduced* would convey more information. Other examples might include such terms as *resource room* for *library, negative patient care outcome* for *death,* and *revenue enhancement* for *increased taxes*.

Still another kind of jargon is the use of technical terms that have escaped their specialized context and are so widely used in other areas that they become almost meaningless. Such a word is *input,* a computer science term for information put into a computer. In a discussion of computer programming, this term may be acceptable; but the person who suggests, "Let's call a meeting and get some input on the question," when opinions are wanted might well be reminded that Jacques Barzun objected to *input* as "vague, since *input* can mean anything from a congressional appropriation to a frankfurter at lunch."

Unfortunately, when a term such as *input* escapes the special context where it is needed, it often spawns a horde of grotesque progeny. An example is seen in the State Department memorandum that stated that a consumer affairs coordinator would "review existing mechanisms of consumer input, thruput, and output." The *Washington Post* concluded that the whole thing was an onput and that the writer deserved a downput.

It may be impossible to get rid of such jargon for it is widely used by people who should know better; but its popularity will probably fade away when its novelty wears off. Meanwhile, the case against such usage was made clear by a retired weather department official who wrote of an editor who used *input,* "The next time he indrops, he shall uppull a chair, downsit, upload a typewriter, and outpound an editorial explaining how onloading a barge is superior to loading it."

All technical fields have produced their share of jargon. You have only to read reports from law, medicine, business, government, the university, and so on, to find examples. Because some of the terms may be correct in certain contexts and wrong in others, no definitive list can be given, and if it could, it would be too long for a book like this one. However, if you keep in mind your readers and their knowledge of the material, you will be less likely to confuse them with "shop talk."

Use of Appropriate Level of Formality

Like the other elements of diction, the level of formality you use is determined by the audience you intend to reach. Language that would be appropriate in one situation would be out of place in another. The following list suggests the difference between formal and informal language. Some of the formal equivalents to informal terms, to be sure, are not the only ones that might have been used, because some of the informal examples are variable or general in meaning.

INFORMAL	FORMAL
raised the roof	objected strenuously
knocked its performance	criticized its performance
out of kilter	in poor condition
O.K.	satisfactory
green operator	inexperience operator
is going to	will
won't stand still for	will oppose
not about to	will not
pretty good record	fairly good record
chance it	take the risk
kick (or gripe)	complain

Just how formal should you be in writing? There is no reason to avoid informality if your relationship with the reader is informal and if the matter is one of only temporary concern. But if a report is likely to go beyond the immediate reader or become a matter for future reference, informality is usually inappropriate. It is usually not advisable to refer to a microphone as a *mike,* to a laboratory as a *lab,* or to television as *TV.* "He lost his temper," is better than "He blew his top." "Not very serious," is better than "not all that serious."

Even contractions are usually avoided when the style of a report is not, in general, informal.

Changing informal to acceptably formal style is not difficult. Suppose, for example, that you wish to replace *you* used informally for *one* in the sentence, "You cannot make the trip in less than four hours." If *one* seems to be stiff, it is possible to write, "The trip takes at least four hours," or "It is impossible to make the trip in less than four hours." In brief, there is no reason that formality should be painfully conspicuous. The only necessity, when informality would not be appropriate, is to write with enough dignity to suit the occasion. And if you engage in professional work, you need to learn how to avoid excessive informality without letting your writing become stiff and awkward.

Avoidance of Stiff, Pompous Language

It is neither necessary nor desirable, however, that in avoiding informality you should go to the other extreme and use language that is stiff or unnatural. The ultimate degree of formality, found for example in legal papers, is never necessary and is seldom even appropriate in technical writing. One of the main reasons that style in technical writing often becomes ineffective is that writers, perhaps subconsciously, try too hard to be impressive. Consequently they sound pompous and affected, producing a style that at best is ostentatious and at worst becomes needlessly hard to translate into concrete meaning.

Style is pompous when a writer uses such expression as *increase the visibility of the incandescent gases* instead of *make the flame easier to see,* or *has a deleterious effect upon* rather than *harms* or *injures* or *damages.* Some writers cannot resist the temptation to use *optimum* when *best* would convey the idea, or *finalize* when they might use *complete.* A practicing engineer once produced the expression *minimize the expenses to an optimum degree.* (Why not just *hold expenses to a minimum?*) The classic example of a rebuke to such ostentation occurred when President Roosevelt, looking over black-out instructions during World War II, crossed out *terminate the illumination* and substituted *turn out the lights.*

Pompous style also results from the tendency of many writers to resort to words like *characteristic, property,* and *condition.* The result is such phraseology as "It possesses the quality of malleability" instead of "It is malleable"; or "It lacked the property of hardness" instead of "It was not hard enough" (or "It was too soft"); or again, "It passed into a solid condition" instead of "It solidified."

A final recommendation concerning diction is this: Remember that standard English is a flexible tool that allows you to say what needs to be said clearly, naturally, and unostentatiously. Your purpose is effective communication, and adapting your diction to the needs of the reader and to the occasion will enable you to accomplish that purpose.

EFFECT OF SENTENCE STRUCTURE ON STYLE

In discussing the effect of sentence structure on style we shall not concern ourselves with the damage done by errors in fundamentals of English. It is true, of course, that such errors often affect style—for example, by reducing clearness—but they are covered in the Handbook. Here, in dealing with the question of how to make sentences not only correct but also effective, we will consider sentence length, simplicity and directness, precision, the choice between active and passive voice, and sentence variety.

Sentence Length

Avoidance of excessive length. Though some long sentences are clear because of their simple construction, there are many times when the use of shorter sentences could make some piece of writing clearer and easier to read. The following example shows how confusion was created by an effort to jam too much into a single sentence:

> High operating speed shall be provided by a high-speed, motor-driven switch operator in order to assure the full inherent mechanical and electrical performance of the circuit-switcher such as power operation under $1\frac{1}{2}$-inch ice formation, fault-closing capability, close interphase simultaneity, long life of fault-closing arcing contacts, and avoidance of excessive switching transients which can accompany prolonged or unstable prestrike arcing.

Note how clearness is improved by use of more sentences:

> High operating speed shall be provided by a high-speed, motor-driven switch operator in order to assure full inherent mechanical and electrical performance of the circuit-switcher. The performance features include power operation under $1\frac{1}{2}$-inch ice formation, fault-closing capability, close interphase simultaneity, and long life of fault-closing arcing contacts. The switch operator also assures avoidance of excessive switching transients which can accompany prolonged or unstable prestrike arcing.

Long sentences are not always so involved as the preceding example. Stringy sentences may not be especially hard to read, but breaking them up will result in a more satisfactory style. Note the following, from a government specifications report:

> Upon completion of the work, the Contractor shall operate the pump in the presence of the Government representative to demonstrate its operation; shall establish accurately the performance of the pump, including efficiencies, through measurements of the discharge rate by use of the flow tube and pressure-differential gauge in the system, the pumping level (by use of gauges furnished with the pump) and horsepower input (by the watt-hour-meter disk-constant method or other equally accurate method); and shall submit a certified report of all such readings, calculations, and measurements.

The content would be easier to grasp if the long sentence were broken into four, as follows:

Upon completion of the work, the Contractor must test the pump in the presence of the Government representative to establish accurately that it meets all performance specifications. Discharge rate shall be determined by the flow tube and pressure-differential gauge in the system. The pumping level shall be determined from gauges furnished with the pump, and the horsepower shall be determined by the watt-hour-meter disk-constant or equally accurate method. All findings shall be submitted in a certified report.

Avoidance of primer style.   In the effort to avoid making your sentences too long you should not go to the other extreme and use "primer style." An excessive number of extremely short, simple sentences causes writing to become jerky and monotonous. Moreover, it does not necessarily result in clearness. Each sentence, it is true, may be clear; but the relationship of the ideas contained in the individual sentences is often hard to perceive. The primer style forces the reader to combine the ideas that the writer should have combined. Also, it necessitates the repetition of words and thus destroys conciseness, as will be pointed out later.

Note how, in reading the following examples, you are slowed up by being forced to come to a full stop too often.

PRIMER STYLE

1. This study will be a continuation of the study carried out on the Glide Ranger District during the summer. Plots will be established in the Tiller Ranger District for study. They will be prepared and located to correspond to conditions encountered in cleared units. In addition, some plots will be altered to determine the effect of conditions which could be economically imposed but do not occur in nature. One example of alteration would be the application of herbicides. Measurement of plant moisture stress (pms) will be taken in all plots. Plant moisture stress readings will be maintained at specific levels in designated plots by the addition of water. Weather observations will be recorded daily. It is hoped that pms readings and weather observations will show a close correlation.

2. Computerized systems can operate in one of three modes. The operating modes are batch processing, remote batch processing, and time-share processing. Conventional batch processing met all but one of the design requirements. The exception was accessibility. Batch processing could not provide output to diversified, remote locations in a timely manner. Remote batch processing, as the name implies, could provide batch processing at remote locations. It could do this through the use of teletype terminals as output devices. Unfortunately, the response time for remote batch processing is unpredictable. It can vary from a few minutes to several hours.

MORE MATURE STYLE

1. The Tiller Ranger District study will be a continuation of the study carried out on the Glide Ranger District during the summer. Plots will be prepared and located to correspond to conditions encountered in cleared units. In addition, some plots will be altered, as with the application of herbicides, to determine

the effect of conditions which could be economically imposed but do not occur naturally. Plant moisture stress (pms) readings will be taken in all plots and maintained at specific levels in designated plots by adding water. Weather observations will be recorded daily to determine whether there is a correlation between weather and pms.

2. Computerized systems can operate in one of three modes: batch processing, remote batch processing, and time-share processing. Except for accessibility, conventional batch processing met all the design requirements; but it could not provide output to diversified, remote locations in a timely manner. Remote batch processing, as the name implies, could provide batch processing at remote locations through teletype terminals. Unfortunately, its response time is unpredictable, varying from a few minutes to several hours.

To sum up the matter of sentence length: Some sentences should be short, some long, some medium. It is very easy, however, to fall into a habit of making all the sentences similar in length and type, and care is necessary to prevent such a pattern from forming.

Simplicity and Directness

Simplicity and directness are closely related—so closely, in fact, that when a passage is simplified it usually becomes more direct as well. The present discussion will not try to draw a line between the two, but since each of them includes certain ideas not included in the other, they will be discussed separately.

Simplicity. Simplicity has already been covered in connection with diction, so here we shall be concerned with it only in connection with sentence structure.

No one would build a complicated mechanism to do a job that a simple mechanism could do equally well, nor analyze a specimen by a complicated method if a simple method would get the same results. By the same token, no one should express an idea by creating a complicated grammatical construction when a simple form would serve the purpose equally well. When you can substitute a word for a phrase or a phrase for a clause or sentence without sacrificing anything of value, you will usually improve your writing by doing so.

The following examples show how sentences that are needlessly complicated can be simplified.

Needlessly complicated: The next step consisted of looking for methods by which the losses that occurred in the bearings might be eliminated.

Simplified structure: The next step consisted of looking for methods to eliminate losses in the bearings.

Needlessly complicated: A series of continuous runs made under conditions that were identical often yielded results that were different.

Simplified structure: A series of runs made under identical conditions often yielded different results.

In each case the sentence was simplified by cutting out clauses that did not make the relationship of ideas one whit clearer than they could be made by a simpler form.

The effort to write simply should not be limited to avoiding needlessly complicated sentences. A passage may consist of relatively simple sentences yet be too elaborate in its general approach. For example, note the following passage: "The process was performed in two steps. The first of these consisted of drying the sample, and the second consisted of weighing it." A simpler style would use only one sentence: "The process consists of two steps—drying the sample and weighing it."

To be sure, if the list of steps were longer and if some or all of the steps could not be expressed briefly and simply, it might be necessary and desirable to introduce the list more elaborately, but for a short list of brief items, a simple approach is adequate.

It is possible, of course, to carry simplicity too far. The writer who substituted *variable elimination* for *elimination of variables* and *concrete information* for *information about concrete* reduced a phrase to a word in each case but in doing so became ambiguous. Similarly, the writer who used *one way of determining system applicability* for *one way of determining the applicability of the system* simplified the grammatical construction but sacrificed readability. Usually, simplifying the structure of a sentence makes it clearer and more readable, but when it produces the opposite effect, it is not desirable.

These examples achieve their simplicity in a manner that calls for additional comment. In each case the phrase was simplified by converting a noun into an adjective. It is true that the English language contains a great many words that can be used as either; but some technical writers go to extremes and make adjectives out of nouns that are not usually used in such a manner. As a result they develop an annoying mannerism that makes it more difficult for them to communicate with the rest of the world, and they irritate even those who understand them.

Directness.   Directness, like simplicity, calls for attention to sentence structure. Consider, for example, the following sentence: "There are parts of the report in which his statements have been influenced by his feelings." The main statement is, "There are parts." The main *fact*, however, is that a writer's feelings have influenced his statements in a report. To make the style direct we might write, "The report shows his biases," or "His feelings have influenced parts of the report." Each of these direct versions, you will notice, has a meaningful subject and a meaningful verb.

You might take the following suggestion as a formula for directness: Ask yourself, "What do I want to make a statement about?" and let the answer to this question tell you what your main subject should be. Ask yourself, "What do I want to say about it?" and let your answer tell you what your main verb should be.

Consider the result of applying this formula to some additional sentences:

It must be expected that there will be complaints from people who live in the neighborhood.

A recent article makes the statement that before 1950 Cuban reserves of manganese had been exploited almost to the point of depletion.

Directness in style would suggest:

Complaints from people who live in the neighborhood must be expected.

Before 1950, according to a recent article, Cuban reserves of manganese had been exploited almost to the point of depletion.

You have probably thought of other versions that would be just as direct. The significance of your being able to do so becomes apparent when two more sentences are considered.

The new alloy is harder than the one we have used in the past.

The alloy we have used in the past is softer than the new one.

These two sentences give us exactly the same fact, and they are equally direct. They differ, however, in one respect: The first makes a direct statement about the new alloy and incidentally gives information about the old. The second makes a direct statement about the old alloy and incidentally gives information about the new. As sentences, they are equally direct; *but they are not equally direct about the same thing*. The choice between them should depend on which alloy the writer is primarily attempting to discuss. In other words, directness is more than just a matter of sentence structure without regard to purpose. It is also a matter of presenting the facts as they bear upon our purpose in writing and the reader's probable reason for reading.

Differences like the one illustrated may seem trivial when we look at only one or two sentences, but many an article or report has extended passages in which the writer seems to have gone off on a tangent because the facts about his real subject are presented only indirectly in sentences that are *direct* statements about other matters. The formula offered above reduces the likelihood that you will obscure your point by this particular kind of indirectness.

Indirectness may result not only from a poorly chosen subject, but also from a poorly chosen verb. By using a weak, general verb you may then be forced to work in the verb idea as a noun or adjective. For example, you might write:

effect a change in instead of *change*
provide information about instead of *inform*
came to the conclusion that instead of *concluded*

accomplish a reduction instead of *reduce*

place more emphasis on instead of *emphasize*

This is not to suggest that you must express *every* idea as a direct statement. If you do, you will make it hard to distinguish between the important and the less important. Consider for example: "We are now using the process described in the enclosed leaflet. It is a continuous-flow process. Last year we used the batch process." Every fact is expressed directly, but it would be better to write, "We are now using the continuous-flow process described in the enclosed leaflet rather than the batch process that we used last year." The revised version is more effective because it is a direct statement of the major idea, which stands out distinctly because lesser matters are referred to in an incidental manner only.

Precision

Though precision involves both diction and sentence structure, the subject is all too often approached as if it concerned no more than the use of words with correct understanding of their meanings. Writers are warned, for example, not to confuse *imply* with *infer*, *liable* with *likely*, *fix* with *repair*, or *unique* with *unusual*. Most of the discussions along this line contain valid and useful information—unfortunately combined with many refinements that are based on textbook traditions rather than on first-hand observation of actual usage.

So far as such questions of general vocabulary are concerned, the Glossary of Usage at the end of the book covers this aspect of precision in enough detail to guard against the most likely errors. It includes the textbook favorites because there are still many people who value these traditional distinctions.

Precision in the use of technical terms is not covered in the Glossary because each line of work has its own extensive technical vocabulary. The most that a general discussion can do is to alert you to the need for mastering the meaning of each term in your field as you encounter it, and to urge that if you do much writing in a field where terminology is rapidly changing and developing, you invest in any special publication that lists and defines the terms you must use. Also, don't underestimate the help that you can get from the latest dictionaries—unabridged or even collegiate. And don't underestimate the extent to which you make yourself look immature if you use technical language incorrectly.

But precision is more than just a matter of words; as pointed out above, it is also a matter of sentence structure. Many a writer who knows the meaning of each word will construct sentences that do not actually say what is intended. Consider for example the statement: *"Test plots using various species give an average range of life of 19 to 25 years, with a probable average of 22 years."* According to this sentence, it is the *plots* that use the species and give the results. But in point of fact, the plots neither use the species nor give the results. It is the tests, not the plots, that give the results and it is the scientists performing the tests who use the species. One way of expressing the idea precisely—certainly not the only way—would be: *"Tests of various species, each performed on a separate plot, indicate that the range of life is 19 to 25 years and that 22 years is the probable average."*

Another example of lack of precision is the sentence: *"The material should be of a hardness and a toughness that will possess good wearing properties."* According to this sentence, the hardness and toughness *possess* wearing properties. In point of fact, they *are* properties. A precise restatement would be: *"The material should be hard enough and tough enough that it will wear well."*

Objecting to lack of precision in sentence structure is not just quibbling about trivial details. A sentence that lacks precision in structure is like a photograph taken by a camera that is not properly focused. We may be able to see the picture, but the blurred lines are annoying and distracting. It is a writer's job to say exactly what is meant rather than just coming close enough to the intended meaning for the reader to figure it out. And the major reason for failure to do this job is writers' failure to check their own work carefully enough to notice what they have said—or in other words, to make sure that the statement made by each verb applies to that verb's grammatical subject.

Choice Between Active and Passive Voice

Many commentators on technical writing devote considerable attention to the frequent superiority of the active voice over the passive. (The difference between them is explained in the Handbook.) Following are typical examples of sentences where the passive reduces effectiveness.

1. It is maintained by the contractor that . . .
2. It is believed that . . .
3. Your letter has been received by us and . . .

The first of these sentences has a meaningless subject. The second not only has a meaningless subject but is also evasive about who does the believing—which may be a fact of considerable importance. The third is awkward and unnatural. The three sentences could be improved by change of voice, as follows:

1. The contractor maintains that . . .
2. I believe that . . . (if not *I*, some other word showing *who* believes)
3. We have received your letter and . . .

Yet in spite of the frequent overuse of the passive voice, it is a legitimate and useful form, and sometimes preferable to the active. For example, "The concrete was damaged by the cold weather" is better than "The cold weather damaged the concrete"—assuming that it is the concrete, not the weather, that we are discussing. Likewise, the passive voice in "The foam should be skimmed from the vats once an hour" is entirely satisfactory, since the skimming is the important fact and the question of who performs the action is not significant.

The passive voice is called for in a sentence such as, "The samples are collected by the supervisors, examined by the inspector, and sent by air to the main office." To use the active voice we would have to write, "The supervisors collect the samples, the inspector examines them, and an airplane carries them to the main office." This sentence, with its three different subjects, obscures the fact that the writer is really trying to tell us about the samples.

It is unfortunate that the terms *active* and *passive* are used to identify the two voices of verbs. Careless thinking makes some people identify *active voice* with *active verb*. But the facts do not justify such an identification, as we can see when we look at two more sentences:

In the darkened room the mother softly hummed a lullaby to her quietly sleeping baby.

The building was shattered by an explosion.

The verb *hummed* in the first sentence is in the active voice but certainly does not convey a picture of vigorous activity. The verb *shattered* in the second sentence is in the passive voice but conveys just as much a sense of activity as if the sentence ran, "An explosion shattered the building." In brief, the extent to which a verb gives an impression of action depends more upon its actual meaning than upon the writer's choice of voice.

A sound guideline in deciding when to use the active voice and when to use the passive is found in the preceding discussion of directness—that is, *use a subject that shows what your sentence actually concerns, and use a verb that says what you really want to say about that subject*. If you will do this there is little danger that your use of voice will damage your style. Yet the warnings against excessive use of the passive are justified, for the passive voice actually *is* overworked by many writers, it often *does* weaken their style, and it is often a symptom that calls attention to lack of directness.

Sentence Variety

One other area of sentence structure requires comment: variety. Textbooks often recommend that writers vary their sentence structure to begin with elements other than the subject. While the advice might be sound if used with restraint, too often inexperienced writers, thinking more is better, torture their prose in an effort to start one sentence with the verb, the next with a predicate adjective, then one with an adverbial clause, and so on, until all continuity and sense is lost. Moreover, an investigation of prose written by professional writers will demonstrate that they begin three out of four sentences with the subject, and when they use some other sentence opener, usually it is an adverb indicating time, such as *then, next,* or *when,* or an adverbial clause serving a similar purpose. In other words, most of their sentences follow the standard English model of subject, verb, complement. This is the pattern used in talking, and the pattern most readers expect, for first it tells the reader what you intend to make a statement about, and then gives the information about that subject matter.

How then do professional writers create variety in their prose? They achieve variety through the use of the elements we have been discussing in this chapter: by using concrete, specific diction, varying sentence length, writing directly and precisely about their subject, using a variety of connectives, such as *of course, to be sure, by the same token,* and through organization, which we will take up in the next chapter. If you decide to vary your sentence structure, make sure that the variety serves the purpose of furthering your communication, rather than calling attention to itself.

CONCISENESS

Lack of conciseness, a defect in the writing of many persons whose work is good in other respects, is a major cause of complaints by busy readers. Therefore, you should learn to trim every word that can be spared when the occasion justifies your taking the time to do so, and should try to avoid wordiness even when writing a paper that you will have little opportunity to revise.

To be sure, you can carry conciseness too far. You will not help your reader by cutting out so many words that you sacrifice clearness, readability, naturalness, precision, and proper application of emphasis. Nothing said in this discussion should be taken as encouraging such a sacrifice.

Since lack of conciseness is sometimes the result of nothing more than a general looseness of style, you can often reduce the length of something that you have written by merely looking it over. You will do a better job, however, if you know what to look for. Therefore, some of the main causes of wordiness are taken up in the following discussion.

Repetition

Repetition for the sake of clearness or emphasis is sometimes desirable, but unintentional repetition is a major cause of wordiness. Many writers use *during the time that,* for example, when the single word *while* would be sufficient, or say that *costs rise to a higher level* when they need only say that *costs rise.* Other examples of such repetition are *a period of three months* (repetitious because three months *is* a period) and *the specimen was placed on the balance and weighed* (repetitious because the fact that it was weighed clearly implies that it must have been placed on the balance).

A slightly different kind of repetition is seen in the sentences, "In one respect, the house was not in good condition. Its roof leaked badly." If the writer had been telling us of several defects, a general statement might have been desirable as a topic sentence; but since there was only one defect to mention, it would have been enough just to say that the roof leaked. Words are wasted when we are told the same fact twice, once in a general way and once in a specific way.

Wordy Phrases

The English language abounds in phrases that convey little meaning but are sometimes necessary to make the flow of words smooth and natural. If your writing is to be concise you should be alert for opportunities to eliminate these phrases. For example, you should avoid *in the event that* when you could substitute *if; due to the fact that* when you could use only *due to* or *because;* and *has proved itself to be* when *has proved* or just *is* would be sufficient. Similarly, *prior to the time that* is a wordy version of *before, at the present time* means only *now,* and *during the years between* could be cut to the single word *between.*

Use of General Rather Than Specific Words

We have already discussed some of the reasons why general words are undesirable; in addition, the damage they cause to conciseness is sometimes ample reason for avoiding them. For example, when writers begin a sentence by saying, "They were fastened together," they then must add "by means of staples," "with rivets," or something of the sort to tell *how.* If they write instead, "They were stapled (or welded, or glued, or riveted) together," the verb would give the full information without the assistance of a modifying phrase or clause.

Other examples of wordiness caused by use of needlessly general words would be: *transported by means of trucks, kept out by the construction of fences,* and *made contact by telephone.* These could be replaced, respectively, by *trucked, fenced out,* and *telephoned.*

Needlessly Complicated Structure

Simplicity has been discussed separately because it is desirable for its own sake as well as for conciseness. You probably noticed in the examples that when a sentence was simplified it usually was shortened. The following illustrations will emphasize this effect further.

Complicated and wordy: It is essential that we take these precautions if our crews are to be safe.

Simple and concise: These precautions are essential to the safety of our crews.

Complicated and wordy: There are four dealers in Kansas City, and any of them can supply us with spare parts.

Simple and concise: Any of four dealers in Kansas City can supply spare parts.

Complicated and wordy: The condition of the machinery that is now located in the basement is not good enough that renovation would be justified.

Simple and concise: The machinery in the basement is not in good enough condition to justify renovating it.

One more fact—often overlooked—should be brought out here. If you simplify your sentences to such an extent that you use primer style, your writing will be wordy even though each sentence may be as short as you can make it. When too many sentences are short and simple, the repetition of certain words and phrases is unavoidable. Thus as you revise for the sake of conciseness, you should watch for opportunities to cut out words and phrases by combining sentences as well as by shortening them—though you will also need to guard against making your sentences too long by such a process. Study of the examples of primer style given earlier will illustrate that the more mature style is also concise.

Superconcentrated Style

One final word of caution: When conciseness is carried too far, the result is a superconcentrated style, as in the sentences that follow.

Preventing catalyst selectivity deterioration is difficult.

The machinery repair cost question demands attention.

When readers encounter sentences like these, they do not grasp the meaning until they stop and figure out the relationship of the words. There simply are not enough functional words (prepositions, conjunctions, and articles) to make the structure clear. A looser style would be better, as for example:

It is difficult to prevent the selectivity of the catalyst from deteriorating.

The cost of repairing the machinery demands attention.

In brief, conciseness is not just a means of saving space. It is also—often primarily—a matter of saving time. And you will defeat your own purpose if you eliminate so many words that you make your material harder to read and force your readers to spend more time on it than they would have spent on a longer version.

FINAL COMMENT ON STYLE

Two possible headings, *clearness* and *readability,* do not appear in the preceding discussion, but their omission does not mean that those qualities are unimportant. On the contrary, it would be impossible to overstate their importance. Whatever else we may do to improve our style, clearness and readability are the end results we hope to achieve. But neither of these qualities can be attained by applying a separate technique. Our style becomes readable when we choose our words intelligently, write sentences of suitable length, develop such qualities as simplicity and directness and conciseness—in fact, do all the things that have been recommended. And clearness, too, is produced by doing all these things and also, of course, by attention to such commonplace matters as reference of pronouns, parallel construction, accurate choice of connectives, punctuation—all the fundamentals of English. Thus clearness and readability have not been overlooked here. Each of them is simply the result of so many different causes that an attempt to discuss it would lead us to the very headings under which the subject of style has been covered.

The subject of readability formulas should perhaps be mentioned here. Since the 1920s, some thirty more or less useful formulas have been devised that purport to show the general readability level of prose. By taking into account such factors as sentence length and complexity of vocabulary, they allow the user to demonstrate that longer sentences and bigger words create prose that is more difficult to read. Unfortunately, the formulas are all flawed to the extent that they do not demonstrate that primer prose is also difficult, nor do they take into account such factors as directness and conciseness, which are also important. What is good in the various readability formulas is not neglected in this book. It is included in the comments on simplicity of diction, length of sentences, and elsewhere. In the end, we feel, the writer is better able to judge the effectiveness of a specific sample of writing than any of the formulas yet devised.

To exhaust the subject of style would be impossible even in a book-length treatment, but it would be wrong to close without pointing out that in regard to each of the desirable qualities discussed, one warning has always been included: Don't try so hard to develop one quality that you sacrifice other qualities equally desirable.

The warning concerns not only the sacrifice of qualities we have discussed but also the sacrifice of other values. Sometimes, for example, if you try to make a sentence or clause as simple and direct as possible, you may prevent it from being parallel to some other sentence or clause that it should resemble. On another occasion, phraseology that looks desirable at first may need to be avoided because it results in unpleasant, accidental rhyme or in excessive repetition of some one sound, word, or phrase. Again, an effort to make a sentence more concise may destroy the intangible grace and rhythm that make writing smooth and pleasant to read.

Our discussion of style should therefore end with one general piece of advice: Pay conscientious attention to the principles that have been presented; but when the result of following them bothers you—when you feel that a revision you are considering simply does not seem right even though you do not quite know why—you will probably do well to trust your instinct and express yourself in a way that does not clash with your convictions about what sounds best.

EXERCISES

Exercise 1

Improve the diction of the following sentences as indicated:

A. Make the following sentences less pompous.
   1. The owner or his representative shall have the right to lawful freedom of ingress to or egress from the premises.
   2. The Supreme Court decision will impact upon the case before the state court.
   3. We hope that this regulation will be effectuated at a sufficiently early date to preserve the navigable character of the river.
   4. In conjunction with the enhancement to our customer-relations system we have developed a procedure to augment our sales staff.
   5. A contract that would, if made, obligate the company to expend funds or convey property must, to be valid, be approved by the financial vice-president.
   6. We are sincerely desirous of working with you in effecting an improvement in this condition.
   7. Because of the cost factors involved with that basing mode, along with the question of whether it was really going to work, maybe even be obsolete by the time it was deployed, and also recognizing the need for some viable defense now, I chose to take opposition to that proposal.
B. Replace the language that is needlessly technical by substituting ordinary language.
   1. The anfractuosity of the wiring between these two points makes the diagram hard to understand.
   2. Much of the shipping from this port consists of cabotage.
   3. His arm was temporarily disabled by a blow where the ulnar nerve rests against a prominence of the humerus.
   4. One defect in this method is the difficulty of preventing the hot water from passing into a gaseous condition.
   5. A sound meter registers differences in volume that are not of sufficient magnitude to be perceived by auricular observation.
   6. Throughout the report his style is damaged by galimatias.

C. Reduce the vagueness of the following sentences. You are free to invent facts, when necessary, in order to be specific.
1. The filters should be changed at reasonable intervals.
2. If carbon builds up very much on the bottom of the burner it should be scraped off to conserve oil.
3. The drain pipes must be buried fairly deep so that they will not freeze in cold weather.
4. Gasoline is wasted rather than saved if a car's engine is allowed to idle very long so as to avoid restarting it.
5. As the sorters gain experience the conveyor belt should be speeded up now and then by moderate increments.
D. Decide for yourself what improvement is needed in the diction of each of the following sentences and make changes that will result in improvement.
1. In high altitudes the cooking time should be considerably longer.
2. Because of the drought, cachexia was widespread among the cattle.
3. There was only a minuscule chance that the tumor would metastasize.
4. We knew that they had recently undergone ecdysis for we discovered the exuviae.
5. Greater lateral dispersion will result when the nozzles from which the droplets are ejected are moved toward the wing tip, when the mean diameter of the droplet spectrum is reduced, and when the airplanes fly at a greater altitude.
6. The administration is attempting several methods of effecting revenue enhancement that would not anger voters.
7. Large numbers of negative patient care outcomes would suggest that treatment of poor circulation by heart transplant is contraindicated.
8. The work is now a bit behind schedule, but we expect to remedy this condition within a reasonable length of time.

Exercise 2

Break each of the following long sentences into shorter sentences. If necessary, change the order of the ideas.

1. As a result of customers' requests for improvement in the method of fault-isolating schematics we shall issue a 300-page loose-leaf book on system schematics, in which we shall utilize the existing wiring Diagram Manual Schematics, which we shall supplement with about 50 new mechanical system schematics as they are developed, and which we shall revise whenever we revise the existing Wiring Diagram Manual.
2. Although the channel shoals severely during storms and is periodically dredged to at least ten feet, over the long term it tends to maintain itself at about the same dimensions, for surveys dating from 1974 show that the channel has migrated east in the vicinity of the breakwater,

and aerial observations show that its cross-sectional area and shape tend to remain constant.

3. Actually plutonium, though it is toxic, is one of a series of similar elements called actívides that includes thorium and uranium, both of which are naturally occuring minerals that abound in some areas and that resemble plutonium in some radioactive properties, the primary radiological difference being that plutonium decays many times faster, and that have not occasioned concern about their presence.

4. The attorney general said that since Jordan somehow obtained a patent from the U.S. government in 1954 he will be in a position, if not required to allege and to prove that the 1905 proceedings whereby the state purchased the property were invalid, or that insufficient money was paid, to have a chance of securing a quiet title to the entire property with all its improvements.

Exercise 3

Rewrite the following passages so as to eliminate their primer style. The order of ideas may be changed, but each revised version should be smooth, natural, and free from stringiness. Your versions should bring closely related ideas together and leave less closely related ideas in different sentences.

1. Each year some 25,000 ships of varying sizes and types sail the oceans and seas of the world. The ships come from every seafaring nation. They haul two billion tons of oil, coal, lumber, and grain. They also carry manufactured products and consumer goods. Nearly all of the world's raw materials are carried by ship. In ten years, the cost for oil to transport these goods has risen ten times. The world's ocean fleets burn some four million barrels of oil each day. The amount of oil burned would supply one-fourth of the daily power needs of all the automobiles, factories, and houses in the United States.

2. The composition of glass is simple. Its main ingredient is silica. Common sand is a source of silica. Another ingredient may be alkali. Primitive glass blowers added alkali to sand to lower the melting point. They also added lime. They got the alkali from ashes of plants or trees. They crushed stone to get the lime. The mixture is heated until it turns to glass. The glass is then cooled. As it is cooled, it may be shaped. It can be rolled, pressed, molded, blown, or pulled. When it cools completely, glass holds whatever shape the glassmaker has given it.

3. Beekeepers have known for centuries that they can move a hive only at night. If the hive is moved during the day, the bees become disoriented. Even a move of a few meters will confuse bees. But if the hive is moved during the night, the bees have no trouble. A move of meters or miles makes no difference. Researchers have discovered that bees relearn the appearance and location of the hive on their first flight each morning. Only at this time is the home-learning programming possible. At no other time will the necessary information imprint on the bee's brain.

4. Most lumber today is kiln dried. It is often dressed as well. Dressing is simply planing. But lumber bought at the yard is not ready for use. The edges of boards are often crooked. A squint along the edge will show hollows and high spots. The sides will have a washboard pattern. The knives of the rotary planing machine leave cross-grain scallops. Such lumber is fine for rough use, such as storage shelves or framing for houses. It is not good enough for cabinet work, such as would be needed in a living room bookcase. Cabinet work requires accuracy and a nice finish. Before boards from a lumber yard can be used for fine work, they must be jointed and smoothed. Jointing means planing the edges straight, square, and true.

5. The intermediate vector boson has been sought by atomic physicists since the 1930s. This subatomic particle may be a cornerstone of the quest for a theory linking four fundamental forces of nature. Physicists call these four forces electromagnetism, gravitation, the strong force, and the weak force. The strong force holds the nucleus of atoms together. The weak force is responsible for radioactive decay. Each force is carried by different particles. The photon, for example, carries electromagnetism. The particle responsible for the weak force has only recently been found.

Exercise 4

Improve the following sentences by making them simpler, more direct, or both. (As you do so you may also make them more concise.) If necessary, study a sentence until you clearly understand what the writer was trying to say, and then rewrite it entirely.

1. It is necessary, if an area is to support a physician-nurse clinic, for it to have a considerably larger population than if it supports the services of only a nurse.
2. A shutdown of one month will have the additional effect of causing production to be lowered for the first week after the plant reopens.
3. The Highway Department issued an announcement to the effect that it will add four new weighing stations.
4. The regulations that have been proposed will put an end to the wasteful expenditure of time on details that are of only minor importance.
5. The failure of the tire occurred because of its not having been properly inflated.
6. We came to a decision against accepting their proposal for two reasons. In the first place, it did not fully meet our specifications; and in the second place, their working schedule appeared impracticable.
7. The Eskimo believed that the wolf sprang from a hole that had occurred in the ice to exercise control over the caribou herds so that weak caribou would not reproduce and that the tundra would consequently be overpopulated with poor herds.
8. Core analysis is of value to the engineer when faced with a decision as to the advisability of testing a formation.

Exercise 5

By using your imagination you should be able to figure out what the writer intended to say in each of the following sentences. Revise each sentence as necessary to achieve precision.

1. When the bypass was ready to be built the crew had to be enlarged.
2. His record consists of operating a bulldozer, repairing agricultural machinery, working in the parts department, and a salesman.
3. The name of his predecessor could not be obtained but was apparently incompetent.
4. The answers, except for one question, are correct.
5. After working for fourteen hours without taking time off the snow still blocked the highway.
6. The supplies needed include paint, paint thinner, brushes, a spray gun, and a ladder.
7. The source of current will be obtained from a generator.
8. The discussion includes the tools used in the process.

Exercise 6

Make each of the following sentences more concise.

1. Sunflower seeds are a crop that it is probable will become profitable to raise in the future.
2. It is the recommendation of the committee that the budget for this program be reduced to one-half of the present amount.
3. It was raised to a height of three feet by use of jacks.
4. We decided against it for only one reason, namely because the cost was excessively high.
5. He should have been taken to the hospital that night and remained there until the next morning to guard against the possibility of complications.
6. I have not as yet made any determination as to whether I shall run for office.
7. A spell of wet weather may be expected, according to the prediction of the weather bureau, beginning on Tuesday and lasting until the end of the week.
8. After the meeting was over Mr. Jones asked that he be given a copy of the report.
9. It is apparent that the regulations which the contractor had put into effect were not achieving the desired result.
10. It can provide a larger proportion of the services than a majority of those who might become patrons believe it has the capability of providing.

4 | Organization

If you are to be effective in technical writing you must do more than just express each separate fact and idea clearly. You must also organize those facts and ideas in a manner that will show how they are related to each other and to the paper as a whole—for it is primarily in their relationships that facts and ideas become truly valuable.

The importance of good organization cannot be stressed too strongly for the simple reason that *all writing is organized in some manner*. It may be organized poorly or well for your purpose and reader, but it will be organized in some manner and your reader will respond to its organization. In fact, as they do to style, readers respond as much to the organization of a piece of writing as to the content. This fact is extremely important for a writer because some methods of organization make material much easier to understand than other methods. Indeed, cognitive psychologists have theorized that memory itself may involve certain forms of organization so that information in other forms must be reorganized in the mind before the information can be mastered. In this chapter, we will discuss forms of organization, methods for writers to use, and the reasons behind the methods.

According to information theorists, there are three general forms in which information can be organized: random, sequential, and hierarchical. The random form is, for example, the order in which cards fall when dealt from a shuffled deck—there being no apparent relation or pattern among them. The second form, the sequence, is any list of items arranged so that each item has the same relationship to each of the other items, such as alphabetical, numerical, or chronological. The third form, the hierarchical, is one that involves more complex, multilevel relationships, as illustrated in the following schematic, sometimes called a tree structure. The numbers indicate the level of the items in the hierarchy.

An example with items as well as numbers is shown in part below:

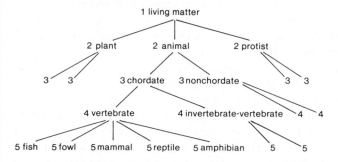

While all three methods of organizing information are useful for certain purposes, such as in computer theory or alphabetical lists, the random and the sequential methods are not as suitable for organizing information to be conveyed to a reader. One reason is the effort required to retrieve information from them. For instance, if you have a list of 1024 items arranged randomly and your readers wished to find a single item on the list, they would have to search on the average through about half the items to find the one wanted. If you organized the items into 32 groups of 32 items, your readers would find the wanted item after looking through an average of about half the classes and half the items within the desired class—a total of about 33 searches instead of about 512.

However, if you organized the items in a hierarchy in groups of four with four levels of classification as shown in abbreviated form below, your readers would have to search 2.5 classes at each of four levels and 2.5 items for an average of 12.5 searches. You will immediately recognize that this example of a hierarchical organization is in the form of an outline. (The asterisks indicate sections omitted to conserve space.)

I. Class 1-256
   A. Subclass 1-64
      1. Subclass 1-16
         a. Subclass 1-4
            i. Item 1
            ii. Item 2
            iii. Item 3
            iv. Item 4
         b. Subclass 5-8
           * * *
      2. Subclass 17-32
         * * *
   B. Subclass 65-128
    * * *
II. Class 257-512
  * * *
III. Class 513-768
  * * *
IV. Class 769-1024

Such an organizational scheme requires the addition of 340 headings, but studies by learning and memory specialists have shown that adding two times as many appropriate headings to a list of random items can improve learning. To be sure, certain mathematical search methods actually make the random and sequential schemes more efficient for mechanical retrieval of information; but since these methods are not readily available to a reader, their efficiency is irrelevant in this discussion. Also, in writing, other factors are involved that make the hierarchical schemes far more effective.

The first factor is perhaps most important: Technical writing is ultimately a teaching process. The writer wishes to convey information to the reader that the latter does not have, and the reader tries to learn what the writer has to say. Because the mind seems to retain information best when the relationships among facts are clear, the hierarchical organization, which shows the relationships, seems best for teaching, learning, and remembering. A study in psychology well illustrates its advantage. A group of subjects was given 112 words to memorize. Half the group was given the words arranged randomly, and half was given the words arranged in a tree structure. Both halves were given equal time to learn the words and were tested to see how many words they could recall. After each test, they were allowed to study the words again, and were retested. The table below shows the results.

Test	1	2	3	4	
random	20	38	52	70	words
tree structure	73	106	112	112	recalled

In a similar study, one group was given the words in random order and told to memorize them. The other group was told to organize the words but was not told that there would be a test of ability to recall them. When the two groups were tested, the group that was requested merely to organize the words remembered the items far better than the group told to memorize.

The differences in the scores appears to be connected to a limitation of the mind. The most recent research indicates that the short-term memory can retain only three to seven items unless the items are organized in some fashion, as they are, for example, in telephone and social security numbers. It is easy to test this assertion. Have a friend or teacher make up a sequence of random digits, each sequence with one more digit than the preceding, such as shown below:

361
8269
10437
593752
2784926

49738693
610839625
8546071637

Have the person read each sequence aloud to you once, and then attempt to write down the digits from memory. The vast majority of people will be unable to reproduce from memory a sequence with more than nine digits, while the average person will be able to recall only seven digits, and even here, the person has unconsciously organized the sequence in groups. The vast majority of people have no difficulty, however, in remembering a number such as 12193824921×732 if it is organized into smaller groups that the mind can comprehend. You will recognize the number when it is properly divided: 1 219 382-4921 × 732. The same principle applies in writing. By properly dividing and organizing the information you wish to convey to your reader, you can make learning much easier. In addition, since most associations made by readers require that they retain in short-term memory what has just been read, it is necessary that you keep the number of points to five or fewer.

The basis for organization, then, is largely psychological, and the best organizational system for the writer appears to be the hierarchy. The primary hierarchical system for the writer is the outline. Three major points are clear:

1. Hierarchical organization—the outline—is best for ease of learning;
2. The outline is best for ease of access to information; and,
3. There should seldom be more than five items at any single level for ease of recall.

Students might take particular note of this for study purposes. An immense amount of information can be made easy to retain by merely organizing it into an outline with no more than five items in any series.

There is one further general point of importance about organization: The organization works best when its nature is clear to the perceiver. This is illustrated by another classic study in psychology: A number of chess masters and novices were presented with chess pieces on a chess board and after a five-second glance, they were asked to reproduce the arrangement on another board. When the pieces were arranged randomly, there was no difference in success between the masters and novices. But when the pieces were arranged at the twentieth move of an actual game, the masters achieved a 90 to 100 percent success, while the novices achieved only about 40 percent success. The difference can be accounted for by the fact that the masters, within five seconds, perceived the organization, while to the novices, the pieces appeared to be randomly placed. Without perception of the organization, therefore, the players—masters and novices alike—could not recall the placement of the pieces. The same principle applies in writing and reading. If readers understand the relationship among the facts and ideas presented, they will be able to understand their significance far more easily. The technical writer provides information about the method of organization by use of such devices as outline-type tables of contents, headings and subheadings, cue words such as *first, second, furthermore,* and so on. The writer's major question should be: Will the reader understand the relationships being discussed?

These are the basic reasons clear organization is needed in writing. There remains the problem of *how* to organize. The remainder of the chapter treats that problem, mostly by discussing the writer's primary organizational device—the outline—and the process of constructing a good one.

Making an outline is advisable not only for the reasons given above but also because the very process of creating a good outline often forces a writer to clarify ideas and to settle questions of relationships that might otherwise go unnoticed. In this chapter, however, the emphasis will be on the outline as a plan that will help you to produce an effective piece of writing rather than as an end in itself. This is no trivial distinction. Many an outline that seems at first glance to be a fine specimen turns out to have flaws when put to use as a writing guide.

Of course, an outline may also be used for other purposes. It may be submitted to the person who will eventually receive your completed paper to see whether you are on the right track, or it may serve as a source of headings for inclusion in the completed paper. However, any outline that turned out to be unsatisfactory as a plan to use while writing would also be unsatisfactory for the other uses. Thus, the emphasis placed here on outlining as a method of planning is justified from every point of view.

To be sure, when a piece of writing is to be short and simple, planning it may be only a matter of deciding what material to include and how to arrange it—jobs that must be done in any writing process. But when a substantial mass of material is to be presented and the relationship of facts and ideas is complicated, the planning involves more than deciding the sequence of presentation. It also calls for deciding on subordination and coordination and for producing a plan of organization that will make these relationships clear.

If an outline is to function effectively, it should meet the following requirements:

1. It should be based not only on your consideration of the subject but also on the knowledge you have about the readers—their reasons for reading, their attitudes, their familiarity with the subject, and anything else that will affect their reception of your ideas.
2. It should cover the material that you will need to include after you have narrowed down your subject by carefully considering the function of your paper.
3. It should be specially designed to accommodate the specific facts and ideas that you want to present.
4. It should give a sense of continuity—of organic unity—rather than being merely a collection of headings that fall within the scope of the subject. (This fourth requirement is less vital in some kinds of writing—for example an instruction manual intended to be used for reference purposes rather than read as a whole—but even when continuity is not essential, the pattern that you create should be systematic, not haphazard.)

Pilot Plant Treatment of Vegetable Wastes

I. Characteristics of raw wastes
   A. Peas
      1. Low volatility
      2. High sodium and chloride
   B. Corn
      1. High volatility
      2. Deficiency of nitrogen
   C. Carrots
      1. Medium volatility
      2. High sodium
      3. Deficiency of nitrogen

II. Continuous operation of treatment system
   A. Acclimatization time and mechanical adjustment
   B. Operation on a continuous basis
      1. Operation with peas
      2. Operation with corn
         a. October corn run
         b. December corn run
      3. Operation with carrots
      4. Accumulation and handling of corn wastes
   C. Hydraulic overflow rates
      1. Overflow after one-day detention
      2. Overflow after six-day detention

III. Nutrient requirements
   A. Results of augmentation of nutrients--October
      corn run
   B. Results of augmentation of nutrients--December
      corn run
   C. Limited significance of results obtained

IV. Clarification
   A. Problems created by changes of vegetables processed
   B. Foaming problems

V. Settling characteristics
   A. Poor characteristics compared with solids from
      substrates of domestic sewage
   B. Possibility of improvement by organic loading

FIGURE 1. **Specimen Outline Showing Correct Physical Form. This specimen illustrates a simple system of indentation and shows the numbers and letters that precede four ranks of points. (For decimal symbols, see Chapter 5.) The specimen is single-spaced except above main points, but double or single spacing throughout would be acceptable.**

THE PROCESS
OF PREPARING AN OUTLINE

When you have analyzed the total writing situation (readers, subject and purpose, and material to be included), you can make your outline by more than
one method.

Sometimes you may best work from the highest level of organization down
to the individual points. That is, to produce an outline illustrating the relationship among all the facts at hand, you first write down a brief statement of
the subject. Next, determine what are the major components of the subject
that you must cover to accomplish your purpose. Then take each component
and decide what elements of that component need to be explained, and repeat
the process for each element.

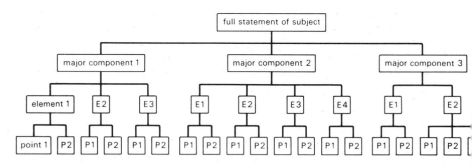

As you decide upon your major components—which will become the
main points of your outline—remember that subjects can have major components with more than one possible basis. For example, a discussion of the
condition of crops in a county may with equal logic be based on the geographical
sections of the county, the kind of crops, or the condition (good, fair, or poor)
of the crops. In such cases you need to decide upon your main points by
considering the total writing situation and adopting the basis that will be easiest
for the reader to understand and to use.

Of course, working from main points downward is possible only when
you already know a great deal about the subject. For example, the diagram
classifying living matter that was shown earlier could not be developed until
scientists learned a great deal about the forms of life. Often, rather than working
from the top down, you may need to work from the bottom up. This will be
necessary when you have a haphazard mass of facts and ideas, or when you do
not even know what material you will need to include.

Coping with this situation may not be easy, but there is a procedure that
makes it easier to get started. It is important, of course, that you regard your
ideas at any stage of the work as merely tentative and subject to constant
revision. To use this method you should:

1. Think about your subject and the function your paper is to perform, and gather information about the subject if necessary. As you do so, jot down your facts and ideas in the form of rough notes. You will probably add or eliminate material later, but so far as possible you should make notes that cover the substance of your paper or at least of the part that you are working on if the total paper will be long.

2. Study these notes and group together material that seems to belong together. As you do so, jot down headings that indicate the common element in each group— headings that may become the basis for main points in your outline. When necessary, postpone decisions about doubtful points.

3. Taking into consideration your purpose, your readers, and your raw material (the latter now grouped under headings that indicate its nature), decide upon the main points of your outline. If necessary, create new headings to cover two or more groups that have something in common. After you decide upon the main points you will probably find that some of your groups of material fall naturally under them as subpoints. In fact, you probably recognized some groups as subdivisions of others when you formed the original groupings. In any event, your main points will not result from just cutting up the subject. They will also result from a process of building up from the actual substance your paper must include.

4. Bearing in mind not only the logic of the subject but also the attitudes and abilities of your readers, arrange your main points in an order suitable to the purpose of your paper.

5. Now apply to the first main point, or to whichever main point you feel best prepared to work with, the same treatment that you applied to the entire subject. Thus work out and arrange the highest rank of subpoints under the main point in question.

6. Organize each main point as far as you consider necessary. (At this stage, the order in which you work on one or another part of your outline will probably depend on the order in which your ideas mature and your facts become available.)

The preceding method should serve as a general guide, not a straightjacket. There is no reason that you should not depart from it when the way in which your ideas develop makes departure natural. For example, you will often have in mind a possible list of main points before you even begin to jot down notes. You may develop ideas for handling some of your material at lower levels before you decide what your main points will be. You should regard such ideas as merely tentative, however, until you see how they fit into the pattern that emerges from the general procedure recommended.

GIVING SUITABLE RANK TO YOUR POINTS

The question of how to decide on the rank to be given to the various points of an outline calls for further comment. Sometimes the decision about the rank of each point is dictated by facts beyond our control. For example, it might be essential to base a discussion of a college or university on its formal organization. Thus it would be unsuitable to make a point covering a department equal in rank to points covering entire colleges, even if the department taught more courses, handled more students, and spent more money than some of the colleges.

Much of the time, however, a writer is free to use personal judgment. Suppose, for example, that in a certain area the acreage and value of each of two kinds of grain equaled or exceeded the combined value of all the fruits. In this case it might be not only permissible but actually desirable to make the point covering each kind of grain equal in rank to the heading *Fruits*.

With a reasonable amount of ingenuity you can usually give each point the rank that its actual importance justifies without producing an illogical result. To be sure, if topics such as "Grain" and "Wheat" were shown as points of equal rank, the equality would not be defensible because one includes the other. But if wheat happened to be especially important to the purpose of your paper, it would be possible to change "Grain" to "Grains of Lesser Importance," and thus justify equal rank for the topic "Wheat." There would then be nothing wrong with taking up "Barley" and "Oats" as subtopics under "Grains of Lesser Importance." Thus the rank of each topic could be made to reflect accurately the importance of the material that it covered.

In brief, when you are not forced to follow a preexisting pattern over which you have no control, you should create a pattern that gives each point the rank its actual importance justifies, but one that is free from such flaws as faulty subordination, faulty coordination, and overlapping.

ARRANGEMENT OF POINTS

Sometimes arrangement of material may be based on a recognized principle—chronological order, order of location, or order of increasing difficulty, for example, and sometimes it may be based on reasons peculiar to the particular occasion. In any event, the arrangement should not be left to chance, but should result from a conscious decision about what is best for the specific task.

Moreover, unless a piece of writing is to be short, no single system is likely to govern the arrangement of *all* the material in the outline. We may often arrange our main points according to one principle and our subdivisions according to another. Indeed the very nature of our subpoints may make it impossible to arrange them on the basis that governs the main points. When the main points, for example, are based on what is to be done at different times during the year and the subpoints are based on what is to be done at certain places or by certain persons, the chronological arrangement used for the main points could not possibly apply to the subpoints.

In addition to the nature of the material itself, there is another important point to consider. When you make an outline you are not just planning how to write about a subject; you are also planning how to communicate with your readers in a manner that accomplishes the result desired. The best arrangement, therefore, is one that is based not only on the logic of the subject but also on the reader's readiness to understand and accept your facts and ideas. Effective arrangement must grow out of the considerations covered in the section on reader adaptation in Chapter 2. A good example of such considerations is seen in the widely used pattern that places conclusions and recommendations of a report before the body. According to logic, the evidence should precede the conclusions, but many organizations reverse the order because the reader wants to know at once what is recommended.

CHECKING FOR PROBLEMS IN ORGANIZATION

However carefully the first draft of an outline is made, problems in organization are likely to appear. The first draft should therefore be checked over with the following points in mind:

1. There should never be a single main point. (Do not count *Introduction* or *Conclusion* as main points.) If there is, further examination will show (1) that this single main point covers the entire subject and hence is not a main *division* of the material, (2) that some of the material placed under this main point does not belong there, or (3) that the full subject has not been covered.
2. There should usually be no more than five or six main points. If an outline must be long, the increased length should result from the use of more subpoints and sub-subpoints rather than from the number of main points.
3. There should rarely be a single subpoint. That is, there should rarely be an A without a B or a 1 without a 2. The only justification of a single subpoint is the need to treat two or more closely related points in the same manner.
4. There should be no faulty subordination. That is, no subpoint should be placed under a main point where it does not belong. (It might be equal in rank or might belong under some other main point.)
5. There should be no faulty coordination. That is, no point should be shown as equal to another when it is logically subordinate.
6. There should be no overlapping among the main points or among subpoints of equal rank under the same main point. This flaw may result from faulty coordination or from other causes.

GENERAL SUGGESTIONS

Classification

The term *classification* here refers to the establishing of divisions. For example, an outline on erosion control in the United States might have four main points:

   I. Erosion control in the East
  II. Erosion control in the South
 III. Erosion control in the Midwest
 IV. Erosion control in the West

Each of these points might be subdivided thus:

A. Erosion control in the past
B. Erosion control in the present
C. Erosion control in the future

Some classification of this sort is often necessary and valuable, especially at the upper levels of the outline. Moreover the outline produced by classification is usually clear, orderly, and apparently logical. There is a real danger, however, that because classification is so easy, it will be carried too far. The resulting outline might contain some headings that are unnecessary because there is nothing much to say about them, and other headings that call for the same material in two or three places. Also, such an outline often fails to provide the headings that would be most useful in organizing the specific facts and ideas to be presented.

Another defect resulting from excessive classification is that it leads to subpoints that always bear the same relationship to their main points. In a sense this relationship does always exist; that is, the discussion under a subpoint is always part of the discussion of the main point. But the relationship can vary. For example, suppose a main point concerns the decline in the production of metals. Mere classification would probably result in such subpoints as ferrous and nonferrous metals, and sub-subpoints such as lead, silver, and zinc. But perhaps there would be no reason for such division into parts. Perhaps, in a discussion that applied to all the metals, we might wish to present first the evidence that a decline is imminent, second, its causes, and finally its probable consequences. Classification is unlikely to provide headings that cover these latter kinds of material.

One virtue of the procedure recommended earlier in this chapter is that by reducing the likelihood of excessive classification it also reduces the likelihood of our producing an outline that appears logical and orderly but later turns out to be unsuitable for our purposes.

Parallel Treatment of Similar Points

Though the desire for parallelism should not lead you to jam any portion of your paper into an unsuitable organization, it is sound practice to organize similar points in a similar manner so far as the subject matter permits. Such a practice enables your reader to become familiar with the pattern and read more efficiently.

When your outline is to be scrutinized by someone else, parallel phraseology as well as parallel organization becomes desirable. Main points should be phrased alike so far as possible, and subpoints of equal rank falling under the same superior point should also be phrased alike. This treatment is recommended not because arbitrary rules demand it, but because it makes an outline easier to grasp. Parallelism is a principle that applies not just in outlining but everywhere in writing.

Avoidance of Undesirable Extremes

It is often necessary in preparing an outline to balance between two extremes, either of which would be undesirable. It is undesirable, for example, to have too many ranks of points; but it is also undesirable to have a long string of subpoints of equal rank—perhaps eight or ten—under the same main point. Such a string might better be divided into two or three parts. Instead of naming in a single list ten advantages of a certain process, for example, you might do well to subdivide them into different *kinds* of advantages and name the specific advantages as points of a lower rank. Whether this would be an improvement, however, would depend on whether this change would create an outline with too many ranks.

Further, it will constantly be necessary to decide whether to make your points long enough so that each is self-sufficient, or whether to keep each point short so it may be grasped at a glance.

Logical Allotment of Space

The space devoted to various points in an outline should be roughly proportionate, if possible, to the amount of space they will occupy in the paper to be written. Thus a reader will not be misled about which points will comprise the bulk of the paper.

Clearness in the Outline

You should take special pains to make your outline a clear enough record of your plan so that a reader can understand it. This should be done even if you yourself are the only probable reader. Many writers, when not able to write from their outlines immediately, find it hard to remember their intentions later on and run into trouble unless the outlines are clear enough to refresh their memories.

Each point should be worded so that its meaning is impossible to misconstrue. It should be broad enough to cover all the material you intend to place under it, but not so broad as to cover more material than you intend to present.

Not only the individual points but also the relationship among different points should be clear—especially the relationship between a main point and its subpoints. Some outlines fall so far short of meeting this requirement that a main point and its subpoints do not even seem to cover the same material. If such a discrepancy exists in your outline, you need first to decide whether you actually *do* have in mind two different bodies of material, or whether your plan is sound but is not clearly recorded. In the former case, either the main point or the subpoints must be moved—or else discarded. In the latter, the phraseology must be changed so that the soundness of your plan becomes apparent. It is advisable to check this kind of defect by asking yourself two questions about each doubtful portion: (1) If someone first read the main point and then the subpoints, would it be clear that the main point could be discussed with the subpoints used? (2) If someone read the subpoints first and then the main point, would it be clear that the main point covered the subpoints without going beyond them?

Sentence or Topic Outlines

In many discussions of outlining, the distinction between the sentence and the topic outline is given considerable attention. When no one except the writer will see the outline, this distinction is obviously unimportant. Indeed, it is unlikely to be considered important in any circumstance. In many excellent outlines, sentences are used for some points and topics or other types of phraseology for others. Unless the points of an outline are to be used as topic headings—in which case they should all be topics—there is no reason to avoid variation in form if the principle of parallelism, already discussed, is complied with.

FINAL CHECK OF AN OUTLINE

When you approach the completion of an outline, you can check your work by considering the following questions:

I. SUBJECT MATTER

1. Does your outline have a desirable number of main points?
2. Do your main points cover the subject, and will they lead to the best main divisions of material in view of your purpose and your readers?
3. Have you provided for the inclusion of the specific facts and ideas that you want to present?
4. Is the outline free from points that are not really called for?

II. ORGANIZATION

1. Is your outline free from classification that serves no useful purpose?
2. So far as the subject matter permitted, have you used parallel organization where it would be expected?
3. Have you balanced properly between the desire to avoid long series of equal points and the desire to avoid points of too many ranks?
4. If you have any single subpoints, can you justify them?
5. Have you avoided faulty subordination?
6. Have you avoided faulty coordination?
7. Have you avoided overlapping points?

III. READERS

1. If your outline is to be examined by someone else, will it create an impression of clearness and continuity?
2. If your outline is to be examined by someone else, will it give a reasonably accurate impression about the relative amount of space the various points will occupy?

It should be apparent by now that these are not arbitrary rules dreamed up by English composition instructors, but that they grow out of the psychological needs of the reader. The successful writer has always followed these points even if unaware of the bases for them.

EXERCISES

Exercise 1

Following are fragments of outlines, each of which contains one or more weaknesses in organization. Point out in one or two sentences whatever weaknesses you find in each.

(1)

A. Areas in need of street repairs
    1. Needs in business district
    2. Needs in Eastgate addition
    3. Needs in Park addition
    4. Satisfactory condition of Selway addition

(2)

I. Mercer sanitary landfill site
  A. Cost
  B. Suitability
    1. Years of potential use
    2. Proximity to residences
    3. Visibility from highway
    4. Possibility of water pollution
  C. Accessibility
    1. Travel mileage on highway
    2. Travel mileage on secondary roads
II. Sand Hollow landfill site
  A. Cost
  B. Years of potential use
  C. Suitability
    1. Proximity to residences
    2. Possibility of water pollution
    3. Visibility from highway
  D. Accessibility
    1. Travel mileage on highway
    2. Travel mileage on secondary roads
    3. New road needed

(3)

A. Errors in hinge-pin retainer installation
1. Interchange of parts 4 and 5 (Parts catalog A-27 p. 17)
2. Installation of part 4 upside down
B. Installation of retainer with tabs offset from pins
C. Results of improper installation
1. Shearing off of pins
D. Remedial measures planned

(4)

A. Plumbing work needed
1. Repair work on showers
2. Replacement of damaged washbasins
B. Additional space needed
1. Need for more storage space
C. Structural work needed
1. Need for new roof
2. Need to construct extra emergency exit

(5)

A. Benefits of closing north drive through the campus
1. Immediate benefits
2. Long-range benefits
3. Improved safety for pedestrians
4. Reduction of noise

(6)

I. Fire protection for residences in undeveloped areas
A. Forest service facilities
1. Response time
2. Primary responsibilities
3. Capabilities and limitations
B. Rural fire protection districts
1. Capabilities and limitations
2. Response time
3. Primary responsibilities

(7)

A. In-flight cooling of X-119 brakes
1. Cooling speed with landing gear retracted
2. Cooling speed with landing gear extended
3. Cooling speed when plane is parked
B. Potential damage from overheated brakes

(8)

A. Problems caused by fall of volcanic ash
1. Control of air pollution
2. Need to obtain face masks
3. Need to obtain more ash-removal equipment
4. Equipment failures caused by ash intake

B. Finding disposal site for ash from streets
C. Cost of coping with ash fall
   1. Cost of equipment rental
   2. Cost of equipment repair
   3. Cost of overtime wages

(9)

B. Damage done by the fire
   1. Damage done by burning
   2. Damage done by water and smoke
   3. Damage to interior of building
   4. Damage to exterior of building

(10)

I. Work needed in motel rooms
  A. Redecoration of eight east-wing rooms
  B. Plumbing repairs in four west-wing rooms
II. Exterior work needed
  A. Repairs needed in the court
    1. Need of resurfacing entry and exit
  B. Repairs needed on roof of west wing
  C. Repairs needed on balcony of east wing

Exercise 2

Following is a list of considerations that might influence a decision about whether to change zoning so as to permit a new shopping center adjacent to a residential area. Make an outline based on this information. You may add other considerations if you wish or may subdivide items listed. The arrangement below is intentionally unsystematic.

1. Effect on tax revenue
2. Effect on traffic
3. The question of protection against fire
4. Compatibility with general plans for city development
5. Availability of adequate water supply
6. Type of landscaping for space not built upon
7. Possibility of excessive noise
8. Kind of architecture proposed
9. Effect on downtown business area
10. The question of providing police protection
11. New jobs that would be created
12. The question of sewage disposal
13. Possibility of attracting business that would otherwise go to other towns
14. Other possible uses for land in the area
15. Possibility of excessive dust and litter

Exercise 3

Make an outline for a paper about the feasibility of establishing a salmon hatchery, using as material the unorganized list of facts provided below.

There should not necessarily be a point in your outline for each item in the list. A single point in your outline may cover two or three items in the list; or there may be items in the list that overlap, just as your own notes will occasionally overlap as you jot them down before making an outline for one of your own papers. In short, the items are not points of an outline. They are merely facts that the paper would include. *You* are to make up for yourself the points that an outline would need to contain in order to accommodate the facts.

It is not necessary for your outline to cover every item on the list, but it should include most of them. It may also include any facts that you need to make up for yourself so that your particular treatment will not contain gaps. Remember that there is no single, correct treatment for the material. The purpose of this exercise is merely to give you a chance to make up an outline without having to spend time looking up information, and to provide a basis for discussion by permitting comparison of the treatments worked out by a number of people.

1. The Bureau of Sports Fisheries initiated an investigation of the feasibility of operating a Federal salmon hatchery on the Makah Indian Reservation.
2. The production of salmon in the Makah Reservation rivers is not capable of maintaining an adequate supply under present ocean fishery demands.
3. In recent years more than 50,000 salmon-angler days were expended annually at Neah Bay in the reservation.
4. The site of the hatchery must have a dependable supply of water.
5. The Sail River is located at the eastern edge of the reservation and runs into the Strait of Juan de Fuca.
6. Low flows of the Sail River, according to the U.S. Geological Survey's water study, would be less than two cubic feet per second.
7. Usable surface-water storage of 425 acre-feet would be necessary to maintain a flow of 3 cfs in Sail River.
8. The Sooes River flow would be 15 cfs one year out of two.
9. The Sooes River has a drainage area of approximately 38 square miles.
10. A preliminary feasibility report suggested two possible site locations: one on Sooes River and one on Sail River.
11. The Makah Reservation is located on the tip of the Olympic Peninsula in Western Washington.
12. The Makah Indians have always relied on fishery resources for their principal livelihood.
13. The hatchery site must be suitable for the construction of the physical facilities for a hatchery development.
14. Most of the Sooes drainage area lies outside the reservation.

15. Test drillings would be required to determine the quantities of ground water that could be developed in the Sooes basin.
16. The Makah Reservation provides an excellent harbor for sport and commercial fishermen.
17. Salmon-fishing fleets from the United States and Canada operate off-shore and in the Strait of Juan de Fuca.
18. The Sail River site has geographic advantages.
19. The Sail River drains an area of five and one-half square miles, 80 percent of which lies within the reservation.
20. A potential site for ground-water development lies along the Sooes River between the reservation boundary and the proposed hatchery site.
21. The hatchery site should be above the flood plain.
22. The Sooes River drains the western side of the reservation and enters the Pacific Ocean at Mukkaw Bay.
23. The Sail River flows through a narrow valley with a steep gradient.
24. One year out of twenty the Sooes flow would be 9 cfs.
25. Sizable quantities of ground water are unlikely in the Sail River basin.
26. The terrain and the foundation material should be such as to hold down the cost of site preparation.
27. The hatchery must have a dependable water avenue downstream so that brood stock can return.
28. The water supply must not contain toxic elements.
29. The Sail River flow will approach 0.2 cfs only one year in twenty.
30. The water must have optimum quality, volume, and temperature.
31. The stream should be free of obstacles to migration.

Exercise 4

The purpose of this exercise is to encourage better organization by causing you to organize the same material in more than one manner. Once you realize that more than one organization is possible, you will be less likely to use the first that pops into your mind and more likely to work out the one that makes it easiest for a reader to find the answers to the questions that are most important for the particular occasion that called for the report.

The following table tells in simplified form the effects of mixing three different substances with soil for the purpose of improving it as a subgrade for highways. So far as the given facts permit (some facts that would affect the results are not given), make an outline for each of the three following occasions:

1. An outline that draws together the facts about each substance.
2. An outline that emphasizes the results desired and shows the effect of using each substance in an attempt to produce each result.
3. An outline that shows which substances merit further testing and which should be dropped, and that provides reasons for the decisions.

Effects of Mixing Other Substances with Silty Clay Soil

EFFECT DESIRED	EFFECT CAUSED BY USING		
	Portland Cement	Calcium Chloride	Dehydrated Lime
Greater ease of compaction	less ease	greater ease	less ease
Greater strength of soil	more strength	no effect	more strength
More resistance to capillary rise	more resistance	lower resistance	more resistance
More resistance to volume change when moisture constant rises	more resistance	more resistance	more resistance

Exercise 5

Following is a list of general subjects to be used as a basis for outlines, either as they stand or as you may adapt them. Choose one or more of them and make an outline containing from 15 to 30 points. These outlines may be topic, sentence, or a combination. They need not be "thought-content" outlines unless the instructor so requests. Supply a title so that your subject will be more specific than those listed.

1. The advisability–advantages and disadvantages—of seeking summer employment in the field you plan to work in after graduation compared to seeking summer employment in the best-paying area you can find.
2. The steps necessary to perform some analysis in your field of specialization. (Business majors, for example, might analyze the feasibility of starting a pancake house near the campus. Civil Engineering majors might consider the flow of traffic through the campus.)
3. The criteria you will consider when you apply for employment after graduation or in seeking a new job.
4. The demands of the curriculum in which you are seeking a degree (specific subjects, kinds of subjects included, types of work called for, differences between courses taken in earlier and later years, and so on).
5. Desirability or undesirability of using artificial turf, or choice between different kinds of artificial turf, for a football or baseball field.

5 | Mechanics

Our discussion of mechanics here will be limited to brief instructions on manuscript form, technical style, hyphenation, decimal numbers for headings, equations, and documentation. Punctuation, capitalization, spelling, and other points that concern all writing are covered in the Handbook.

MANUSCRIPT FORM

Much of the time, technical writing does not demand any special manuscript form. In such cases the following general instructions should prove sufficient.

1. Ordinary manuscript should be typed on one side of $8\frac{1}{2}$-by-11-inch white paper of good quality, i.e. nonerasable, 25 percent cotton or rag, 16 or 20 pound weight. Double spacing is usually preferable to single, though single spacing with double spacing between paragraphs is standard in letters and often desirable in reports.

2. The margin at the top should be 2 to $2\frac{1}{2}$ inches on the first page and 1 inch on other pages. Other margins should be: left, 1 inch; right, $\frac{3}{4}$ of an inch to 1 inch; bottom, 1 inch. If a manuscript is to be bound at the side, the left margin should be increased by $\frac{1}{2}$ or $\frac{3}{4}$ of an inch. If it is to be bound at the top, the top margin should be increased to 2 inches.

3. The beginning of each paragraph should be indented five spaces.

4. Unless a separate title page is used, the title of a paper should be placed on the first page, centered on the first line. It may be entirely in capital letters, or, if the author's name does not follow, may be underlined and written in upper and lower case (capital letters to begin each important word, small letters elsewhere).

5. If the name of an author accompanies the title, it should be centered a double space below the title and should be upper and lower case. There should be three or four blank spaces between it and the text.

6. The page numbers may be placed in the upper right corner of each page except the first, from which it is omitted. Alternately, page numbers may be centered at the bottom of all pages, especially if a paper is bound (not clipped) at the top.

7. Long quotations (75 words or more) should be single-spaced except for double spacing between paragraphs. Margins adjoining such quotations should be in-

creased by $1/2$ to $3/4$ of an inch on each side. In manuscripts sent to printers, however, all material is double-spaced.

8. Ordinarily, manuscript should be fastened together by means of paper clips or not at all, and should not be folded. Occasionally, however, especially when it is placed in final, permanent form, it should be semipermanently fastened in a cover that opens at the side, or provided with backing paper and stapled together at the top.

9. Manuscripts to be submitted for printing should never be fastened permanently. In such a manuscript, illustrations should not be attached to the copy. Rather, an identifying number should be written on the figure's back and a note should be inserted in the manuscript to indicate where the figure belongs.

TECHNICAL STYLE

Technical style, in the present context, concerns the form in which numbers are written and the use of abbreviations. In this discussion, suggestions will be offered to help you determine when to use technical style and how to use it correctly.

Use of Figures or Words for Numbers

Numbers in ordinary style. Before the use of numbers in technical style can be explained, it is first necessary to recall the conventions that govern their use in ordinary writing. *The basic rule in ordinary writing is: If a number can be expressed in no more than two words, it should be written out. Otherwise, it should be expressed in figures.* (Examples: *four, seventeen, twenty-seven, one hundred, one thousand;* but *114, 1198, 14,456.*) There are many modifications to this rule, the most important of which follow:

1. Figures are never used at the beginning of a sentence. Such numbers must be written out, or else the sentence must be changed so it does not open with a number.

2. All numbers in a series are written in the same form—preferably in figures if any number in the series is long enough to call for use of figures.

3. There are many special uses in which figures may or should be used regardless of the size of the number. These include degrees of latitude and longitude or of temperature, prices, scores, time of day, dates, and tabular statistics.

4. For extremely large numbers a mixed form is widely used which is extremely easy to write correctly and to read accurately. (Examples: *50 billion, 125 million, 6.4 billion.*)

Numbers in technical style. In technical style, numbers are more likely to be written as figures. The main differences between ordinary and technical style are as follows:

1. In technical style, 10 and all numbers above are expressed as figures. Any number below 10 is written out, except as mentioned below.
2. In technical style, a number that precedes a unit of measurement is written as a figure even if it is below 10. (Examples: *6 inches, 4 hours, 8 cubic yards;* but *six hoes, three stories, eight gusset plates, four arches.*) Writing out a number as a word is especially undesirable when it precedes a unit of measurement that is abbreviated.
3. In a passage where numbers are especially frequent, all numbers may be expressed as figures. (Example: *He used a crew of 3 carpenters, 1 plumber, 6 laborers, 1 foreman, and 1 timekeeper.*) This is particularly desirable when statistical information is being presented.
4. When one number appears immediately after another as part of the same phrase, one of the numbers is spelled out. (Examples: *7 six-inch timbers, two 7-man crews.*) It is preferable that the shortest number be the one spelled out; but when two such terms are close together, the same form should be used for both.
5. Numbers that are merely approximations are often written out, regardless of their size, unless the result would be cumbersome. (Examples: *The company has enough timber to operate for twenty years. The building should stand for fifty years.*)
6. Sums of money are expressed in figures. (Examples: *5 dollars or $5, $7.95, $0.80,* or perhaps *80 cents.*)
7. Technical style tends to use decimals rather than ordinary fractions because they make it possible to indicate a greater degree of precision. In a decimal fraction with a value less than one, a zero is placed before the decimal point (0.719). Also, it may at times be desirable to add a zero after the decimal point to indicate the degree of precision (0.6840).
8. Decimal fractions should not be used to express information for which ordinary fractions are customary, especially when accuracy would thus be misrepresented. For example, *2.812 inches* should not replace *2 13/16 inches,* since this measurement is not accurate to the thousandth of an inch.

The preceding rules do not answer every possible question; if they did, they would be too long and complicated to be useful. Moreover, there are times when two rules may conflict with each other. Hence there are problems you will have to solve by using your own judgment, taking into account the abilities and expectations of your readers. Most of the time, however, these rules will make it possible for you to choose between figures and words with confidence that the form you decide upon will be acceptable.

Use of Abbreviations in Technical Style

As any reader may observe, abbreviations are used more frequently in technical writing than in writing in general—so frequently, in fact, that their use is clearly the result of a special set of conventions. If you are to decide intelligently when to follow these special conventions, you will need to know the underlying reason for their existence.

Abbreviations are justified not because they make the writer's work easier—though they may have that effect—but because they assist the reader. In technical writing, certain terms related to the field of discussion may be used over and over again. To write them out in full would take more time and space; but more important, it would make what is written harder to read. This is especially true when the terms are phrases rather than single words. It might be necessary, for example, to write *parts per million, feet per second, revolutions per minute,* or *board feet* eight or ten times in a single brief paragraph. Written out in full, these terms might occupy so much space that the remainder of the material would be overwhelmed; yet each of them expresses only a single concept and would actually be clearer and more compact if expressed by a single symbol. Hence, abbreviations that shorten such phrases to a few letters have come to be recognized and may be used when they serve a useful purpose.

If terms for which there are abbreviations do not appear often enough to be a problem, however, nothing is gained by abbreviating them. In fact, unless they are numerous, abbreviations are distracting and annoying; they look peculiar, and they hinder rather than assist communication. Therefore it is advisable to refrain from using the technical abbreviations unless the terms appear frequently. When you decide, after considering the points mentioned, to abbreviate in accordance with technical style, you will have sound authority for what you do if you comply with the following rules:

1. To establish for your paper an abbreviation of a term not necessarily familiar to your readers, first write the words in full and then place the abbreviation in parentheses after the words; for example: "Veterinarians belong to the American Veterinary Medical Association (AVMA)." In subsequent references, use just the abbreviation. If you do not plan to use the term often, however, do not abbreviate the term and do not include the abbreviation.

2. A term denoting a unit of measurement is abbreviated when it follows a figure. Examples are *inch, yard, pound, ounce, gallon, cubic yard, revolution per second, watt, board foot,* and *horsepower*. Unless it follows a figure, however, none of these terms is abbreviated. One would write *63 ft, 2300 rpm, 435 ppm, 125 hp, 50 cc;* but, "It is expressed in *horsepower*" or "The measurement is converted into *cubic centimeters.*"

3. An abbreviation *for a unit of measurement* is always shown as singular. You should use *lb,* not *lbs; bbl,* not *bbls; gal,* not *gals*.

4. A few extremely short terms denoting units of measurement are not abbreviated. Among these are day, mile, and acre. Since usage is not consistent, no exhaustive list can be given. Systematic personal observation is the only way to be sure about the customs in your own profession.

5. In many professions there are terms in addition to units of measurement that are used with extreme frequency and consequently are abbreviated when it is reasonable to believe that the readers will grasp the meaning instantly. Some examples are *a-c* for *alternating-current* used as an adjective, *F* and *C* for *Fahrenheit* and *centigrade, cp* for *chemically pure, el* for *elevation, emf* for *electromotive force*.

6. Even when you are using the technical style of abbreviation, there are many terms that you may abbreviate or not according to the dictates of your own judgment, provided you handle each term consistently.

7. The fact that technical style permits the use of abbreviations does not mean that it is desirable to use arbitrary signs for words. You should write *8 in.*, not *8"; 12 by 15 ft*, not *12' × 15'; percent*, not *%*.

There are a few exceptions to this rule. It is correct to use the dollar sign, and in appropriate context the use of *¢* would be permissible. Degrees are another exception. Three forms are widely used to indicate degrees of temperature: *84°C, 84 C*, and *84 deg C*. (The same forms could be used for degrees Fahrenheit.) Degrees, minutes, and seconds of angles or degrees of latitude and longitude may be expressed by signs. It would be correct, for example, to write *21° 55' 15"*.

8. Capitalization need not be affected by abbreviation. Abbreviations are not ordinarily capitalized unless the terms they stand for are capitalized.

9. In some professions, notably engineering, it is customary to omit the period after many abbreviations. In deciding whether or not to use periods, you should follow the practice of books and magazines in your own field. Even in engineering, however, the period should be used if omitting it could cause confusion, for example in the abbreviation for *inch*. Also, the period should always be used after abbreviations that do not result from technical style but would be used in writing in general—for examples: *a.m., p.m., c.o.d., B.C., Fig.*, and abbreviations used in footnotes and bibliographies, such as *ibid., op. cit., Vol., p.*, and *ff.*

10. One final rule about abbreviation will settle many questions: When in doubt write the word out.

HYPHENATION
OF COMPOUND TERMS

The question of whether some term is or is not a compound, and if so whether it should be hyphenated, arises with unusual frequency in technical writing. Consequently, though the use of hyphens to divide words at the end of lines is covered in the section on punctuation in the Handbook, hyphenation of compounds calls for treatment in this chapter.

Actually, usage is so far from uniform that it is impossible to draw up a definitive set of rules; and even if it were possible, exceptions would be so numerous as to make it advisable to look up some terms in an up-to-date dictionary. When you consult a dictionary, however, a certain amount of caution is necessary. First, you should make sure that the dictionary is not out of date, for some terms that are hyphenated in one edition have come to be treated as solids by the time another edition is published. Second, be sure that you do not mistake a symbol used to indicate the division of a word into syllables for a hyphen. Third, note well what part of speech is shown after the term in the dictionary entry. A term, for example, that is hyphenated when used as a noun might be treated as two words when used as a verb.

Helpful as a dictionary may be, it will not solve all your problems, for as a technical writer you will often create compounds that it does not contain. The discussion that follows will be helpful on such occasions. It is not a set of absolute rules but rather an explanation of tendencies, based on study of authorities. If you apply its principles, you may feel confident that the form you use will be acceptable, even when some other form would not necessarily be an error.

1. In writing a term formed by placing a prefix in front of a word, you will almost always be safe if you use the solid form. Such a term may be hyphenated, however, to permit internal capitalization *(pre-Cambrian)* or to clarify pronunciation or meaning *(reform, re-form; re-employ; re-anneal; intra-atomic)*. *Self* is an exception to the general rule; when used as a prefix it is always set off by a hyphen *(self-sufficient)*.

2. A compound adjective is usually hyphenated, especially when it precedes and directly modifies a noun *(time-consuming method, all-inclusive statement; 60-horsepower motor)*. It is somewhat less likely to be hyphenated if used as a predicate adjective *(the man was hard-hearted; the rules were all-inclusive;* but *the job was half completed; the method was up to date)*. A chemical term, for example *carbon monoxide* or *calcium chloride*, is not ordinarily hyphenated even when used as a compound adjective. Also, one should not hyphenate terms when one of the words is an adverb ending in *-ly*, such as *easily answered*, even when they appear in phrases such as *easily answered question*.

3. Compound nouns are more likely to be written as separate words or as solids than to be hyphenated. A few special types that are hyphenated are exemplified by the following terms: *cave-in, motor-generator, weigher-in, I-beam, foot-pound*.

4. Compound verbs vary, but the type shown in the following sentences is hyphenated: *They double-tracked the railroad. They dry-cleaned the canvas*.

5. Usage in hyphenating compound adverbs varies widely, but a hyphen is always used in a compound adverb formed by adding *ly* to a hyphenated compound adjective *(half-heartedly, quick-wittedly)*.

6. Try to avoid writing sentences in which you would be forced, if you followed the customs of hyphenation, to create a form that is unclear or illogical. Many problems are best settled not by consulting rules for hyphenation but by revising the sentence. (Undesirable: *a mercuric chloride-activated compound; a 7 by 16-foot area; the North Dakota-South Dakota boundary*. Improved: *a compound activated by mercuric chloride; an area of 7 by 16 feet; the boundary between North Dakota and South Dakota*.)

7. Long, unwieldy compounds are poor style. Instead of trying to decide how to hyphenate them it is better to refrain from using them. (Undesirable: *internal-combustion-gasoline-engine cylinders; a piece of 1½-inch-inside-diameter pipe*. Better: *cylinders for internal-combustion gasoline engines; a piece of pipe with an inside diameter of 1½ inches*.)

8. It is important to remember that the same term may be hyphenated in one usage and not in another. *(The blast-furnace crew worked overtime,* but *A new blast furnace is being designed. The blast-off occurred as planned,* but *It will blast off at 1:15 p.m. We plan to hard-surface the highway,* but *The highway has a hard surface*.)

THE DECIMAL SYSTEM
OF NUMBERING HEADINGS

You have probably seen the decimal system of numbering headings in instruction manuals, specifications, and perhaps in some of your textbooks. It is useful when frequent reference to some specific section of the material will be necessary. In this system, a number is placed before each heading in the text, and also before each heading in the table of contents if one is used.

The following examples show how this system, by use of Arabic numerals and decimal points, indicates the rank of each point and identifies the superior point under which the lesser points appear.

1 main point
1.1 subpoint under first main point
1.2 second subpoint under first main point
1.2.1 subpoint under 1.2
1.2.2 second subpoint under 1.2
2 second main point

Unlike the numbers used in an ordinary outline, each number in this system is complete and self-sufficient. That is, the number preceding a point includes the numbers of points of higher rank under which it is taken up. Because of this fact, and because its numbers are compact and easy to include in text, the decimal system does its job extremely well. When there is no necessity of frequently referring to some specific section of the text, however, it is undesirable because it gives the text a cluttered appearance and serves no added purpose.

EQUATIONS

When equations appear frequently in technical writing they are set apart from ordinary text, and sometimes they are numbered for easy reference. Following is a typical example.

. . . The formula for heat flow under these conditions is

$$(7) \qquad Q = \frac{k(T_1 - T_2)At}{d}$$

where

Q = heat flow,
k = coefficient of thermal conductivity for the refractory material,
$T_1 - T_2$ = temperature drop from hot face to cold face,
A = area of the wall,
t = time,
d = thickness of the wall.

Note that the alignment is based on the "equal" sign. Note also that punctuation is just as it would be if the material were placed in regular lines of text rather than set apart.

DOCUMENTATION

As a technical writer you will often use information obtained from printed sources. Consequently, you will need to acknowledge your indebtedness, partly as a matter of courtesy and honesty, and partly because you serve your readers' best interests by doing so. This acknowledgment of sources is referred to as documentation.

By acknowledging sources you may perform various functions: You may show the source of quotations or the authority for statements that might arouse skepticism; you may indicate the extent of your study, and thus build confidence in your remarks; or you may show the reader where additional information can be obtained.

Beginning writers often have questions about what information to document. Ignorance about this matter can lead the unwary into problems of copyright infringement or charges of plagiarism, so every writer needs to understand the requirements. The basic point to remember is:

If you use another person's words or ideas, give that person credit as the source.

Even some textbooks are confusing on this point, implying that if you change the words, you are not guilty of plagiarism. The fact is, however, that if you use a writer's words, you are using both words and ideas; if you paraphrase, i.e., change the words, you are still using the writer's ideas. Consider the problem for yourself: Which do you value more, your specific words or your ideas?

The foregoing does not mean you should never paraphrase. The contrary is true. Rarely do one person's words fit easily into the context of another paper, for the readers and purpose are almost certain to be different. Therefore, the writer of the second paper changes the words to suit the new situation, but of course gives the first person credit for the ideas.

Once you understand the general principle involved, there are several situations where the rule is relaxed. Information generally known and accepted in your field is not documented. For instance, you do not cite a source for E = mc^2 or "The coefficient of expansion of steel is 0.000010." Sources are not cited for information that your readers are unlikely to question or can easily verify, such as "The stock market crashed in 1929." If you are writing as part of your regular job and wish to use material produced for your company by another employee as part of that person's job, you do not need to credit the source, because under current copyright law, the material is part of a "work made for hire" and is owned by the company for whatever use it wishes. And finally, short quotations used for specific wording, such as the phrase "work made for hire" above, need not be documented.

The following discussion should assist you in selecting an appropriate system of documentation; it will cover the general conventions and will illustrate forms that ordinarily are safe to follow. Many fields, however, have their own special conventions; and on any particular occasion you should not hesitate to use whatever system the readers whom you address are probably accustomed to. And if you think your work might be published, you should use the system of documentation you find in the publication where you think it most likely to appear.

Four methods of documentation are common in technical writing: (1) footnotes only; (2) an alphabetical list of references (with or without citations in the text); (3) a numbered list of references, and citations in the text referring to these numbers; and (4) footnotes plus a bibliography.

In choosing among these methods you should take into consideration both the functions you need to perform and the customs that prevail in your particular field.

Footnotes

If you do not expect to cite sources frequently and if your total indebtedness to references is not great, footnotes alone will often be the most satisfactory system of documentation.

A footnote may be used for various purposes. It may merely explain something in the text that will be clear to most readers but not to all; it may provide additional comments or facts that would destroy continuity if placed in the text. These uses, however, are unrelated to documentation and may occur even in undocumented work. Footnotes used for documentation usually perform one or more of the following functions: (1) identify the source of a quotation; (2) acknowledge indebtedness for facts or ideas; (3) cite authority for some statement that might arouse skepticism; and (4) indicate sources for additional information.

Footnotes used for documentation should be numbered. Sometimes all the notes applying to each page are placed at the bottom of the page itself. In this event, each page may begin with footnote 1, or there may be a single sequence of numbers for all the footnotes in a paper. Sometimes all the footnotes are placed at the end of the paper, sometimes at the end of each section, and are called endnotes. In this event, a single sequence is used for numbering all the notes at any one location. Placing the footnotes on the pages to which they apply makes it easier for the reader to consult them. Placing them all in one location eases the work of the writer, and of the printer if the manuscript is to be printed. Also, it permits the reader to see in one place all the references that have been cited.

The presence of each footnote is indicated by inserting its number in the text at the appropriate place. If the footnote identifies the source of a quotation, the number is placed at the end of the quotation. If the note applies to material that is not directly quoted, the number is placed at the end of the material that the note concerns. (The text must of course be phrased so that it shows where the application of the footnote begins.) Some forms permit a writer to place this number in parentheses on line with the regular text, but more frequently the number is raised half a space above the line, thus: ". . . Using this system, it was possible to produce about 3,000,000 lumens of light."[2] As can be seen, the number identifying the footnote follows all other punctuation and is not preceded by a blank space.

Regardless of their location, footnotes are ordinarily separated from the text by a solid line beginning at the left margin and extending 1½ inches (15 or 18 spaces) toward the center of the page. This line is a double space below the last line of text and a double space above the first footnote. A double space is left between two footnotes, but single spacing is used within the footnote itself except in manuscripts to be set in type.

The form of footnotes varies in minor respects, but most forms are similar in general arrangement and content. A footnote usually answers the following four questions, so far as possible, in the order indicated: (1) Who said it? (2) What was the title of the piece of writing in which it was said? (3) Where or by whom was this writing made available? (4) On what page or pages may the material cited be found? The following examples show widely used forms for footnotes citing various types of material. They are numbered as if they appeared together, but it is unlikely that so many footnotes would be found on any single page.

Forms for Footnotes

For a book with one author:

[1]Niels Jensen, *Optical and Photographic Reconnaissance Systems* (New York: John Wiley & Sons, Inc., 1968), p. 144.

For a book with two or more authors:

[2]Harold S. Graves and Lyne S. S. Hoffman, *Report Writing*, 4th ed. (Englewood Cliffs, N.J.: Prentice-Hall, Inc., 1965), p. 317.

For a book with no author given:

[3]*Preparation of Technical Publications* (Indianapolis: U.S. Naval Avionics Facility, 1966), p. 34.

For an edited book:

[4]William K. Tobold and Larry Johnson, eds., *Automotive Encyclopedia*, (South Holland, Ill.: The Goodheart-Willcox Co., Inc., 1977), p. 345.

For an article in a book:

[5]Donald W. Head, "Chemical Cycles in the Sea," in *Oceanography*, ed. Richard C. Vetter (New York: Basic Books, Inc., 1973), pp. 344–46.

For a magazine article (signed):

> [6]Robert M. Pirsig, "How To Avoid Gumption Traps," *Handyman*, January, 1980, p. 42.

For a magazine article (unsigned):

> [7]"A Bacterial Wash," *Time*, December 14, 1981, p. 100.

For an article in an encyclopedia:

> [8]"Telautograph," *The Encyclopedia Americana* (1969), 26, p. 327.

For an article in a newspaper (unsigned):

> [9]*The Washington Post*, July 29, 1983, p. 8.

For a numbered report, technical paper, or any similar citation:

> [10]V. P. Puzinauskas, *Gussasphalt or Pourable Asphaltic Mixtures*, Research Report 70-2, February, 1970, The Asphalt Institute, College Park, Maryland, p. 5.

For a reference to a secondary source:

> [11]John Diebold, "The World and Doomsday Fads," *New York Times*, February 15, 1973, as cited in M. J. Osborne, "Water Lilies and the Lombard Steersman," *Tappi*, May, 1973, p. 84.

For a thesis or dissertation:

> [12]Kenneth V. Erie, *John Ruskin as a Prophet of the Conservationist Movement* (Master's thesis, Channing State University, 1972).

For an address:

> [13]John R. Stokes, *Problems of Protecting Lakeshore Quality*, unpublished address at a meeting of the Northwestern Conservation League, Yakima, Washington, May 6, 1983.

For a personal interview:

> [14]Harvey Forester, Biologist, U.S. Forest Service, in a personal interview in June, 1982.

As mentioned earlier, the form of footnotes varies in detail. Different systems of punctuation are used. Some forms indicate the volume and issue of magazines, and some omit this information. Indentation is sometimes hanging (first line extending to the margin and the other lines indented) rather than in paragraph form. And some highly specialized scientific books and periodicals use abbreviation far more extensively than is customary elsewhere.

In many respects, however, usage is well established and should be complied with. The title of a book, magazine, or other independently printed reference should be underlined (the equivalent of italics in print), but quotation marks should be used for articles in a magazine, separate units of a collective work, and other writings that are not separately printed. Since footnotes are not alphabetized, the name of an author is arranged in normal order and should be shown exactly as it appears in the original. A publisher's name should be used exactly as it appears on the title page of a book. If it is known that a publisher's imprint has changed since publication of the book, it is proper to give the new imprint provided the form used is precisely as sanctioned by the publisher. A title, the first time it is mentioned, should appear exactly as in the original, although it may be shortened in later citations. That is, a footnote referring to a book may be reduced to the name of the author, the title, shortened if necessary, and the page number; and the name of a periodical or organization, if long, may be sharply abbreviated.

Since a footnote need not repeat information already given in the text, if an author's full name appears in the text, it may be omitted from the footnote. If only an author's last name appears in the text, however, the footnote should provide the full name. The question as to whether the name of an author or any other information belongs in the footnote or in the text can be settled by applying the following principle: Whatever is actually part of the message belongs in the text; what is provided merely as documentation should be in the footnote.

Certain abbreviations were once widely used in footnotes, the most common being *ibid.* and *op. cit. Ibid.* is the abbreviation for *ibidem,* Latin for "the same." *Op. cit.* is the abbreviation for *opere citato,* Latin for "work cited." (Italics are used because the terms are foreign; in a typed paper they would be underlined.) *Ibid.* without a page number indicates that a reference is exactly the same as the one immediately preceding it. *Ibid.* followed by a page number indicates that the two references are the same except for the page. *Op. cit.,* which is preceded by the name of the author unless named in the text, is used when a reference has already been named but other references have intervened. Obviously, *op. cit.* cannot be used if more than one work by the same author has been named. In order to spare the reader the necessity of turning back to preceding pages, it is best to use *ibid.* and *op. cit.* only to refer to a footnote on the same page. *Loc. cit.,* Latin for *loco citato,* place cited, is sometimes used in much the same manner as *ibid.* but is less definite. *Passim,* Latin for here and there, is used to indicate that ideas have been gathered from various places in a reference, perhaps within a stated succession of pages. *Cf.,* Latin for *confer,* compare, is used just as *compare with* might be used.

It seems desirable, however, in the interests of effective communication, to hold the use of a foreign language to a minimum; more and more technical writers are coming to the conviction that *ibid., op. cit.,* and *cf.* are the only Latin abbreviations that are really necessary, and some are even omitting the underline or italics that identify *ibid.* and *op. cit.* as foreign. Many journals, rather than use even these few terms, simply repeat the author's name in the second footnote and the page number of the new reference.

Alphabetical List of References

Often, in technical writing, documentation is limited to a list of materials headed *References.* Such a list may include only the sources from which you drew specific information (which is then sometimes labeled *Literature Cited*), may be broadened to cover material that contributed to your general background, or may even include references that would merely be helpful to a reader who wants additional information.

Obviously, a list of references not cited in the text conveys only a limited amount of information. It does not show your authority for specific facts but only indicates whether you have examined a satisfactory number of authoritative, up-to-date references before writing. Still, the evidence that you have consulted good authorities makes your statements more convincing, and the use of this system does not prevent you from mentioning, in the text, the sources of specific facts when you think such information is important. This may be done by merely mentioning the authority in a regular sentence, as for example, "This theory, as Arthur Miller points out. . . . "

The form used for the individual entries in such a list of references closely resembles the form used for footnotes. The main difference is that the last name of the author is placed first to facilitate alphabetical arrangement. When more than one title by the same author is listed, a solid line one-half inch long is often substituted for the author's name in the second and later entries. Unsigned references are alphabetized by title. Titles are indicated as in footnotes: quotation marks for a title such as that of an article in a magazine, and an underline, the equivalent of italics, for the title of an independent publication. Two or more references by the same author are alphabetized by title.

Details may vary, but the following alphabetical list illustrates a form that is widely used and illustrates the principles that apply to most types of entry.

References

BROWN, JAMES, *Casebook for Technical Writers*. Belmont, Calif.: Wadsworth Publishing Co., Inc., 1961.

———, *Cases in Business Communications*. Belmont, Calif.: Wadsworth Publishing Co., Inc., 1962.

ECKENFELDER, W. WESLEY, JR., *Industrial Water Pollution Control*. New York: McGraw-Hill Book Company, 1966.

Effects of Studded Tires, A Research Progress Report. Minnesota Department of Highways, December, 1970.

FASAL, JOHN, "Forced Decisions for Value," *Product Engineering*, April 12, 1965.

GIBBY, J. C., *Technical Illustration*. 3rd ed., Chicago: American Technical Society, 1970.

JONES, E. L., and PHILIP DURHAM, eds., *Readings in Science and Engineering*, New York: Holt, Rinehart and Winston, Inc., 1961.

MANDELL, D. A., G. W. HINMAN, and others, *The Gas-Cooled Breeder Reactor with Gas Turbines and Dry-Cooling Towers*. Washington State University College of Engineering Bulletin 332, Pullman, Wash., 1973.

"Micromanipulation," *Encyclopaedia Britannica* (1970), 15, 376.

MONTARI, F. W., "Legislative and Administrative Measures Needed," *Proceedings of the Fifth Annual Conference of State and Federal Water Officials*, Washington, D.C.: Water Resources Council, 1971.

"Productivity: Key to Our Economic Health," *Changing Times*, February, 1974.

A variation of this method, widely used in some fields, is to insert the name of the author and the date of the reference in parentheses, as for example, "A recent study of the effect (Miller, 1983) points out. . . ." When two works by the same author were printed in the same year, they are differentiated by adding a small letter after the date, such as (Gibby, 1980a) and (Gibby, 1980b). Sometimes the page number is included within the parentheses as well: (Miller, 1983:235). When this method is used, the date follows the author's name in the list of references and in many fields greater use of abbreviation is made.

Literature Cited

Brown, James. 1961. *Casebook for Technical Writers*. Belmont, CA: Wadsworth.
Eckenfelder, W. Wesley, Jr. 1966. *Industrial Water Pollution Control*. N.Y.: McGraw-Hill.

Numbered List of References

To make documentation more explicit without the use of footnotes a numbered list of references is often used in place of an alphabetical list. Thus a reference can be cited by merely inserting its number into the text. If this method is used, the list of references is limited to cited materials and the entries are arranged in the order in which they are first cited. The form of individual entries is identical with the form used in an alphabetical list except that authors' names are arranged in the normal order. The following brief list should be a sufficient illustration of form.

References

1. J. C. Gibby, *Technical Illustration*, 3rd ed., Chicago: American Technical Society, 1970.
2. John Fasal, "Forced Decisions for Value," *Product Engineering*, April 12, 1965.
3. James Brown, *Casebook for Technical Writers*. Belmont, Calif.: Wadsworth Publishing Co., Inc., 1961.
4. E. L. Jones and Philip Durham, eds., *Readings in Science and Engineering*. New York: Holt, Rinehart and Winston, Inc., 1961.

To insert the number of a reference into the text of a paper, either of the following forms may be used:

"The temperature loss rarely exceeded 2.9 deg F for 100 lb of ore (7). This information. . . ."

"The tests performed by L. V. Smith /2/ showed the form of curves to be those reported for austenite steel."

A numbered list of references should not be confused with the use of notes at the end of a paper. When endnotes are used, each has a separate number even though it may refer to a reference that has been cited previously.

When a numbered list of references is used, however, each reference is listed only once and the same number is used each time the reference is cited. Also, as can be seen from the examples, the page number of the reference from which material was drawn is not indicated when a numbered list of references is cited, for a single listing may be cited more than once and the page number might be different on each occasion. Some publications overcome this difficulty, however, by including the page number within the parentheses in the text: (7. p 73) or (7:73).

Like the alphabetical list of references, the list that is numbered and cited by number is widely used—perhaps more widely in technical writing than any other form.

Footnotes Plus Bibliography

The most detailed system of documentation is that in which footnotes are supplemented by a bibliography. This system is used more frequently in scholarly work than in technical writing. It is suitable when references to sources are numerous and when a writer wishes not only to give the exact page number of every citation but also to show in one place all the references that were consulted, regardless of whether all were cited specifically.

When footnotes are supplemented by a bibliography, their form is the same as that already illustrated, except that the publisher and place of publication of a book are not included. The form used for the bibliography is the same form that is used when there are no footnotes.

Final Comment on Documentation

It is impossible to discuss the subject of documentation without presenting a great many rules. Omission of any of these rules would have meant failure to answer questions that constantly arise; and the writer who does not know or does not look up the answers to such questions is likely to use forms that impress the reader as strange and awkward. Also, rightly or wrongly, a reader may form the impression that writers who have not learned how sources are usually cited may be inexperienced. And of course the conventional forms are preferable also because a reader will grasp the content faster if the form is familiar.

None of this should make us lose sight, however, of the fact that the purpose of documentation is to give readers information that they want or need. Anyone who does much documentation will occasionally need to present information that does not fit into the conventional forms. When this is your predicament, there is no reason to be disconcerted. Following the conventional forms only so far as they fit the occasion, merely tell the reader the facts in the simplest, most concise manner that will be clear, arranging them, so far as possible, in the order shown near the beginning of the discussion on documentation. One of the comforting facts about knowing the conventional forms is that you can feel free to improvise when necessary, undisturbed by the fear that you are betraying ignorance.

Finally, you should be aware that different fields and different publications within a single field sometimes use slightly different forms. Our best advice is: Check what is done in your field. Once you have studied the forms we have illustrated, you will be able to detect what elements vary in your specialty, and then you may write with confidence.

EXERCISES

Exercise 1

Change the following sentences so that numbers and abbreviations are written in accordance with technical style. Do not make unnecessary corrections.

This exercise does not involve hyphenation, although hyphenation sometimes occurs, so you should not introduce errors in hyphenation in your corrections.

1. The volume will be 6.5 cu ft per sq ft of surface.
2. A weight of 15-20 lbs will be acceptable.
3. The size of the base plate, 11" × 14", is too large.
4. Power is supplied by 4 100-hp diesel engines.
5. A contamination level of more than seven ppm is considered dangerous.
6. Use of the new machines resulted in a $1/4$ increase in production.
7. 4 inches of rain fell in two days, causing approximately 148 homes to be flooded.
8. The leakage was 0.75 gallons per minute, which was unacceptably high.
9. Proper adjustment increased the performance to seventeen mpg.
10. The clearance, which formerly had been under $1/2$ in., was increased to 1.25 in.
11. When the temperature dropped to minus two degs. F, work was discontinued.
12. The speed of the conveyor belt, 20' per min., was too high to permit accurate work.
13. The theory was originated in 125 B C and was accepted for several centuries.
14. The estimate of possible rpms proved to be too high.
15. We will need to requisition 15 shovels, 15 rakes, four wheelbarrows, and 12 axes.
16. An increase of 22% in production followed the change.
17. A thickness of 2 in will be adequate.
18. Showing the elevation in both ft and meters is a recent change.

19. The plant shown in the fig. was raised in soil containing 150 to 160 PPM of nitrate nitrogen.
20. It will be necessary to hire a crew of about 30 men.

Exercise 2

A. The following sentences contain expressions that a writer might feel should be hyphenated or written as single words. For the purpose of this exercise they are all written as separate words. Rewrite in correct form each expression that needs to be changed. Although the treatment of compounds is far from uniform, it is suggested that in this exercise you follow the principles recommended in this chapter.

1. The machine washing process is ineffective.
2. The hand woven tapestry covered the entire wall.
3. Enlargement of the sewage treatment plant is essential.
4. The instructions say that we must hand wash the material.
5. The self cleaning oven will be well worth the extra cost.
6. The labor saving potential was exaggerated.
7. The 40 pound sacks were easy to handle.
8. They tried to cover up the shortage but the cover up was easily detected.
9. The semi conscious patient was taken to the hospital.
10. It will be necessary to buy dry cleaning machinery.
11. It is time to re assess the merit of the claim.
12. It is a mistake to use hit or miss methods.
13. It is protected by a double layer of cotton or polyester fabric.
14. The company shut down the plant in May but the shut down was only temporary.
15. Their non conforming attitude is causing considerable trouble.
16. The illegal U turn almost caused an accident.
17. The death was caused by a quickly absorbed poison.
18. To test the machinery they used no load runs. (runs without loads)
19. The haulage is estimated in ton miles.
20. The use of a hyphen permits discrimination between such terms as a man eating tiger and a man eating chicken.

B. Improve each italicized expression in the following sentences either by inserting hyphens, changing the construction, or both.

1. *Monocoque and unitized construction heavy gauge* sheet steel was used for the enclosure.
2. The *externally operable three pole* switch established a new standard of security for high voltage transformers.
3. *Adverse soil condition* findings indicate a need for a pile foundation.
4. The *scenic waterways systems* analysis was completed satisfactorily.
5. The study concluded by recommending the purchase of *two propeller type 1000 gallon per minute* pumping units.

6. A *bi directional pawl and ratchet manual switching* mechanism provides *trouble free* operation in areas where positive switching is important.

Exercise 3

Tell whether each of the following statements calls for a citation of its source, assuming that it appears in a work that is documented.

1. When a nuclear bomb came loose from a B-52 and fell into a swamp, four out of five interlocking safety devices failed.
2. In 1981, when Canada was complaining about the damage done by acid rain from the United States, it had not exerted on its own industries the kind of controls it demanded be placed in industries south of the border.
3. Residents of the state of New York were warned by the state health department not to eat more than one meal per week of fish taken from any body of water in the state because the fish were contaminated by a cancer-causing agent.
4. Some day, solar wind may propel huge space sails across space like old-fashioned clipper ships.
5. Few of the 21,000 offshore oil wells have caused environmental contamination problems, and around some oil platforms commercial and sports fishing have actually improved.
6. It is far safer to swim with sharks than to drive on the average city street or highway.
7. It costs American railroads 34 cents out of every dollar of revenue for tracks and right of way compared with five cents paid by trucks and .0003 cents by barges. The current aid to railroads consists of loans that are repaid with interest while trucks and barges have long received federal grants.
8. Attaching this device, which converts alternating current to pulsating current, results in a cooler burning bulb and thus extends the filament up to 100 times, at the same time providing the same amount of illumination with a 30 percent saving in energy.
9. By 1996, the market for agribusiness applications of genetic engineering should be ten times greater than the potential medical-pharmaceutical applications.
10. "This plastic-insulated wire is a highly fire-resistant alternative to costly metal-encased wiring . . . and can reduce the cost of wiring buildings by up to 50 percent."
11. The new parallel-linkage system eliminates the tendency of the rear wheels of a frontwheel drive car to toe out when, for example, it goes over a bump.

Exercise 4

A. Assume that you have written a paper that you are documenting by means of footnotes only and that you must include, either on one page or at the end of the paper, citations of the material provided below. Rewrite each item in correct form. If the form for an item was not provided in the text, write it in accordance with the general principles that govern documentation.

1. A book entitled The Physical Chemistry of Surfaces published in 1976 by John Wiley and Sons of New York. The author was A. W. Adamson. The passage cited was on page 76.
2. An unsigned article entitled Optical Determination of Linear Expansions and Shrinkage of Wood. The passage cited was on page 39 of the June, 1980, issue of Forest Products Journal.
3. The same as number 2.
4. The same as number 1 except that the page was 84.
5. A doctoral dissertation entitled Some Logical Properties of the English Component System, written by Thomas Hukari at the University of Washington in 1982. It has not been published. The passage cited appeared on page 14.
6. The third edition of a book entitled Manual of Color Aerial Photography, edited by John T. Smith and Abraham Anson. The book was published by the American Society of Photogrammetry of Falls Church, Virginia in 1968. The passage cited was on pages 67 and 68.
7. A paper identified as Technical Paper 517 of the United States Steel Corporation, written in 1982. The citation was to page 9. The paper was entitled Percent Diversion Versus Measurement Limits. The author was not named.
8. A book entitled Theory of Elasticity published by McGraw-Hill Book Co. of New York in 1959. The page cited was 132. It was written by S. Timoshenko and J. N. Goodier.
9. A book entitled Theory of Games and Economic Behavior written by John Von Neumann and Oskar Morganstern. You found this book quoted in an article by Howard F. Emerson in the Journal of Engineering Education for April, 1982, and entitled A Mathematics Foundation for Industrial Engineers.

Exercise 5

A. Assume that you have written a paper that you are documenting by providing a numbered list of references and citing them in the text. The facts about each reference are provided below. Arrange the material in each item in correct form.

1. A book entitled Irradiation Effects in Nuclear Fuels published by Gordon and Breach of New York in 1969. The author is J. A. L. Robertson.
2. An article by Robert Runde published in the magazine Money in October, 1981. Its title is Coping With the Crisis.
3. An unsigned magazine article entitled Herpes, the Misery and the Hope, published in Changing Times in October, 1981.
4. The second edition of a book by Warren J. Luzadder published in 1980 by Prentice-Hall, Inc. of Englewood Cliffs, New Jersey. The title is Basic Graphics.

5. An experiment station bulletin of the Washington State University College of Engineering Research Division. The bulletin is number 530. The date is March, 1981. The author is Howard Capp and the title is More Responsive Water Planning is Possible.
6. A book by James E. Souther and Myron L. White. It was published in 1977 by John Wiley and Sons of New York. Its title is Technical Report Writing.
7. An unsigned article entitled Major Decisions of the U.S. Supreme Court, 1978–79. It appeared in the World Almanac for 1980 of which George E. Delure was editor. It was published by the Newspaper Enterprise Association, Inc. of New York.
8. An article from the Encyclopaedia Britannica, 14th edition, Volume XIX, dated 1949. The title is Rivers and River Engineering.

B. Assume that you have written a paper that you wish to document only by means of an unnumbered list of references. Using the material in A of this exercise, make such a list of references. Provide a suitable heading, arrange the items in proper order, and place the material in each item in the right form.

6 | Special Areas

The special areas discussed in this chapter are those related to definitions, technical description, the explanation of processes, instructions, analyses, and technical or semitechnical papers and articles. Except for articles and instructions, you will find that the kinds of writing discussed do not usually appear as complete, independent papers, but as component parts of reports or other longer pieces. Nevertheless, each of them calls for special skills that may best be acquired by examining it separately.

DEFINITIONS

In writing about technical subjects you will continually find that to express your ideas clearly you must use a term that calls for definition, especially if you feel that a number of your readers will not understand the way you are using the term. When you discuss key terms, a definition is particularly important if a lack of understanding on the part of your readers could cause them to have difficulty following your argument. The term may be one you are using in a special way, it may be a term most of your readers are unacquainted with, or you may wish to remind your readers of the exact meaning of a term often used more loosely. Sometimes simply explaining informally what a term means will suffice, often a sentence definition will be required, and sometimes an expanded definition is necessary.

The Informal Definition

Perhaps used most often in popular writing, the informal definition does not call attention to itself, but fits smoothly into the flow of the discussion. It is usually incomplete in that it does not really define a term but simply identifies the use being made in a specific context. Often the definition consists merely of a word, phrase, or clause substituted for the term in question, but sometimes it may take the form of an entire sentence. When a sentence is used, it may superficially resemble the formal sentence definition but does not meet the requirements as will be explained when we discuss that topic.

You can find examples of informal definition in almost anything you read. The following examples demonstrate some of the forms you might find:

The British often prepare kippered (dried and smoked) herring for breakfast.

The punt (indentation) in bottles allows for greater pressures than would a flat bottom.

Mercuric oxide decomposes, i.e., breaks up, into mercury and oxygen when heated.

The enceinte in military construction is the main enclosure of a fort.

It is the curse of the twentieth century that totalitarianism—the one-party state that allows no opposing political or social ideas—still appeals to many societies as a way to solve social problems.

Contemporary metallurgists have created metallic mixtures or alloys that would have made medieval alchemists envious.

The common element in all the examples above is their casual nature—the definition is provided as a minor point to the main discussion. In technical writing, however, clear definition of terms often takes on greater importance. The writer, in fact, may want to call attention to the meaning of the terms. Then, a formal sentence definition may be required.

The Sentence Definition

A sentence definition is a statement that contains the bare minimum necessary to cover the meaning of a term and thus to reveal the essential nature of the thing that the term stands for. Usually it consists of a single sentence; but if a single sentence would be long, involved, and hard to read, there is no reason that two or three sentences should be considered objectionable, in spite of the name "sentence definition." In addition to the term itself, this type of definition contains two other parts, the genus and the differentia. The genus indicates a classification or group that includes the term, and the differentia discriminates between the term and whatever else the genus includes. These parts are seen in the definition of a microscope as "an optical instrument (genus) consisting of a lens or combination of lenses for making enlarged or magnified images of objects too small to be observed easily with the unaided eye (differentia)."

The genus must be chosen carefully. It must be accurate and should be as narrow as possible so that the differentia will not have to include an excessive amount of information. It would be better to identify asbestos, for example, as a nonmetallic mineral than merely as a substance. The differentia should contain enough information to draw the line between the term being defined and everything else that the genus might include without overwhelming the reader with detail.

A good sentence definition must be based on a clear understanding of the essential nature of the object for which the term stands. The definition of a pearl, for example, would need to emphasize the pearl's origin. Color, beauty, and value might also be mentioned; but unless the definition were based on the fact that a pearl is an abnormal growth within the shell of certain types of mollusk, it would not serve its main purpose.

The following sentence definitions may be studied as additional examples:

A capillary is a thin-walled tube that carries blood between any of the smallest arteries and its corresponding vein.

Isothermal compression is the compression of a gas under such conditions that the heat generated by compression is removed as fast as it is generated, so that the temperature of the gas does not change.

The flash point of oil is the lowest temperature at which the mixture of oil vapor and air above the surface of the oil will flash up if ignited. The fire point is the lowest temperature at which the surface of the oil will catch fire and burn.

The Expanded Definition

Though a well-written sentence definition is logically complete, it often must be expanded if the reader is to realize all its implications, find answers to all the questions that may arise, and see how it does or does not apply in specific cases. Sometimes it must be expanded by means of additional definitions that make clear the meanings of words it contains or the derivations of the terms involved. For instance, the definition of isothermal compression above might be expanded to include the derivation of *isothermal:* "*Iso* is a combining form meaning equal and *thermal* refers to heat." Usually, however, definitions are expanded by one or more of the following methods: use of illustrative examples, comparison and contrast, listing of component parts, and elimination. Other methods might include telling the history of the term, explaining the operation, or any of a number of techniques that give the reader a clear idea of how the writer is using the term in a particular context.

Illustrative examples make a definition more concrete and sometimes, also, clarify its scope. A definition of *parasite* might be expanded by mention of tapeworms, sheep ticks, and other specific parasites that live in or on some host from which they obtain food or shelter. Illustrative examples are helpful in that they make a definition concrete. There is danger, however, that the examples chosen will not be representative and will therefore create a one-sided picture. If all parasites listed were types that live inside the host, a reader might not notice that the term *parasite* includes types that live outside, even though the sentence definition had been broad enough to include them.

Comparison or contrast is a useful method of expanding a sentence definition because it facilitates discrimination between the term defined and other terms with which that term might be confused. The full meaning and exact limitations of *contagious disease,* for example, might not be realized until that term was compared with *infectious disease.*

A list of parts or divisions encompassed by a term will often help a reader to realize the full scope of the term. For example, a definition of *physics* might read, "Physics is the science that deals with those phenomena of matter involving no change of chemical composition." This statement may be complete in itself, but it would be possible to help a reader by pointing out that under this definition physics includes the science of matter and motion, mechanics, heat, light, sound, electricity, and the branches of science devoted to radiation and atomic structure.

Naming the parts that comprise something has a different purpose when done in a definition than when done for the sake of analysis. When we analyze something, it is assumed that the reader knows what the object is but needs to be told what its parts are; but when we name the parts for the sake of definition, the purpose of naming them is to let the reader know what the whole object is. Naming the parts of something for the sake of analysis is comparable to listing the ingredients contained in a cake—it being assumed that the reader knows what a cake is. Naming them for the sake of definition is comparable to listing the counties that comprise an electoral district so that people will know what is meant when the district is mentioned.

Expanding a sentence definition by elimination is a process in which a writer clarifies our understanding of a term by pointing out what it might seem to include but does not. In defining *insanity,* for example, it might be necessary to mention that the condition called insanity does not include feeblemindedness, imbecility, or any other condition of mental deficiency as contrasted with mental derangement. Elimination can never by itself serve as a complete definition, for essentially, definition is a process of telling what something is rather than what it is not. But by specifying some of the meanings that a term does not include, a definition makes it easier to see the *limits* of its meaning.

The foregoing methods of expanding a sentence definition may be used separately or in any combination. Any kind of material may be used, in fact, if it will help the reader to absorb and remember the sentence definition. The following example shows how a sentence definition has been made easier to grasp by being amplified.

> The term *median* signifies the particular value among three or more which has an equal number of other values above it and below it. For example, if annual incomes for five individuals were $6000, $7000, $18,000, $19,000, and $20,000, the median income would be $18,000. The term with which *median* is most likely to be confused is *mean,* which refers to a point midway between the

lowest and the highest values in question. In the list of figures above, the *mean* amount would be $13,000. The mean, that is, depends on only two figures and is unaffected by anything else, whereas the median results from taking into consideration not only the highest and the lowest but everything in between. The term *average*, of course, differs from both—referring, as is generally known, to the amount arrived at by finding the sum of all the figures in the list and dividing it by the number of figures.

Though the difference between these three values may often be slight, there are times when it is highly significant. For example, if all but one of twenty people who took a test made scores between 80 and 98, but the single exception scored only 20, the mean score would be only 59, whereas the average and the median would be much higher—the exact amount depending in either case upon the distribution of all the scores concerned.

Point of View in a Definition

Many terms vary in meaning according to the context in which they are used. When lawyers use the term *insanity* in a courtroom, they refer to legal insanity and intend to convey a meaning that is not usually in the mind of the average person who uses the term. If a psychologist were to define *normal human being*, the definition would differ from the one that would be offered by a physiologist or sociologist. Thus the point of view from which a definition is written may affect its contents.

There is nothing objectionable about a definition written from a special point of view. However, it is important that, unless the special point of view is obvious from the context, the definition itself mention it.

In technical writing, concerned as it is with specialized subject matter, the need to mention a special point of view occurs with unusual frequency. In particular, the difference between the scientific and the popular meaning of many terms should not be overlooked. A zoological definition of *insect*, for example, would include butterflies but not spiders; but to a lay reader the term would probably include spiders but not butterflies.

The popular meanings of words are valid in ordinary circumstances. In technical work, however, precision is expected, and words that have exact scientific meanings should not be used or defined in a loose, casual manner. Consequently, when as a technical writer you address lay readers, you should provide a scientific definition for any term that the reader might otherwise misunderstand.

The following specimen shows how a writer has recognized the obligation to define a term from a special point of view.

The term *plant disease,* as used in the science of plant pathology, refers to the plant's being in such condition, regardless of the cause, that it cannot perform its vital functions in a normal manner.

Sometimes a point of view is special to the extent that a definition may become arbitrary. When government authorities, for example, establish grades of livestock and produce, each grade must be defined as exactly as possible. Though there may be a real difference between beef that is *prime* and beef that is merely *good*, the exact point where the distinction between the two grades is made is an arbitrary point. Similarly, to cite a long-established example, the exact amount of alcohol a beverage may contain before it is legally considered intoxicating must be set arbitrarily.

Definitions that draw arbitrary lines are especially common in specifications. Indeed, definitions are so important in specifications that a special section is often devoted to them. A typical example would be the definitions of *coarse aggregate* and *fine aggregate* that might be included in specifications in a contract that involved the pouring of concrete.

Even an individual is sometimes justified in writing a definition that is to some extent arbitrary. If there is disagreement about the meaning of some term, a writer may define it arbitrarily—not with the intention of settling the general controversy, but merely to let the reader know which of the possible meanings is being used in the current discussion. Following is an arbitrary definition:

An Intermediate Care Home is an establishment that provides protection, physical assistance in meeting daily needs, social services, rehabilitation, remotivation, and recreation for no more than 40 individuals of moderate to high dependency but does not provide medication that must be administered by a doctor or registered nurse.

The Effect of Purpose and Readers on Definitions

Like any other writing, a definition is written because we are trying to communicate with readers, as discussed in Chapter 2. Consequently, we face the question, "Who are the readers and why do they need a definition of this term?" Whether a definition is good or bad depends in part upon whether these questions receive enough attention.

For example, you may need to define a term so that readers who are entirely unfamiliar with it, or who at best have a vague idea about it, may grasp its meaning well enough to meet their needs. Or, you may need to define a term so that readers who already understand its general meaning, or who at least have special knowledge in some field where the term is used, may use it as a basis for making fine discriminations.

The basic facts in a definition and the language used should both be affected by the question of which of these two conditions exists. Suppose we defined butanone as "a highly flammable liquid used as a solvent." On many occasions, nothing more would be necessary to accomplish our purposes; but if chemical matters were under discussion and chemists were among the readers, it would probably be desirable also to say that butanone is derived from acetone and to include its chemical formula. The point is that facts of more than one kind can be used to draw the line between what you are defining and all else with which the thing you are defining might be confused. And the question of what facts to include depends on whom you are addressing and on the specific purpose for which your definition will be used.

As for language, the advice in the chapter on style applies to definitions as it applies to other writing. If you want a particular definition to be useful to anyone who is not a specialist in the subject, it is best to avoid the use of words that are more technical than the term you are trying to define. A lay reader who did not know what is meant by *mumps* would learn very little from one dictionary's definition, "a specific, febrile disorder characterized by a nonsuppurative inflammation of the parotid and sometimes other salivary glands." Such a person would be better served by, "a contagious disease marked by inflammation and painful swelling of the salivary glands below and in front of the ear." Yet if one were giving instructions to medical students, the language as well as the additional facts would make the former definition the better of the two.

Most instructions on how to write definitions warn against the use of highly technical words. This warning should often be heeded, but it is based on the assumption that the purpose of a definition is to give the reader a first introduction to the term defined, and that the reader is unfamiliar with the technical vocabulary of the field involved. Sometimes this is true, but not always. When a definition is written to enable a scientist to pin down the meaning of a term with extreme precision, it may well be expressed in different language than would be used to give a lay reader a general understanding of the term. There are times when the very reason that the definition is brought into existence is the need for an exactness that ordinary language does not possess.

Still another condition may exist. A definition may be written to establish a special meaning of an ordinary word that science has adopted because no single word in its vocabulary will serve the purpose. An example would be the use of *doping* to mean, "adding minute, carefully controlled amounts of impurities to growing crystals of silicon or germanium." It is sheer nonsense to object to scientific language in a definition that exists in order to tell the special scientific meaning of an ordinary term.

In general, the point to remember about the effect of purpose on a definition is this: In writing a definition, as in writing anything else, always ask yourself, "Who will my readers be, and how can I write so as to meet their needs?" In any kind of writing, it is bad to concentrate on substance alone and disregard the reader and the function of the written material.

GENERAL SUGGESTIONS

1. Do not base your definition of a term on any word that comes from the same root as the term itself. To define *permeability* as the quality of being *permeable* does no more than tell the reader that permeability is a quality. (This rule will be modified in rules 2 and 3 that follow.)

2.   Always phrase a sentence definition to match the grammatical form of the term defined. (This requirement rules out such sloppy phraseology as "Osmosis is when . . ." Even if one were defining a noun that means a time, it would be better to use such phraseology as ". . . is the time when . . . ," thus avoiding the use of an adverbial clause when a noun is demanded by grammar.) Sometimes the need for grammatical consistency delays the moment when a writer comes to grips with his real problem. For example, the noun *fertility* is derived from the adjective *fertile,* so the first statement one might make in defining *fertility* is that fertility is the quality of being fertile. The real problem of definition would then consist of defining the adjective *fertile.*

3.   In defining a term that consists of more than one part, concentrate your attention at the point where attention is needed. For example, in defining *soil physics,* your main problem would probably be to show how soil physics differs from physics in general. Or in defining *biochemistry,* the main problem might be to differentiate between biochemistry and other chemistry. Of course, if you felt it necessary, you might first define the basic word (*physics* or *chemistry*) and then go on to the discrimination indicated above.

4.   Be sure that a definition includes everything that should be included. If *bird* were defined as "a warm-blooded animal that flies through the air," the definition would exclude the ostrich, which is a bird even though it does not fly. There is no reason, however, that you should not resort to a device constantly used in dictionaries and insert the word *usually,* so that you can utilize a helpful fact even though it may not apply to every individual thing that the definition includes.

5.   Be sure that a definition excludes everything that should be excluded. (The foregoing definition of *bird* did not exclude bats and hence was inadequate.)

6.   In writing a sentence definition try to avoid language with which the reader is unlikely to be familiar. The acceptability of whatever language is necessary for the sake of precision has been mentioned; but in a definition, as elsewhere, ordinary language is better than technical language if it will convey your ideas as effectively.

7.   In an expanded definition, be sure to point out how the term you are defining differs in meaning from any other term with which it might be confused. For example, if you were defining *toxin* you would probably need to discriminate between *toxin* and the more general term *poison,* and also, perhaps, between *toxin* and the more limited term *venom.* A sentence definition might cover all the facts, but the expanded definition would be written so that all the facts would not only be covered, but would actually reach the consciousness of the reader.

8.   Do not forget that you can sometimes make a definition easier to grasp and remember if you mention the root or roots from which the term is derived. For example, *isobar* might be fixed in a reader's mind by a comment on the fact that it derives from *iso,* meaning equal, and *baros,* meaning weight and in this instance referring to the barometer.

TECHNICAL DESCRIPTIONS

To understand how technical description differs from "literary" description, we need only imagine two descriptions of a room that has been ruined by fire. The literary description would enable us to imagine ourselves in the room— impressed by the fierceness of the fire that had wrought the damage, moved by the half-burned remains of personal belongings, startled to see the sky through a hole in the roof, conscious of the smell of wet, charred wood and the feeling of shattered plaster under foot. A technical description, however, would make no effort to create an imaginary experience nor to arouse our emotions. It would merely tell us, in an objective manner, the facts about the condition of floors, wall, ceiling, and contents—perhaps with the intention of enabling us to consider the need for repairs or to judge what might have been the cause of the fire.

A technical description is sometimes complete in itself, but it is more likely to appear as part of a longer paper. Such a description might deal with the damage done by an accident or a flood, or with the construction, facilities, and condition of a building. It might be written to help us judge the possibilities of remodeling or to decide whether safe working conditions prevailed.

A technical description might also be written to tell us about some newly developed machine, some device that had been used in an investigation, or some piece of equipment the purchase of which was contemplated. Our discussion will deal mainly with subjects of this latter type because such subjects will make the greatest demands on your skill as a writer.

The description of an object may concern the type of object in general or one particular example, but in either event, the basic method would be the same. Indeed, it is often best, in describing some device in general, to describe a single characteristic specimen and then mention the respects in which variation from the chosen example is most common.

The following technical description illustrates many of the points made in this section:

The Champion Scrubber-Polisher

The Champion electric floor scrubber-polisher is a machine designed for domestic or light-duty janitorial service. It scrubs floors by means of two circular brushes driven by an electric motor while a cleaning fluid is fed to the floor from a dispenser mounted on the handle. It can also be used as a polisher, in which case felt pads are attached that cover the surface of the brushes. It weighs 18 pounds, is 12 by 6 inches in size at the base, and stands 38 inches high with its handle erect. It consists of five main parts: the brushes, motor, hood, handle, and dispenser.

The brushes are 5 inches in diameter and are positioned side by side, separated by two inches at the inside. They have $1\frac{1}{2}$-inch plastic bristles fastened to stiff plastic plates as wide as the brushes. In the center of each brush is a metal peg, short enough that it clears the floor by an inch, to which the felt pads are attached when the machine is used for polishing.

The motor runs on 110- to 120-volt current and can be operated at 1200 or 2400 rpm. It is riveted to an aluminum base above the brushes. The electric cord runs down through the handle, which is hollow. The cord is 20 feet long and can be wound onto brackets on the underside of the handle.

The motor and brushes are covered by a rigid plastic hood that is 8 inches high and clears the floor by $1\frac{1}{2}$ inches. Its horizontal dimensions are 5 by 8 inches where it covers the motor, but it widens to 6 by 12 inches where it covers the top of the brushes. A rubber strip around it protects walls and furniture.

The handle has an outside diameter of 1 inch and is 34 inches long. At its bottom, two steel strips are riveted to it and extend through slots in the hood down to the base that supports the motor. They are hinged at this base so that the handle can stand erect for compact storage or can be held at a convenient angle, when the machine is in use, without tilting the brushes and motor. The lower 12-inch section of the handle is made of the same plastic used for the hood, and the rest is polished metal. The top 8 inches are covered with rubber for insulation. A bend 6 inches from the top allows the user to hold the handle at a natural angle. On the lower side of the handle, 8 inches from the top, is the switch that controls the motor. On the upper side, just above the bend, is a lever that controls the flow of the cleaning liquid.

The cleaning-liquid dispenser is mounted on the upper side of the handle. The reservoir containing the liquid is made of semirigid plastic, transparent enough to make the height of the contents easily visible. It holds 1 gallon, is graduated in 1-cup units, and has a capped opening at the top for filling. Its length, which extends along the handle, is 13½ inches. Its other dimensions are 2½ by 6½ at the top and 2½ by 4½ at the bottom. The bottom is supported and held steady by a base 2 inches deep made of the same plastic used for the hood. The top is held firm by a cover of the same plastic, fastened to the handle by a ring that can be moved up and down to facilitate removing or attaching the reservoir proper.

Through the length of the reservoir runs a rod that opens and closes a valve and thus controls the flow of the cleaning liquid, which is carried down through a plastic tube to two openings, one near each of the two brushes. The rod itself is controlled by another rod running inside the handle to the lever mentioned above. (The use of two rods is necessary in order to permit removal of the reservoir.) The valve referred to is located in a small protuberance at the bottom of the reservoir that fits against a rubber seal in the base to prevent leakage.

The Champion scrubber-polisher is a light-duty machine, so its main usefulness is for small-area jobs such as in motel rooms, offices, or private homes. It also serves well as an auxiliary machine for use in restricted areas of larger buildings.

Description of the Object as a Whole

Though most of the information in a technical description is likely to concern some single part of the object described, the opening section deals with the object as a whole. In this section you should include any of the following materials that are needed: (1) a definition of the object to be described; (2) an explanation of the general manner in which it performs its functions; (3) a general description of the object; and (4) a list of its major component parts, each being preferably a part that performs some particular function.

Your definition of the object, if you decide that one is needed, will be sufficient if it merely explains what the reader might be doubtful about. For example, consider the definition, "A demagnetizer is a device for removing the magnetism from hardened tool-steel parts that have been held on a magnetic chuck and thus have become permanently magnetized." The term *demagnetizer* has an obvious meaning, but the writer considered it necessary to indicate that a demagnetizer is used in a certain kind of work and for the purpose of demagnetizing a certain kind of object. The definition was adequate to the occasion and performed a useful function. On the other hand, it would be pointless to write such a definition as, "An electric brake is a brake that operates electrically."

Sometimes the reader will know without being told what an object is used for, and sometimes the use of the object will be indicated in the definition. There are times, however, when the use of an object will not be apparent unless you make a point of stating it. Such might be the case, for example, if you were writing about a permeameter, a vane borer, or an autoclave. It will always be part of your job, when you write a technical description, to pass judgment on whether the use of an object is obvious or whether it calls for comment, and to include such information if you feel that the reader needs it.

The explanation in the opening section about how a device works may not need to be extensive, but if your readers are to understand a detailed description, they must not be left entirely in the dark about how the object performs its function. They would need to know that an autoclave sterilizes or cooks by means of superheated steam under pressure; that a centrifuge separates different materials by means of centrifugal force; or that the air brakes of a train function when the air pressure between different cars is lowered or released so that the compressed air stored below each car can enter the cylinders and apply the brakes. Even a minimum of such information can make all that follows far more intelligible.

When you become familiar with some object by inspecting it personally, you notice the object in general before you notice the details. Similarly, it is desirable in a description to give the reader an overall look at the object before presenting information about its parts. Readers should be told early in the description about the shape and size of the object and also, perhaps, about its finish, its color, and the material of which it consists, if these are important.

Obviously, however, this information cannot be given about some objects because they are not sufficiently visible. For example, no one could effectively describe the appearance of the hydraulic-braking system of an automobile; and in this instance the appearance is unimportant anyway.

When readers understand what an object is, what it is used for, how it functions, and perhaps what it looks like, they are ready for a list of its functional parts. This list should not be excessively long. Rather, it should consist of the main functional units, each of which may be broken down into *its* parts, if necessary, when the time comes to discuss it. For example, you might divide an automatic washer into the tub and agitator, the motor and timer, the frame and housing, and the pump and tubing. The parts that make up each of these main units would not be mentioned until later.

The items in a list of the main parts of an object may be numbered if numbering them seems helpful, and should be arranged in the order in which they are to be discussed. One possible arrangement would be the order in which the parts would be noticed by a person who was looking at the object. A second would be the order in which they perform their respective functions. The latter is usually the best order when the parts function in succession rather than simultaneously. A third possibility might be the order of decreasing importance, for it might be desirable to tell about the important portions of the object at once rather than delaying them while trivial matters are presented.

Description of the Functional Parts

When the main parts or divisions of an object are described, the treatment of each main part is somewhat the same as the treatment of the entire object. That is, it is defined if definition is needed. Its purpose and the general method by which it accomplishes this purpose are made clear. Its general characteristics such as size, shape, and material are indicated. Its position in relation to the object as a whole and to the other main parts of the object is explained. And if it is complicated enough to merit such treatment, it is in turn broken down into its component parts, which are then discussed in the order in which they are listed. This continues until all the functional parts have been described in as much detail as seems necessary.

The Conclusion

Unless the object described is extremely simple, a brief conclusion is desirable, so that the reader's final impression is not limited to the last small detail. No single formula can be prescribed for the conclusion, but its general purpose will be to make the reader visualize the object as a whole, serving its purpose and functioning as it is intended to function. Often, if the object is one that performs some process, the best approach is to review one cycle of operation so that the reader can imagine the object in action.

Adaptation to the Reader and Occasion

The reader's knowledge and the demands of the occasion should affect both the contents and the language of a description. The fact that a description of some article is needed does not necessarily mean that all parts of the article must be described in equal detail. For example, electric motors are fairly well known and standardized, and if such a motor is part of a certain mechanism, it should be sufficient to tell the type, the power, and the kind of current it uses.

GENERAL SUGGESTIONS

1. A technical description, as we have seen, is intended to help the reader to understand an object as well as to picture its appearance. Thus it must include

information about parts that may not be visible, and the emphasis is constantly directed to the manner in which the object functions.

2. In actual practice, a technical description is often accompanied by one or more figures. Instructions on the use of figures are given in Chapter 7.

3. It is often possible to make a description clearer by comparing an object with something that the reader is familiar with. For example, by saying that an object had the approximate shape of a light bulb with the socket end down, you might make shape and position clearer than you could ever make them by direct description. Comparison to the shape of letters is especially likely to be helpful, as seen in the phrases *L-shaped, shaped like an H, somewhat resembling an inverted U,* or *having an S curve.*

4. In telling about size, try to avoid general words such as *large* and *small*. To illustrate their vagueness: a cupel, used in assaying, has been called a *small* cup; likewise a crucible, used in the steel industry, has been called a pot for melting a *small* amount of metal. Yet a cupel would hardly hold a teaspoonful of material, whereas a crucible might hold as much as 200 pounds of steel.

5. The positions of various parts of an object must be indicated with care unless the use of a figure makes the position of each part clear. Words such as *above, below, behind,* and *beside* must be used constantly. Information on position must include not only location but also information on such points as whether a cylinder is horizontal or vertical, or whether a hole in the center of a rod is parallel to the axis or across its diameter. Trivial facts should of course be omitted, but whatever is worth presenting at all is worth presenting clearly.

The preceding discussion has dealt with objects that function actively. A technical description that deals with such a subject as the characteristics of a building or the condition of a piece of equipment is simpler but not essentially different. First it covers the subject in general and then breaks the subject into its main divisions, continuing this treatment until it provides all the information necessary.

EXPLANATION
OF A PROCESS

In discussing the explanation of a process we shall deal mostly with processes that are performed mainly by human action. Some of the suggestions will also apply to processes that occur in nature and processes performed mainly by machinery, but for reasons that will be obvious during the discussion, not all the techniques of explanation are usable for the two latter types.

First, it is necessary to differentiate between explaining a process and giving instructions for performing it. Instructions are written to enable those who may perform a process to perform it properly. The *explanation* of a process, on the contrary, is addressed to readers who may be unlikely to perform the process themselves but want to understand it so that they can judge its reliability, practicality, or efficiency. The manner in which certain tests were performed might be explained, for example, so the reader can judge whether the results were valid. Or again, a method of doing some job might be explained so the reader can judge whether it would be more efficient or less expensive than some other possible method. In such cases the typical reader is a supervisor or executive rather than the worker who performs the process, and the explanation appears in a report, article, or proposal rather than in an instruction manual. In contrast to a set of instructions, it is not worded in such a manner that it gives commands; it tells what actions are performed so that readers may understand what is, was, or is to be done rather than so that they may do anything.

When you explain a process, you will usually find it desirable to follow a fairly well standardized pattern—one that has become standard not because of arbitrary considerations but because it leads to explanations that are easy to understand. This pattern calls for an introduction, an overall picture including a list of the main steps that make up the process, an explanation of each of the steps listed, and a conclusion if one would be helpful.

The Introduction

The introductions should be limited to information that readers really need, and it is unlikely that on any single occasion they will need information on all the points to be mentioned. This qualification should be held in mind as you read the following suggestions. Also, if any of the introductory information applies only to one part of the process, you should feel free to wait until you come to that part before presenting it.

Sometimes the introduction opens with a definition of the process. It indicates why, where, when, and by whom the process is performed. It includes information on the materials, tools, and apparatus needed in performing the process; and if any important apparatus or materials are likely to be unfamiliar to the reader, it identifies them. It tells, if necessary, whether those who perform the process need special skills or training. Sometimes, too, it mentions special requirements about the time when the process must be performed, or special conditions that must exist, such as temperature, humidity, freedom from dust, or ventilation. And finally, it sometimes tells about preparations that must be made before the process is performed.

The Overall Picture

The introduction is followed by any comment necessary about the process as a whole—for example, the theory on which it is based—and by a list of the main steps that make up the process. An effort should be made to hold the main steps to five or six, for if they are too numerous it will be extremely hard for the reader to grasp and retain an overall picture. The main difference between the treatment of a long, complicated process and a simple one is that the former is subdivided more extensively, not that it includes more main divisions.

The main divisions of the process should be named in chronological order so far as possible. They should be expressed in parallel form and may be numbered if numbering seems likely to be helpful.

Each main division should be based on the completion of a stage of the work rather than on some arbitrary consideration such as place or time. It might be undesirable, for example, to divide a process into work done in the field and work done in the laboratory. Even if the work done in the field comprised one specific task and the work done in the laboratory comprised another, you would give your reader a better picture of the process by identifying that task rather than merely identifying the place where it was done.

The Explanation of Successive Steps

After listing the main steps, you should next take up these steps one by one and treat each of them somewhat as you treated the process as a whole. For each division, a definition is provided if needed, and the facts about time, conditions, apparatus, personnel, and preparations are made clear. Then, if necessary, the parts of which the step consists should be listed.

When nothing would be gained by further subdivision, you can begin to tell what is done when the process is performed. In telling what is done, you should emphasize the results that the acts performed are supposed to accomplish rather than the actions themselves. The effect of this emphasis may be seen by comparing the following examples:

First, the form is filled with mix and then the mix is tamped with a ½-inch rod 25 times, which leaves room for more mix. Again, and finally a third time, the process is repeated, the result being that all samples are compacted to the same degree.

In order to compact all the samples to the same degree, each is tamped in the same manner. First, the form is filled with mix, and then the mix is tamped. . . .

The important fact here was that the samples had to be uniformly compacted. The tamping was performed in a certain manner only because it produced that result. Yet in the first example, the result aimed at seemed added as an afterthought, and the reader was expected to follow a series of actions the purpose of which had not been indicated. The difference between the two methods may seem small when observed in a brief specimen, but in a longer paper it can mean the difference between understanding and confusion. When reading about an action, a reader should know what that action is intended to accomplish; and when several successive actions are all performed in order to accomplish a single result, that result should be indicated at the beginning of the passage. Otherwise, the reader will soon be lost in a maze of details.

As your explanation progresses from step to step, you should keep the reader aware of the progress. When you take up a new step you should point out that you are doing so, using the phraseology or numbers that you employed when you originally listed the steps. Special care is necessary when the steps do not follow one another in a regular, chronological order, as when two steps are performed at the same time by different persons. By careful use of transitional material, however, you should be able to keep the reader oriented and constantly aware of the process as a whole.

In the explanation of a process, as in any other writing, there is no reason that all parts of a process must be covered in equal detail. Suppose that one step of a process consists of some test which the reader already understands or which is a recognized, standard procedure. It would probably be sufficient merely to say that this test is performed. Yet some other part of the process might involve unusual procedures that would call for detailed explanation. Thus, in regard to each portion of the process as well as in regard to the process as a whole, it is part of your job as writer to include the material that serves your purpose and to omit what can be omitted without loss.

The Conclusion

There are many occasions when the explanation of a process does not need a conclusion—especially when the process was performed on some specific occasion and the explanation is part of a longer paper. When you feel that a conclusion is desirable, however, you need not hesitate to add one. Your conclusion might summarize the process, perhaps restating the main steps so that the reader's final impression will include the process as a whole rather than only one small part. It might evaluate the process or the results of the process. It might comment on why the process is important or indicate how it fits into some larger process. But unless such material fits the occasion, your explanation should end with the final remarks about the last action performed.

Following is an example of an explanation of a process performed by human action.

Crossing Strawberries[2]

Crossing is the process by which the geneticist or plant breeder combines the genetic characteristics of one plant with those of another. The breeder does this in the hope of obtaining a new individual that will maintain as many desirable qualities of both parents as possible and will be free from a maximum number of undesirable qualities. The person who performs the process must have access to a greenhouse in which to grow the breeding stock and to a plot of ground in which to raise the seedlings. The only equipment needed in addition to what is ordinarily available in a greenhouse and garden is a pair of very sharp long-nosed tweezers for use in emasculation. Also needed, of course, is a supply of the desired varieties of plants for use as parents. The

[2]Courtesy of DeLance Franklin, University of Idaho Experiment Station, Parma, Idaho.

crossing involves three fundamental operations: (1) cross-pollinating, (2) growing the progeny, and (3) selecting the desirable individuals from the progeny.

In the cross-pollinating the breeder's first task is to prevent self-fertilization by emasculating, in the greenhouse where they have been raised, a number of flowers from all the plants that are to be used as female parents. With this purpose in mind, the breeder selects primary and secondary flowers that are just beginning to unfold and uses the tweezers to remove all the stamens of each flower selected. Then, in order that the flowers chosen may be easily distinguished from any flowers that may open later, the corolla and calyx are removed from each of them.

When all the flowers to be used as female parents have been thus emasculated, the breeder picks from the desired male parents a suitable number of flowers that have unopened, pollen-bearing stamens. One such flower is usually sufficient for four to six pollinations. The picked flowers are allowed to wilt until anthers dehisce (burst open). The breeder then does the actual pollinating by holding a male parent against the pistils of a female parent and twirling it, making sure that it touches the stamens so that the pollen will reach the stigmas. Since casual cross-pollination by insects is not a danger when the work is done in a greenhouse, it is not necessary to bag the pollinated flowers as would be done if the operation were performed out-of-doors.

When the berries have matured on the crosses, the breeder can grow the progeny. The first step is to separate the seeds from the pulp. This is accomplished by mashing the berries into a container of fine, clean sand. Later, the breeder sows the mixture into pots or flats of good soil, covering the seeds with a one-eighth inch layer of the sand. The seedlings are raised at a temperature that is held between 65 and 85 degrees, and the seedbed is never allowed to dry out. When the plants are large enough that they can be handled conveniently, they are transplanted into thumb pots or flats—in the latter case being placed 2 inches apart. Within two to four weeks they are again transplanted, this time into the outdoor plot.

When the berries first show signs of ripening, the breeder begins the process of selecting desirable individuals by first pulling up and discarding all the plants that show disease or weak growth. As the berries ripen further, all plants that are obviously below standard are discarded. These two steps result in the elimination of all but about five percent of the plants that have been raised. The breeder compares the plants that remain with the best of the already existing varieties, and saves only the choicest individuals. From these, new plants are propagated asexually (usually from runner plants) to obtain enough specimens of each hybrid that can be tested more critically throughout the range of conditions that may be required for proving or disproving the merits of each.

By means of this process, the breeder may accomplish the following objectives: resistance to disease, insects, drought, or extremes of temperature; qualities that are especially desirable for a specific use such as canning or freezing; or berries that are firmer or taste better than other varieties.

Explaining a Process Performed by Machinery

When a process is performed mainly by machinery that functions to some extent automatically, explaining it involves telling how the machinery functions. This creates a problem very much like the problem of writing a technical description. The difference is mainly in the facts that are emphasized and about which the details are numerous. The technical description emphasizes the facts about the machine and tells, incidentally, how the machine operates. The explanation of a process emphasizes what the machine does as it operates, and makes the actual description of the machine incidental.

One possibility of handling the job is to tell first what actually is done by the machinery, organizing the information if possible on the basis of stages of the work, as we would do if the work were performed by human beings. The next step is to tell what it is necessary to know—no less and no more—about the machinery that does the job—what it is like and how it functions.

Sometimes this treatment can be varied. If what is said about the machinery falls naturally into divisions that are basically the same as the main divisions of the process—that is, if each separate and distinct part of the process is performed by a separate and distinct part of the machinery—it may be advisable to open with an overall picture of the process and then combine the discussion of each part of the process and the necessary information about the machinery that performs that part. Discussion of human action can be limited to information about how many people are involved and what abilities they need.

One might summarize by recommending the following general method. First, use an introduction consisting, so far as it would be appropriate, of the same kind of information you would use in explaining a process performed by human action. Next, give an overall picture of the process by telling its main divisions. Then tell what is needed about each of the divisions, placing emphasis in descending order on *what* is done, *how* it is done, and the kind of machinery that does it.

Some of the processes that this technique would clearly apply to are:

Feeding cattle by automated machinery
The manufacture of paper
Crushing ore
The extraction of minerals by the flotation process
Manufacture of steel by the oxygen process
Harvesting by means of a combine
Cracking petroleum
Treating water with various kinds of water-softening equipment
Treating water for purification

Explanation of a Natural Process

The introductory information for the explanation of a natural process obviously cannot include some kinds of information used to introduce a process performed by human action, but it might call for the following: a definition which may either tell the meaning of the term that identifies the process or clarify the exact nature of the process; a brief statement of the scientific facts, principles, or theories that cause the process; and possibly the conditions under which the process occurs.

This introduction may best be followed by an overall picture of the process indicating the stages in which the process occurs. If possible, the stages should be so presented that there will be no more than five or six main divisions, just as is the case with processes of other types; but such a limitation may be prevented by the facts.

When the general picture of the process has been completed, each stage should be explained, in chronological order. The explanation of a stage consists merely of telling what happens, and, if possible, why it happens, unless the reason is common knowledge among the readers addressed or has been covered in the introduction. The suggestion made above (in the discussion of processes performed by human action) that we should emphasize purpose does not apply here, but writers should treat single incidents as significant mainly because they show how major elements develop. That is, a natural process in which twenty or thirty happenings occur need not be treated as twenty or thirty distinct happenings. Rather, the happenings can be grouped together as they combine to produce the major developments, so that we conceive the process as occurring in a reasonable number of stages.

The explanation of a natural process may or may not call for a conclusion. The conclusion might include a summary of what has preceded, information about the significance or importance of the process, or both. In the last analysis, however, the use or omission of a conclusion should probably depend upon whether the explanation of the process is an independent piece of work that for psychological reasons should not end abruptly, or whether it is a part of some larger piece of writing and leads naturally into whatever follows.

Some examples of natural processes you might wish to describe include:

cell division
plate tectonics
formation of oil
childbirth
operation of the nervous system
volcanic action

INSTRUCTIONS

A piece of writing giving instructions has much in common with the explanation of a process, but it is addressed to a different audience and serves a different purpose. Readers may be expected to perform the process, and the purpose is to enable them to perform it properly rather than just to understand it. Instructions may sound like commands; but there is a difference between ordering a person to do something and telling *how* to do something. The two possibilities are not mutually exclusive, and both may be called for on some occasions. In our context, however, giving instructions means giving information rather than giving orders, even though such information may sometimes be expressed in the imperative form.

In technical writing giving instructions is usually a matter of telling how to perform some physical process such as assembling, using, inspecting, or repairing equipment, performing a test, or doing some job in a laboratory. Such instructions are given by manufacturers to dealers who sell and service their products and to consumers who use them; they comprise the bulk of the contents of laboratory manuals; they are produced, in the form of instruction manuals, as a result of enormous labor, by the multimillion-dollar industries that serve the armed forces. But in spite of the variation in complexity that these extremes represent, it is possible to offer many general suggestions that should be helpful.

Introductory Material

The same kind of introductory material used in explaining a process may be called for in giving a reader instructions for performing a process. That is, the term used to identify the process *may* call for definition. The purpose of the process may need to be stated. Other useful information might concern equipment needed, conditions that must exist, skills necessary, and preparations to be made before the job itself is begun. It would be wrong to assume, however, that every item in the list is always to be covered. On the contrary, the introductory information should be limited to what will be useful for your purpose and readers.

How Much to Include

When the introduction is complete and the actual instructions begin, one question that always arises is what to include and what to omit. For example, should purposes be explained or should the contents be limited to the actions called for? It is certainly true that instructions are grasped more easily, remembered better, and carried out more intelligently if the reader understands the reason things are done in a certain manner. But such explanations are a means to an end and should be kept distinct from the instructions proper so that the latter will stand out sharply and be easy to identify. This result can often be accomplished by a simple change in manner, as shown by the following examples.

POOR

1. An accurate record shall be kept of the time the machine has been used on each job that has a separate account number.
2. The job account shall be charged with the amount arrived at by applying the rental rate.
3. Minor repairs and service shall be charged . . . etc.

BETTER

. . . To charge the expense of a machine against production for a short period would be extremely complicated. Therefore an hourly rental rate period has been set up for each type of machine. It is to be applied as follows:

1. Keep an accurate record of the time a machine is used on each job that has a separate account number.
2. Charge the job account with the amount arrived at by applying the rental rate.
3. Charge the time lost for minor repairs and service to the job on which the machine is working.
4. Charge major repairs or overhaul against operating accounts 700 to 799.

Use of Numbers and Other Mechanical Devices

Numbering the items is almost always useful in giving instructions—though it is no substitute for careful, systematic arrangement and organization. Sometimes a single sequence of numbers is adequate, but if the subject matter is at all complicated it may call for a system of numbers and letters comparable to the system used in an outline. That is, a major order (A. Assemble the blower unit) might be followed by a numbered list of specific acts involved in the assembling. The basis for treating several actions as a single larger unit should practically always be the fact that they combine to produce a single result, or that they concern a common problem.

In addition to numbers, other mechanical devices are useful in giving instructions. Illustrations may be almost essential. Other possibilities are extra-wide margins, parentheses, explanatory notes—either footnotes or notes labeled as such and inserted in the text—and variation in the style of lettering (different typefaces in printed matter, and the use of capital letters and underlines in typewritten copy).

These devices are useful mainly because they permit the writer to use inserts that clarify the meaning of terms and give warnings, yet still make the instructions proper stand out so that they cannot be missed. It is possible, of course, to overdo these devices, and thus produce a cluttered-looking paper; but judiciously used they make it easier for the reader to focus attention mainly on the parts that tell the essential action and to recognize the other material for what it is.

Special Elements

In writing instructions it is often necessary to emphasize something that is not part of the instructions proper. For example, in giving instructions on the use of an insecticide it may be necessary to warn against letting it get onto the skin or into the eyes. In giving instructions about using a poisonous substance it may be necessary to tell what antidote is most effective. Such material should be placed under conspicuous headings and inserted wherever it applies.

These special headings not only guarantee attention, but also make the instructions proper easier to follow.

Warnings may be ranked under three levels of severity: those things or actions that may result in a less than perfect product, those that may cause damage to the equipment performing the process, and those that may injure or maim the operator. In each case, the kind of heading or attention-getting device must reflect the relative importance of the danger involved. Thus, in instructions for running a plywood press, the danger about keeping hands and clothing clear must be presented in larger type, or some other eye-catching way, than the warning about allowing the press to build up to its full load.

Less vital than warnings against damage or danger are special instructions that apply in exceptional circumstances. It might be necessary, perhaps, to tell about a special procedure to use in extremely cold weather, or what substitute can be used if material that is normally used is unavailable. Here again the main consideration is to make the information easily available without letting the instructions proper become so cluttered with *if*'s and *when's* that it is needlessly hard to discover what to do under normal circumstances. The exceptional circumstance can sometimes be provided for by the simple expedient of a parenthetical insertion at the appropriate place. Much of the time, however, the best method is to present the instructions that apply under normal circumstances without interruption and then cope with the variations by adding what is necessary, suitably headed or introduced. A suitable introduction might be somewhat as follows: "Though the preceding instructions can ordinarily be followed as they have been given, it may be necessary, under special circumstances, to modify them as follows."

Style and Tone

Because anyone who writes instructions tries to make them easy to follow, and because this involves breaking down whatever is complex into its simplest elements, the sentences used in instructions are likely to be extremely short. And since a person who reads instructions expects them to consist of a series of separate actions, preservation of continuity is less important than in most kinds of writing. This does not mean that gaps in thought are permissible, but merely that connective words and phrases may be used sparingly because the relationship among the items is already understood to be that of separate elements in a series.

In keeping your sentences short and simple, you should not ordinarily try to save words by departing from normal language. Unless space is artificially limited, as, for example, on a package, a cookbook style is not desirable. (A characteristic example of the cookbook style is the sentence, "Place 8 ounces water and three pellets in jar and shake until dissolved.") Not enough space is ordinarily saved by the omission of a few words to compensate for the loss of naturalness and readability.

Tone, sometimes, as in a maintenance manual or a laboratory manual, calls for little consideration because the situation is entirely impersonal. There are times, however, when a pleasant tone is an asset. If you compare the booklet that an automobile manufacturer prepares for a purchaser with the repair manual prepared for mechanics, you will see that the automobile industry recognizes the desirability of addressing the user of its products in a manner that develops good will. Often, instructions addressed to subordinates also will get better results if they are given pleasantly and courteously with a view to securing willing cooperation. Most of the effort to create a pleasant tone is expended in introductory and explanatory passages rather than in the instructions proper, which can thus be kept clear of everything except the bare minimum necessary to cover the action.

Conclusion

Writing instructions that will be clear to the reader involves, first of all, asking yourself who your readers will be, what they already know, how much detail they must be given about the action, and how much explanation about reasons they must be given in order to persuade them to do things properly. You will need to use your imagination in order to foresee the possibilities of errors and misunderstandings, and to provide for every likely contingency. You will need to use good judgment in deciding how simple your overall organization can be. You may usually feel free to sacrifice some of the niceties of style (for example, primer style is often effective) when clarity demands such a sacrifice. And you should often use illustrations and mechanical devices such as headings, notes, and variations in lettering to help your readers see how everything is related to what they are expected to do. If you handle your instructions in this way, you will probably find that your readers can understand your instructions well enough to comply with them.

The following specimen shows how the recommendations for writing instructions work out in practice.

Etching Concrete Substrates

As an alternative to sandblasting, portland cement concrete substrates may be prepared for priming by etching them with muriatic acid to remove the powdery residue on the surface. If such treatment is used, the surface must be in proper condition. Plaster, mortar, loose aggregate, and other such materials must be removed. Grease must be washed off and paint must be removed by any method that will get rid of it. When the surface has been prepared, the etching shall be done as follows:

1. Add one volume of 20° Baume muriatic (hydrochloric) acid to two volumes of water.
2. Pour this solution onto the surface from a plastic watering can, using one gallon of solution for 50 to 100 square feet of surface. (The

amount necessary varies with the condition of the surface.) Work the acid into the surface with a stiff-bristle broom to make sure that no part of it remains dry.

3. Brush the entire surface vigorously.
4. Remove the spent acid from the concrete by flushing the surface with water and going over it with a squeegee or by using a commercial vacuum cleaner. (WARNING: If a vacuum cleaner is used, make sure that it is equipped with an acid-resistant tank.)
5. Wet the entire area again with fresh water and brush it vigorously.
6. Remove the excess water as you did before—by squeegee or vacuum.
7. Examine the surface to see whether the results of the etch are satisfactory. (The surface should have a finely pebbled texture.)

If the surface does not appear pebbled, repeat steps 2 through 6. Otherwise, it is ready for application of the primer.

ANALYSIS

Analysis, as viewed in the context of technical writing, consists of the process of examining something to distinguish its component parts or elements either separately or in their relation to the whole. Situations in which it is necessary to analyze something and present the results in writing are likely to confront anyone in a position where technical writing is called for. It would be impossible to name everything that those who use this book will be called upon to analyze, but the list would include conditions, causes, results, trends, and problems of every kind.

The nature of a written analysis can be made clearer by recalling some of the analyses that all of us encounter in our reading. We read analyses of business conditions or the conditions of some particular industry; analyses of the causes of increase or decrease in production; analyses of the manner in which funds have been distributed or the areas in which expenses have increased; analyses of the results of new technology or legislation; analyses of the shift of industry from one region to another; analyses of problems such as the detection or prevention of defects in materials or the design of better products.

For a concrete example, let us see what was involved when an analysis was made of the suggestion that a college introduce a program designed to prepare students for technical writing as a full-time occupation. Some of the questions that arose were as follows:

Course Work

What existing courses would the program include?
Could these courses accommodate the increased enrollment?
Are new courses needed?

Administration

Would qualified instructors and acceptable textbooks be available?
Would any existing department be prepared to administer the program?
If not, would a new department of technical writing offer enough work of its own to justify departmental status?

Students

Would students enter such a program if it existed?
If so, from what existing programs would they be diverted?
If students completed the program, would they enter the vocation it had trained them to enter?
Would the program assist industry if those who entered it were diverted from other fields where there is as acute a shortage of workers?

Budget

How much would it cost?
Where would the funding be generated?

This list could be expanded, but it is long enough to show that analysis of the suggestion demanded consideration of the content, staff, administrative organization, students, value, and expense of the program. In short, analysis showed the numerous parts into which an apparently simple suggestion had to be broken down before it was possible to see all that was involved. The major point to remember is:

> An analysis (as with any other problem) becomes manageable when the major problem is divided to the degree that each separate part can be easily grasped.

In actual practice, of course, many an analysis can follow a pattern established in previous analyses of the same or similar matters. But when there is no established pattern to rely on, one must often use considerable imagination in order to discover what facts might be significant. An analysis of the unemployment situation, for example, would obviously include the total number and percentage of employable persons out of work, but would also need to tell such facts as the geographical distribution of the unemployed; their distribution in towns and cities of different sizes; their distribution as to age, sex, and race; their education and their training in vocational skills; the number in various brackets based on length of time since last employment; the changes since previous figures were released; the effect of the season or of special causes (strikes, for example) on increase or decrease; and the specific industries affected. All these facts and of course many more might affect the validity of any generalizations upon such points as the seriousness, the causes, and the possible means of improving the situation.

Unemployment is of course a vast and complicated subject, and an analysis of it on the scale described above would probably be produced in order to make available all the facts anyone might need for any special purpose. Most of the time, however, the person who writes an analysis, like the person who writes anything else, should not exhaust the subject but should bear in mind the specific purpose of the analysis by taking into consideration the question of who the readers will be and why they need an analysis of the subject. The entire approach to the job of analyzing, the organization, and the amount of detail and emphasis on various aspects of the subject can be determined more intelligently if the reader and purpose as well as the subject receive constant attention.

This brings us to the question of how one goes about the task of writing an analysis. The following method is suggested.

First, concentrate on the question of who will read the finished product and why. Will it be read by those who need information only so as to understand the subject, or will it form a basis for decisions and action? To illustrate the latter possibility—a reader might want: an analysis of competing products in order to decide which one to buy; an analysis of expenses so as to decide upon the areas in which to economize; an analysis of the probable side effects of a change in the design of some product in order to decide whether these would offset the benefits; an analysis of the demand for heat and power that would result from some enlargement of operations so as to know whether existing sources would be sufficient.

Next, bearing the function of the analysis firmly in mind, begin to list the points, large and small, on which information will be needed; and as information becomes available, decide upon an appropriate pattern for presenting it. Perhaps the key question in organization is whether to weave together the facts on different aspects of the subject or whether to keep each aspect separate from the others. In an analysis of industrial accidents, for example, facts might be presented on such different matters as the number of accidents in different kinds of jobs, the number in different shifts, and the relative number for workers with different amounts of experience. It would be possible to create a pattern in which these matters comprised respectively, main points, subpoints, and sub-subpoints; but it would also be possible to handle each of them separately.

When an analysis is made for what might be called random use rather than for specific application, the general method suggested for making an outline (Chapter 4) would be hard to improve upon as a method of starting the job. This consists basically of jotting down rough notes and creating a plan by grouping together related material. When an analysis is made for a specific application—that is, because it is needed for a predetermined purpose—the writer's understanding of what is needed permits a narrowing of approach to concentrate attention on what is significant in view of the particular occasion. For example, an analysis of the quality of some curriculum in a specific university might be different if made for the general purpose of accreditation than if made by the personnel department of a corporation as a basis for recruitment.

The actual writing of an analysis poses no unusual problems. A brief introduction is usually sufficient to cover, so far as circumstances demand, such points as the purpose for which the analysis has been made, the basis on which the major divisions rest and perhaps the reasons for their choice, the source of facts included and the scope and limitations of whatever investigation was made, and the assumptions, if any, on which the interpretation of the facts is based. To these points may be added individual facts of special importance and general conclusions that are specially significant.

When the bulk of the material in the body of an analysis consists merely of objective facts for their own sake, the main problem in presenting it is the avoidance of monotony. Unless a writer has a natural gift for interesting style, there is not much that can be done about this except to break the prose with informal tables, which would probably be desirable for the sake of clearness even if monotony were not involved.

When the facts call for interpretation, the main consideration is to make the relationship of facts to each other entirely clear and to explain the processes of reasoning involved in deciding upon their significance.

An analysis may or may not call for a general conclusion. Sometimes, merely breaking down the total subject into its component parts is all that is necessary. Yet, as mentioned earlier, an analysis indicates parts of something either separately or in their relationship to the whole; and when their relationship to the whole is important—when the analysis is made to provide a basis for decisions and action—a recapitulation of the major facts, a reinforcement of whatever has been said about their significance, and an application of the analysis to the unsettled issues becomes a practical necessity.

Many people who write analyses do not think of them as being a distinct type of writing, but most of us can obtain better results if we recognize what we are doing. And skill in writing an analysis is well worthwhile because an analysis may not only exist independently but is often an important part of a report, a proposal, an article, or any other kind of technical writing.

Following is an analysis of the amount of electrical power used for various purposes.

The growing demand for electricity presents an extremely complicated problem—a problem that is constantly increasing because the demand for electric power has been doubling each ten years. The solution will not be found by urging reduction in its use for so-called frivolous and unnecessary purposes such as, for example, the electric toothbrush. This becomes obvious when the actual amount of power used for various purposes is considered.

In the area served by one power company, for example, 58 percent of the electrical power generated is used not in homes but by such customers as factories, aluminum plants, agriculture, business establishments, and governmental services—street lights included. An excessive cutback in this use would not only deprive the public of needed services but would also wreak havoc on the economy of the region by damaging industry and creating unemployment.

The rest of the power, 42 percent, goes to residential users. This amount would not be substantially reduced even if all the "frivolous" uses—toothbrushes, can openers, toasters, and the like—were eliminated. The fact is that these "gadgets" account for only about 3 percent of the power used in residences—about 1¼ percent of the total for all purposes. Practically all the electricity used in residences goes to perform such heavy-duty jobs as heating, cooking, and heating water.

The problem has another aspect that may disturb those who oppose new efforts to build new power plants because most kinds of power production create pollution and environmental problems. The fact is that increased power production is essential to pollution control itself.

It is generally recognized, for example, that thousands of sewage-treatment plants are needed in order to reduce water pollution. A secondary sewage-treatment plant serving a population of 100,000 requires over 500,000 kilowatt-hours of electricity a year. Also, recycling plants to handle paper, plastics, or aluminum cans require large amounts of power—as do plants that compress or cut up old automobile bodies. The same is true of mass-transit arrangements, which are so strongly urged to reduce air pollution and the consumption of gasoline. When the Bay Area Rapid Transit System in the San Francisco region was under construction, calculations indicated that it would require about the amount of power used by 43,000 residences.

(The analysis from which this was adapted goes on to discuss methods suggested for increasing power production.)

TECHNICAL ARTICLES

Although most students may think it unlikely that they will ever engage in writing articles, the chances that they will do so—or at least might profit from doing so—are greater than might be imagined. Hundreds of periodicals dealing with technical and scientific subject matter roll steadily from the presses, and the students of today will eventually be the authors who fill their columns.

To be sure, articles will not be written just to keep the periodicals in existence. On the contrary, the articles must be written because in every aspect of science and technology there is a great deal of information to be disseminated, and the periodicals are needed as a place to publish the articles written to disseminate it. To cite a single example: Medical journals exist because without them doctors would not have access to the information that keeps them abreast of developments in their profession.

Scope and Importance

The term *technical article* is, of course, an oversimplification. Articles vary in the extent to which they are technical because the publications that print them perform different functions and are intended for different kinds of readers. At one extreme are the highly specialized scientific journals such as those dealing with nuclear physics or higher mathematics. Articles in these journals are far beyond the grasp of any readers except specialists. Not quite in this rarefied atmosphere are other publications whose readers must still be able to comprehend formulas, mathematical calculations, and technical terminology that call for extensive knowledge of the field concerned as well as general grounding in science and technology.

Somewhat farther down the scale of technical and scientific specialization are periodicals containing articles in numerous technical or scientific fields rather than in only one. Also, articles on such subjects appear in many periodicals that are not limited to science and technology in their interests. The widespread public interest in such matters as pollution, ecology, power shortages, and long-range energy problems results in a steady stream of articles by authoritative authors, written for intelligent, well-educated readers who have no more reason for reading than a natural interest in the world that they live in. The publications carrying such articles include even the better newspapers; and a newspaper article by such an author as the science editor of a worldwide press service is entitled to respect. Some of these articles in periodicals with general circulation may contain generally accepted facts; some may attempt to give a balanced picture of a controversial subject; some may take one or the other side of a disputed question; but all of them are the work of well-informed authors, some of whom are recognized authorities.

Not entitled to inclusion as technical articles are the purely popular articles that also appear in great numbers. These are a different breed. Whereas the technical article is written by an author with special qualifications whose purpose is to inform, the popular article is likely to be the work of an author whose main occupation is writing and who may have neither an enduring interest nor a claim to expertise in the field where the article lies. Such work is read not only for information but for its entertainment value. It represents an attempt to be colorful and may be inclined toward sensationalism. To be sure, it may serve its purpose; but it is less likely than a bona fide technical article to be organized and written with enough objectivity and restraint to bear up under the scrutiny of well-informed readers.

The above kinds of publications are numerous, and the writers who will eventually fill their pages will include students who are studying this book. An even greater number of today's students will contribute, however, to publications that fall into none of the preceding categories. Rather, they will write for periodicals concerned with specific industries or technologies. Some of these, in fact, are published by one or another of the nation's major corporations and deal almost entirely with its own activities.

The number of these industrial periodicals is immense. Every industry, from the automotive to the beverage-dispensing, has one or more. Their importance would be hard to overstate. Though reports are perhaps the main source of information in industry concerning matters that fall in the area of professional workers' personal responsibilities, it is articles in industrial publications that on a broad scale keep professionals abreast of general developments, perhaps in their own companies and certainly in the industry where they are employed. And it is in such publications that those who use this book will probably find an outlet for the articles that they are eventually in a position to write.

Finding Subjects for Articles

Finding subjects to write about will not be a problem for a person holding a professional position rather than preparing to hold one. When a person holds such a position, subjects grow naturally out of the work itself. That is, such people will write articles because they have developed some information or ideas that are worth sharing, rather than hunting subjects in order to write articles. They may describe some newly designed machine or process, for example, or a modification or improvement of one that is already in use. They may explain some new technique or evaluate alternative techniques. They may tell about some method to solve a chronic problem that has long plagued their industry. They may describe the manner in which some project has been carried to completion—the difficulties encountered and how they were overcome. Examples of these and other types of subjects can be found by looking through industrial periodicals in any field.

Often, the subject of an article develops by stages. Work done in the line of research, development, or production will first result in a report. Then a key figure in this work may have an opportunity to read a paper about it at a convention or conference. Finally, someone may work over the paper for publication. All in all, the time is likely to come when any person in a professional position will have a ready-made subject for an article if he or she is alert enough to recognize it.

For the student, the problem of finding a subject is obviously different. Instead of writing an article because you have the subject matter for one, you may need to look for a subject because you are asked to write an article. It should assist you in this quest if you bear in mind that most published articles are of one of four types: (1) the article that concerns a genuinely new subject, (2) the article that presents new information or a new attitude about a subject which has been written about before, (3) the article that concerns a subject made timely by some new trend or event, and (4) the article that is written to instill a certain conviction about a particular question.

To find a subject for an article about either of the first two types you might start by considering your own school. If it is one where research is carried out, you may be able to learn enough from an instructor to write about one of the research projects. Or again, you might explain some new course, program, or curriculum. You might tell about unusual features of some new building, or the construction of a roof over a stadium, or the means by which artificial turf for an athletic field can be rolled up and stored between seasons.

Looking beyond school, you may find a subject in the community where it is located, for in almost any community improvement projects are usually being carried out or studied. Occasionally, too, a subject may be found that involves both school and community—as, for example, student participation in a survey of the attitudes of residents in an area toward a land-use plan.

Articles of the third type—dealing with a matter brought into attention by some recent event—are also a good possibility. For example, in a year when tornadoes have been especially frequent, even a writer who is not an authority on them can acquire enough information, old and new, to use them as the subject of an article. Most of the information in such articles may not be new, but a minimum of new information may provide an up-to-date flavor. Not to be overlooked are the possibilities of writing articles made timely by the passage of legislation or by the announcement of a decision by some regulatory board. Accordingly, when you are looking for a subject, you can do worse than to give some attention to recent news developments. There is always a chance that they may revive interest in a subject that has been dormant.

The fourth type of article, one that aims to support some particular conviction, may deal either with a new subject or an old one. If your subject is old, however, your idea for an article about it needs to be evaluated with considerable care. Sometimes, to be sure, though your subject may not be new, your conviction about it may be new or perhaps unorthodox—in which case you are fortunate. But if your conviction is no longer disputed, the chances that you can write a worthwhile article about it are low. If your conviction is one that many other people share but that is still controversial, you will need to have something to add to arguments that have been previously advanced. This last statement should perhaps be modified. If the particular periodical that you have in mind as an outlet for your article has not previously published anything of the sort, yours might meet a need. But even in this event you should try hard to find something fresh to say rather than just offering a rehash of what others have said already.

All in all, perhaps the best way to find a subject of the type in question is to ask yourself, "Do I agree with every attitude that is conventional in my intended profession? Do I have some ideas of my own?" Many people have within themselves the germ of an article if they realize that it lies in the unorthodox remarks they make in private conversation.

The extent to which students can make an article technical in content obviously depends on how far they have progressed in a professional curriculum. There is no reason, however, that a student who has not yet taken any highly technical courses should be unable to gain experience and skill by writing on a subject. Moreover, students can gain such skill without necessarily thinking in terms of an article for publication in a periodical with nationwide circulation. Many publications circulate almost entirely within a single state or region, and it is probably most practical to think of some such periodical as a potential outlet.

Adapting the Article
to the Periodical

As the preceding paragraph implies, when you write an article you should have in mind at least one periodical where it might be published. This is sound advice even when you write an article as a classroom assignment. After all, when you write an article during your professional career you will certainly have publication in mind, so you should consider now, so far as possible, the matters that you will have to consider then. And any experienced writer will advise against writing an article and then trying to figure out who might publish it. Rather, you should give some thought to what periodical or periodicals might be interested while the article is still taking shape in your mind. Having reached a decision, you should then adapt your product to the one you have singled out.

Both of these steps—choosing a potential outlet and adapting your article to it—call for carefully examining any target you think is likely. During your examination you should bear in mind the questions below, which apply to your study of any publication you consider either as a student or during your professional career.

Will the article that you have in mind concern a subject that the publication is interested in, to judge from the articles you find it to contain?

What kinds of readers does the publication seem intended for? How much knowledge do they possess? Closely related to this is the question: To what extent should you use technical language (where it might be useful) if your article is to be like those you examine?

How far should you go in the use of formulas, equations, and mathematical calculations?

Should your style be entirely formal? If not, just how informal should you make it? Should you use any jargon?

How prevalent is the use of tables and figures; and if it is frequent, how sophisticated are those used?

How long are the previously published articles? (If you find none outside the range of 500 to 2000 words, for example, and you submit one that runs to 4000 words, the best you can hope for is to have it returned and be invited to cut its length in half and submit it again.)

How long are the paragraphs in most of the articles?

Do the articles seem to be written with restraint or do their authors sometimes give free rein to their enthusiasm?

What kinds of openings seem to be preferred? Have the authors attempted to arouse interest by imaginative openings or do they depend upon the subjects to arouse interest without such assistance?

What kinds of titles are customary? Are they intended to arouse curiosity or are they matter-of-fact announcements of the subjects?

Are the articles documented when there might have been a reason for documentation? If so, what form does the documentation take?

The preceding list is long, but each of the items is well worth attention. It is unfortunate when an article fails to be published simply because it needlessly differs in too many respects from the kind of material a periodical customarily uses.

Writing the Article

The process of actually writing an article is not essentially different from the process of writing anything else, so you should utilize the instructions in the preceding chapters. Since a technical article should be well-organized, the suggestions in the chapter on organization should be useful. And though some matters of style may be affected by the adaptation of your article to the prospective outlet, the general recommendations in the chapter on style should be borne in mind—especially after you have written the first draft and are polishing it. The chapter on mechanics will answer many questions that may confront you. (Don't overlook what is said near the opening of the chapter about manuscript form.)

Also, bear in mind the earlier passages of the present chapter. Many an article-writer needs to define terms, explain a process, offer a technical description, or analyze a problem or situation. When these matters were discussed, it was pointed out that the problems in question crop up constantly in such places as reports and articles.

It may be desirable to place headings in your article. If you do so, they should almost certainly be limited to the main points of your outline. Not all the headings that you see in articles were placed there by the authors. In many publications, headings are inserted by an editor where they are needed to break up long, formidable-looking columns of type. These are not to be confused with headings used systematically to show divisions and subdivisions of books and reports.

Use of tables and figures was mentioned in connection with adapting your article to the periodical. If you decide to use such material, you should know what it will be before you attempt to put your written material into its final form. The use of tables and figures will be covered in more detail in the next chapter. For the moment, it is sufficient to point out that except for the possible use of photographs to arouse interest, the figures and tables in a technical article are an integral part of it and are used because they are the best means of conveying some portion of the information it should contain.

Benefits of Writing Articles

The value of technical articles to industry has already been pointed out, but you are naturally interested, also, in whether you will personally benefit from writing them. Most assuredly they can make a substantial contribution to your professional progress. By writing articles that are published in scientific or technical periodicals you can secure favorable attention from the management of the organization you work for, acquire a reputation among your colleagues, and make a great many people outside your own organization aware of your existence and your professional competence. Consequently, by developing the ability to write technical articles, you acquire a skill that can advance your career. An example of a technical article follows. The article appears on the right-hand pages and an analysis of it is placed on the left facing the part under discussion. We suggest that you read the article completely before proceeding with the analysis.

SPECIMEN TECHNICAL ARTICLE

One of the primary differences between articles in magazines and those in professional journals is the necessity of capturing the reader's interest. Unlike a technical report, which people read to gather information necessary to their jobs, an article provides entertainment as well as information. True, professional people read articles to keep up with their fields; nevertheless, the profusion of reading material in every area gives great latitude for people in choosing which articles to read. The title alone of this article would be enough to attract most readers interested in science, so the authors devote only the first sentence to capturing the reader's interest. In an article on a less exciting subject, or aimed at a more general audience, the authors might have spent far more time at capturing reader attention, or in general, explaining why the article should be read.

The second sentence gives a classical definition of the subject, despite the fact that probably no reader of *Science* magazine, in which this article first appeared, would be unfamiliar with the information that the dolphin was not a fish. The sentence serves two purposes, however: It reminds readers of specific information they may have forgotten, and it firmly delimits the subject of the article. The rest of the paragraph serves as a summary or abstract of the article, and could easily be used without change in a report on the same subject. It gives a brief description of the planned investigation, the problems that hindered it, and the results of the experiments aimed at solving the problem.

The second paragraph provides a history of study in this area, starting with the first known investigations. In a study of an area with a greater number of investigations, the authors might have simply dealt with the more recent or perhaps the most relevant studies in order to save space. In this case, there is enough space to give a brief description of the actual experimental method and its results. Terms such as *intraperitoneally, asphyxia,* and *nasopharyngeal sphincter* are used without definition, since the authors expect that any scientifically sophisticated audience, such as the readers of *Science,* would be familiar with them. For a more general audience, the authors would have defined such terms or attempted to avoid using them.

A transition sentence opens paragraph 3 and moves the reader from general history to consideration of the authors' investigations. Paragraphs 3, 4, 5, and 6 provide a detailed description of the development of equipment and methodology. Exact measurements and the use of trade names of equipment give the kind of specificity required in a scientific report. In a report on the same subject, however, design drawings would probably be included as well, but since the authors do not expect anyone to attempt to duplicate their work from this article alone, they provide less information.

Anesthesia for the Bottlenose Dolphin: Tursiops Truncatus*

1. The great whales, dolphins, and porpoises have recently attracted much interest, particularly because of their apparent intelligence and large, well-developed brains. These animals belong to the order Cetacea and are air-breathing, warm-blooded mammals which spend their entire lives in an aquatic environment. Our investigations have dealt primarily with the neuroanatomy of this animal and, especially, with the construction of a brain atlas from several thousand wholemount brain slides. The application of classical neurophysiological techniques to the massive dolphin brain as planned by this laboratory has been hindered by two major problems: (i) no known safe anesthetics for this animal are available, and (ii) there are no adequate methods for artificially supporting respiration. These difficulties are related in part to the specialized central nervous control of the blowhole mechanisms associated with the complex engagement of the larynx into the nasopharyngeal sphincter and with the unique pattern of respiration seen in these diving mammals. Our experiments have led to a method of anesthesia for the dolphin which has proved safe and reliable.

2. The first attempts at general anesthesia in the dolphin were made by Lilly et al., in 1955. These investigators injected intraperitoneally 30 mg of pentobarbital per kilogram of body weight, the usual primate dose, and found that respiration became uncoordinated and death resulted from asphyxia. When the dose was reduced to 10 mg/kg, air leaked from the mouth, the lungs gradually deflated, and respiratory efforts became ineffectual. In a series of animals receiving various doses of pentobarbital or paraldehyde, leakage of air through the nasopharyngeal sphincter always presaged eventual respiratory failure and death.

3. In October 1963 we began experiments designed to develop a safe method of anesthesia for the dolphin. First we had to perfect suitable endotracheal and respiratory equipment for maintenance of adequate ventilation. Dissections were carried out to study the anatomical relationships of the oral, nasopharyngeal, and respiratory passages. Manual palpation of the laryngeal-nasopharyngeal area was performed in living animals, a specially constructed surgical tank being used in which the animal was firmly secured and kept moist by hosing frequently with sea-water. Four wooden restraints were constructed to fit the contours of the body, and were placed posterior to the blowhole, anterior and posterior to the dorsal fin, and anterior to the flukes.

*E. L. Nagel, et al., "Anesthesia for the Bottlenose Dolphin: Tursiops Truncatus," *Science*, 146 (December 18, 1964), 1591–93. Copyright 1964 by the American Association for the Advancement of Science. Reprinted by permission of the author and publisher. [Paragraphs were not numbered in the article, but numbers have been supplied to facilitate reference.]

Despite the generally clear language of the article, the authors sometimes lapse into what could be considered jargon: "gross observations" means simply that they watched the animal. "Extubation" in paragraph four verges on the same fault. Both terms, however, are found frequently in scientific writing.

The discussion of the first set of experiments begins with the seventh paragraph—about halfway through the article. All of the preceding is essentially preliminary information necessary to make clear the significance of the actual work. In the following discussion of the experiments, the authors explain not only what they did, but why—an important factor in science writing, all too often omitted. Thus it is clear not only why they attempted a particular method (using the intraperitoneal route) but why they did not use another (intravenous channels). By explaining both, they forestall objections that might rise in the mind of a knowledgeable reader, and provide for all readers confidence in their techniques.

Halfway through the paragraph the authors give the results of the first experiment: The animal died. The rest of the paragraph is devoted to a discussion of the conclusions they drew from this result.

This rather long discussion of an unsuccessful experiment is included in the article for the same reason that the discussion of using intravenous channels is discussed—to indicate that the successful methods were developed for good reasons, and that other methods were either not considered or were considered and rejected for good reasons. This paragraph illustrates the fact that in science, we learn from failures as well as success. Unless the failures are reported, however, others may unknowingly repeat them.

Wet, heavy, foam padding protected the animal's body from the tank and the restraints. The use of a padded, oral speculum (Sands type for horses) allowed the dolphin complete range of jaw motion and provided adequate access to the animal's oral cavity.

4. For the endotracheal catheter we used a clear Tygon tube (2.5 cm outside diameter, 50 cm in length) having a short Murphy bevel and a highly distensible cuff (80 to 100 ml of air for full inflation). A heavy stainless steel coil inside the tube prevented its kinking and collapse. The final design of the respirator was based on a large-animal ventilator, the Bird mark 9X. Preparation for placement of the endotracheal tube was accomplished by insertion of the arm into the animal's oral cavity and through the oropharyngeal sphincter, so that two fingers were hooked around the vertically placed larynx. This structure was then pulled anteroventrally and a finger placed into the glottis for guidance of the endotracheal tube into the larynx. The animals were then maintained on air by means of the respirator for periods of up to 3 hours. Following extubation they swam, vocalized, and fed immediately on return to the home tank. As a precautionary measure, all animals were given three daily administrations intramuscularly of 1,500,000 units of benzathine penicillin G and 10 ml of vitamin B complex, with vitamin C as supportive post-anesthesia care.

5. The respirator was later modified by the addition of an apneustic plateau control unit which permitted variations to be made in the duration of the inflated stage. With this device, the animal was ventilated in a manner mimicking its natural pattern. Thus, the respirator rapidly inflated the animal, maintained it in an inflated state for a preset period of time, and then deflated and inflated the animal in rapid succession.

6. At this time the state of the animal was monitored by: (i) taking the rectal temperature by means of a telethermometer (Yellow Springs model 43) and a thermistor probe; (ii) an electrocardiogram recorded by means of a multi-channel recorder (Grass model 7 polygraph); (iii) gross observations of the eye, melon, blowhole, and general body movements; (iv) observing the color of the tongue as a crude, but effective, measure of the degree of oxygen saturation of hemoglobin.

7. We then began experiments with various anesthetic drugs. Since we planned to map the brain in one of the studies, we hoped that barbiturate anesthesia could be utilized to enable adequate comparison with results seen in other animals. Intravenous channels were difficult to obtain in the dolphin, however, other than by means of a percutaneous puncture of the great vessels in the abdominal region. In the first experiment with barbiturate we therefore utilized the intraperitoneal route and short-acting drugs. Thirteen miligrams of thiopental per kilogram, in two doses 5 minutes apart, followed in 10 minutes

In paragraph 8 the authors discuss other anesthetics that might be used, based on the conclusions drawn from the first experiment. Once again, they are careful to explain why they chose one agent rather than another.

In paragraph 9 the preliminary results of the next set of experiments are discussed. These successful results led to a further development of the method, which is described next, and the further results are detailed. Note especially how the authors use the techniques of technical description of equipment and the process description of their experiments—techniques that were discussed in this chapter.

by 5 mg of methohexital per kilogram in two doses 5 minutes apart, produced a loss of all apparent reflex activity for the next 3 hours. Spontaneous movements of the eyes and flukes began 5 hours after drug administration, and voluntary respiration was first noted approximately 2 hours later. The animal was returned to the home tank 9 hours after drug administration. Leakage of air from the mouth proved to be an ominous sign, since death followed in another 1½ hours. However, the animal's lack of buoyancy in air and the increased possibility of developing hypostatic pneumonia presented an increasing risk to the animal with further maintenance on the respirator. This experiment delineated certain major problems concerned with administering anesthesia to the dolphin, namely, that any significant degree of central nervous system depression after anesthesia would probably result in the animal's death. Short of intravenous administration of drugs, further experimentation with rectal, intramuscular, intrapleural, or intraperitoneal injection of hypnotics seemed contraindicated because of uncertain absorption, metabolism, and excretion of these agents in the dolphin.

8. Inhalational anesthetics were tried next. Flammable agents were ruled out because of the explosion hazard in the laboratory. Potent agents were also excluded because of insufficient monitoring to insure safe, controlled administration. Halothane, therefore, was not considered because this agent depresses the cardiovascular system even in light planes of anesthesia. In seeking an agent apparently free of such side effects, nitrous oxide containing varying concentrations of oxygen was then used. Nitrous oxide is a good analgesic and amnesic agent, but a weak anesthetic. It was chosen because of the lack of side effects seen in other species of animals, including man, and because of rapid and complete recovery following its use. The depth of anesthesia is known to be dependent on the species, individual, and the premedication given.

9. In preliminary experiments 50 percent nitrous oxide in oxygen was administered for periods of up to 3 hours, during which time it was well tolerated. This mixture produced a definite degree of analgesia as evidenced by an elevated threshold to noxious stimuli. An uneventful recovery of the animal followed, with resumption of normal breathing pattern, swimming, and feeding activity. To secure deeper planes of anesthesia, the concentration of nitrous oxide in oxygen was then increased to 70 percent, and the following effects were noted in several experiments. Instead of passive acceptance of the respirator, the animal now attempted to breathe around or dislodge the endotracheal tube. These movements seemed uncoordinated and exaggerated during the induction phase of anesthesia. Following this brief excitation phase the animal became largely insensitive to noxious stimuli, such as subcutaneous insertions of needle electrodes or penetration of the blowhole to the nares with the fingers. Such stimuli, in our experience, have always evoked strong evasive reactions in the awake animal.

The last paragraph in the article, paragraph 10, gives the final results of the experiment and discusses directions that future work might take. The authors identify exactly what they have accomplished: They have developed a method that has allowed them, on *four* occasions, with *two* animals, to insert electrode sleeve guides and record brain waves. Since success in such work more than once is adequate to prove the value of the method, they have reported their results at this time. In a later article, they may report what they found from the recordings of the electroencephalographic activity.

Paragraph 10 is a suitable conclusion. The authors had done the work described in the preceding material in order that they might perform certain experiments that would be possible only after anesthetics had been administered. In their last paragraph they show that their methods had been successful by telling about the performance of such experiments. Also, they point out that their methods not only made possible the kinds of experiments they themselves wished to perform, but have also opened up the possibility of performing a wide range of other work—major surgery in general. Thus the authors end their article by making sure that their readers realize the full significance of their work.

This article is remarkable for a number of reasons. It deals with a highly technical subject in a manner that is interesting even to readers not trained in the field. The organization of material is clear, proceeding from early failures to final success. But perhaps most remarkable of all is the calm, restrained tone of the article—so ordinary that we barely note it at all. Because the tone is exactly what we have come to expect in science writing, the casual reader may be tempted to pass over the very real significance of the work reported. This article is the first mention in print of a major scientific breakthrough, of the kind that would have rated major headlines in the nation's newspapers had it occurred in baseball or film, for instance. Prior to this work, surgery in dolphins was impossible—anatomical studies could be made only on dead animals. Since the development of these methods, the authors and their colleagues at the University of Miami have done considerable work in this area and have published numerous articles in various journals. Nevertheless, in keeping with the long tradition of scientific objectivity, the authors reported simply, "These methods open the possibility of performing major surgery in this species for the first time." Students would do well to study this article carefully. They could learn more about technical writing from it than from dozens of lectures on the subject.

10. In the last three experiments we extended the monitoring procedures by utilizing the infrared CO_2 gas analyzer (Beckman) in order to measure $_pCO_2$ in the expired air. Two penetrations of blood vessels were performed percutaneously with a 20-cm hypodermic needle. In the last such experiment, 0.25 mg of thiopental per kilogram was injected directly into the abdominal *vena cava*. This small dose of barbiturate did not appear to affect the animal adversely in any perceivable way. Monitoring devices which measure arterial $_pO_2$, $_pH$, and pressure will probably be necessary, however, to test safely such agents as the barbiturates or halothane when given in effective anesthetic doses. Using nitrous oxide anesthesia, we have been able to insert electrode sleeve guides into the skull on four occasions in two animals, and to record successfully cortical electroencephalographic activity. These methods open the possibility of performing major surgery in this species for the first time. However, for some surgical procedures supplementation of nitrous oxide anesthesia with other agents may be necessary.

EXERCISES

Exercise 1

Indicate the genus under which you would place each of the following terms if you were defining it in a sentence definition. Be sure that your decisions are accurate, and try in each case to choose a genus that limits the term as much as possible so that the differentia need not be unnecessarily long and complicated.

(Note: In this and other exercises concerned with definition most of the terms will be familiar to you so that heavy reliance on a dictionary will not cramp your independent thought. The exercises are not intended to test your knowledge; they are designed to demonstrate the need to avoid accidentally omitting essential facts that you know but do not think about, and the need to express all essential facts in a precise and orderly manner.)

octane	electrolysis	polyester	cellophane
brandy	volcano	autoclave	butterfly
pediatrician	comet	butterfly	petroleum
galvanize	atmosphere	glass	
cold front	transit	submarine	
hybrid	metabolism		

Exercise 2

It is often necessary, in writing a definition, to discriminate with special care between the term to be defined or the referent that the term stands for and one or more other terms or referents. For example, if you were defining *tree* it would be necessary to discriminate between *tree* and *bush*. If you were defining the following terms, what other terms would call for such special discrimination? (Choose five to answer.)

fog	peat	brazing	caterpillar
psychoanalysis	polyester	jelly	detergent
whale	silt	ocean	calculus

Exercise 3

Indicate the faults that you find in the following sentence definitions.

Note: The function of a single definition is to tell what the term means in one legitimate use rather than to cover all of its possible meanings. For example, the definition of the term *lock* would not need to cover the meaning of the term as used in *lock of hair, lock in a canal,* and *lock on a door.* The definitions below are faulty not because they fail to cover all the possible meanings that the term may convey but because they fail to express satisfactorily the single meaning obviously intended on the particular occasion.

1. A bee is a flying insect that stings.
2. A bridge is a structure supported by piers or hanging from cables and built in order to make it possible to cross a river.
3. An outboard motor is an internal combustion engine used to propel a boat.
4. A lock is a device operated by a key or combination and used to prevent a door from being opened.
5. Welding is when metallic bodies are caused to unite by heating them and allowing the heated parts to flow together or by hammering or compressing the bodies.
6. Glue is a substance with which materials are glued together.
7. Water is an element each molecule of which is composed of two hydrogen atoms and one oxygen atom.
8. Hail is a form of snow consisting of small, roundish lumps of ice.
9. Silk is a strong, lustrous fiber or the fabric made of that fiber.
10. Sleep is a condition in which sensation, thought, and feeling are reduced to the extent that the person or animal concerned is unconscious.

Exercise 4

Write a sentence definition for each of five terms from the following list. None of the definitions needs to cover more than one meaning of the term concerned, and each may be limited, if necessary, to a special point of view.

yeast	pasteurization	solar wind
planet	horsepower	northern lights
heat	sand	catalytic converter
cybernetics	inversion (as	geyser
diving bell	a condition of weather)	ecology
propjet	windmill	wet suit
soap	calculus	dredge (noun)
computer		

ASSIGNMENTS

The following assignments cover the writing problems discussed in the chapter. In almost every case a list of possible subjects is provided. These lists call for comment. *First:* Some of the subjects may impress you as not being especially technical, but there are valid reasons for their inclusion. This book is intended for the use of people in a wide range of fields; and the more technical and specialized a subject is, the smaller would be the percentage of readers who could use it. Because it would be impossible to suggest several highly technical subjects in every conceivable field of specialization, a good many subjects have been included that can be handled without specialized knowledge.

It is not necessary to discuss a highly technical subject in order to profit from doing what an assignment calls for. It is entirely practical to learn the technique of handling a technical subject by writing about a subject that is not technical. For example, you could improve your ability to describe any machine by attempting to describe an automatic dishwasher, even though you do not need a technical education to learn the facts about one.

Second: In choosing a subject for any assignment you may find it helpful to look over the lists for other assignments also. A certain mechanism, for example, might be listed as the subject for a technical description; but it might also be possible to base an explanation of a process on the way that it functions or to write a set of instructions for using it or repairing it.

Third: As you look over the lists, use your imagination. The lists may be helpful in stimulating you to think of additional subjects for yourself.

Assignment 1

Write one or more expanded definitions of terms listed in Exercise 4, or of other terms if permitted by your instructor. Each definition should be from 100 to 400 words long. (Some of the terms, of course, may not present enough problems to call for an expanded definition.)

Assignment 2

Write a technical description of one of the following objects or of some other relatively simple object with which you are familiar. Aim at a length of 400 to 800 words, adjusting the degree of detail so that the paper falls within these limits. Do not attempt to describe anything so complicated that you cannot do justice to it. (This assignment may or may not be accompanied by one or more figures; if figures are used, it may best be postponed until after the chapter on Tables and Figures has been studied.)

a wind-speed indicator	a spinning reel
a musical instrument	an automobile muffler
a fire extinguisher	an automatic dishwasher
a hydraulic jack	a kerosene lantern
an incubator	a safety razor
a retractable pen	a pencil sharpener
a stapler	a hand operated egg beater
the eye	a safety-release ski binding
a laser generator	a bicycle brake
a lawn mower	a simple steam engine

Assignment 3

Write an explanation of one or more of the following processes or of some other relatively simple process with which you are familiar. It is suggested that you explain a process that you can cover in 400 to 800 words. If you wish to explain a process that cannot be covered within this length, you may be able to reduce its size by covering only one part of it. List A consists of processes performed mainly by persons, and list B consists of processes performed mainly by machines.

LIST A

1. Tuning an automobile engine
2. Grafting a branch onto a tree
3. Preparing a slide for microscopic examination
4. Performing a specific welding or brazing job
5. Producing a casting using a sand mold process
6. Sewing a specific garment
7. Making a picnic table
8. Cruising timber
9. Making bread
10. Taking and analyzing a soil sample

LIST B

1. How an automatic dishwasher operates
2. How a furnace thermostat functions
3. How self-developing film works
4. How a specific type of duplicating machine operates
5. How a nerve cell works
6. How an oscillating electric fan works
7. How a seed germinates and grows into a plant
8. How binoculars work
9. How a movie projector works
10. How a transistor works

Assignment 4

Write a set of instructions for performing one of the following processes, one of the processes mentioned in list A of Assignment 3, or a process of your own choice that you know how to perform.

1. Pruning a fruit tree
2. Developing and printing film
3. Starting a new lawn
4. Making a concrete walkway
5. Bleeding automobile brakes
6. Operating a specific type of office equipment, such as a duplicating machine
7. Truing a bicycle wheel
8. Tying a specific kind of knot, such as a bowline or a butterfly
9. Decorating a cake
10. Setting up a filing system

Assignment 5

Write an analysis of one of the subjects listed below or of a comparable subject about which you can obtain information.

1. Analyze the contents of this book: the kinds of material—such as discussion, exercises, examples—and the subjects covered.
2. Analyze the use of space in a specific area, such as a building, a city street, a park, or a farm. If you choose to analyze a street, for example, you might consider the number and kinds of businesses, the flow of pedestrian traffic, parking, the changes during a 24-hour period, space devoted to landscaping and traffic control, and any other features that affect the use of the space.
3. Analyze the effect of a new regulation or a change in an existing regulation. An example might be a change in the draft law, drinking age, speed limits on city streets, visiting hours in campus dormitories, or similar regulations.
4. Analyze a periodical—for example, a technical journal in your field—in regard to subjects it deals with, length and number of articles, space devoted to advertisements, and the kinds of advertisements.
5. Analyze the expenses involved in owning a house compared to renting.
6. Analyze the composition of a specific group of people, such as your student body, your business's employees or customers, or your club's membership.
7. Analyze the expenses of attending school.
8. Analyze the expense of owning and operating an automobile or truck compared with leasing it.
9. Analyze parking needs for your school or business. Include changes during the day, the week, and the year.
10. Analyze the benefits and shortcomings of a particular career.

Assignment 6

Make an outline for a technical or semitechnical article. After the outline has been approved by the instructor, write the article. When you submit the outline, you should name at least one periodical to which the article might be submitted. Your choice of a subject and the nature of the article should result from previously acquired familiarity with the periodical. The length of your article should be approved by the instructor and appropriate for the periodical.

7 | Tables and Figures

It is almost impossible to imagine technical writing that does not make frequent use of figures and tables. This is true because some kinds of information—for example, statistical data or the shape, arrangement, and size of objects—are difficult if not impossible to convey by the use of words alone. Hence it is important to know when and how to supplement words by using figures and tables, and how to use words in order to point up the significant features of the figures and tables used.

This chapter will describe the main conventions that govern the use of these nonverbal forms, but there is only one way in which you can really learn to use them to the best advantage: Study what you see done in the books and articles related to your field. Not that everything done is desirable, but systematic observation will increase your resourcefulness and will help you to form an intelligent opinion about whether something you are contemplating is done often enough that it will not strike your readers as strange.

TABLES

Of all the nonverbal devices used in technical writing, tables are used most frequently because they can present pure data without interpretation; that is, tables allow the presentation of exact results of experiments, surveys, or analyses without bias, and allow readers to provide their own interpretations. Usually tables function to enable readers to compare statistical information. Tables often take considerable time to compile, and you may be tempted to take the easy way out and present your statistics in ordinary text, but there are times when it will be almost hopeless for readers to grasp your information unless they receive it in tabular form.

Though tables usually consist of numerical data, resourceful writers use them to present other kinds of information also. A consumers' magazine uses them, for example, to present the facts about different makes of automobiles.

It names the different makes of car in the column headings, lists the feature being compared—such as brakes, visibility, steering mechanism—at the left, and indicates the facts about each car in the column under its name. This enables readers, with equal ease, to check all the features for any car covered or to compare each make to the others in any particular feature. This same type of table can be used for numerous other types of comparison—the characteristics, for example, of different kinds of apples; the identification of insects that damage crops and the measures for controlling them; the results of using different quantities of fluoride in water to reduce tooth decay.

Because tables can be used for so many purposes, technical writers should always be on the alert for opportunities to utilize them. Also, writers should realize that when the construction of a table seems extremely complicated, their readers' interests as well as their own may often be served best by dividing the material into two or more simple tables.

For your assistance in using tables, the following suggestions are offered:

CONSTRUCTION OF TABLES

1. Every table should have a title, usually placed at the top. If there is more than one table, each should be numbered, using either Roman or Arabic numbers.
2. If all data in the table are in the same unit of measurement, that unit should be listed in the title, or in parentheses under the title.
3. A table should indicate all the factors that affect the data it contains. For example, the size of pipe used would affect data on the performance of a pump.
4. Each line and column should have headings that show accurately the nature of the contents. When necessary, the headings should designate the units in which quantities are expressed. If there is not enough room for the essential information in a heading, part of it may be added to the table in notes.
5. Figures in columns are usually aligned under similar digits—usually the right-hand digit or the decimal point. Fractions should be expressed as decimals unless decimals are not customary for the type of data or unless decimals would misrepresent the degree of accuracy achieved.
6. Do not use needless lines, especially solid lines, between columns of data, between lines, or at the bottom of the table.
7. Standard symbols and abbreviations may be used to save space.
8. When a note is needed to explain some part of a table, its presence may be indicated by a lowercase *letter* (rather than a number) raised half a space above the line at the point where the note applies. The notes to a table are placed at the bottom of the table rather than at the bottom of the page, where they might be confused with ordinary footnotes.

GENERAL SUGGESTIONS

1. All tables should be referred to in the text so readers will know when to give them attention. The only exception is tables placed in the appendix. Reference to such tables may also be desirable to indicate the reason for their inclusion.

2. Each table (unless relegated to the appendix) should be placed shortly after the point where it is first referred to in the text, and on the same page if possible. Under no circumstances should a table be placed in advance of the point where it is referred to or discussed.
3. A table should not break into the normal margins of the page. If necessary, it may be placed sidewise, with the top of the table toward the left.
4. Tables should be complete on one page. When long tables are unavoidable, *continued* or *cont.* should be placed at the bottom of the first page to indicate

1		Table 2. Average Daily Emission (in tons) of Atmospheric Pol-				
2	lutants					
3						
4						
5		: Hydrocarbons and Hydro- :		: Sulfur	: Nitrogen
6	Source	: carbon Derivatives	: Aerosols	: Oxides	: Oxides
7		: Hydro- : Hydrocarbon :		:	:
8		: carbons : Derivatives :		:	:
9						
10	Automotive[a]	1100	150	35	35	775
11						
12	Petroleum Marketing	70	0	0	0	0
13						
14	Petroleum Refining	160	2	5	45	45
15	and Production					
16	Solvents and	210	150	0	0	0
17	Coatings					
18						
19	Totals	1540	302	40	80	820
20						
21	[a]These figures are now being revised to incorporate later data.					

1		Table II				
2		Average Daily Emission (in tons) of Atmospheric Pollutants				
3						
4						
5		: Hydrocarbons and Hydro- :		: Sulfur	: Nitrogen
6	Source	: carbon Derivatives	: Aerosols	: Oxides	: Oxides
7		: Hydro- : Hydrocarbon :		:	:
8		: carbons : Derivatives :		:	:
9		: :	:	:	:
10	Automotive[a]	: 1100 :	150	: 35	: 35	: 775
11		: :	:	:	:
12	Petroleum Marketing:	70 :	0	: 0	: 0	: 0
13		: :	:	:	:
14	Petroleum Refining :	160 :	2	: 5	: 45	: 45
15	and Production :	:	:	:	:
16	Solvents and :	210 :	150	: 0	: 0	: 0
17	Coatings	: :	:	:	:
18		: :	:	:	:
19	Totals	: 1540 :	302	: 40	: 80	:, 820
20						
21						
22	[a]These figures are now being revised to incorporate later data.					

FIGURE 2. Specimen Tables. The forms shown, unlike most forms seen in print, can be produced on a typewriter. They may be altered as necessary to accommodate different materials. The numbers at the left are not part of the tables but are added in order to indicate vertical spacing.

that the table has not been completed, and at the top of the second page to indicate that part of the table has preceded. Column headings must be repeated on the second page. If totals are included at the bottoms of columns, the subtotals should be at the bottom of the first page and at the top of the second page and the word *forward* should be used at the left side of the subtotals to show that they are not final totals.

5. A table from an outside source must be acknowledged, as one would acknowledge any other borrowed material. This may be done by naming the source in parentheses after or under the title, or by use of a footnote. If the number and title are on one line, the footnote number follows the title; if they are on separate lines, the footnote number follows the number of the table.

The preceding suggestions are modified when a writer uses a form that strictly speaking is not a table. The term *table* means something more than a list, or two or three lists placed side by side. If you examine the work of experienced writers you will find that they rarely call anything a table unless it not only has headings at the top of its columns but has a left-hand column that also consists of headings, as can be seen in the specimens.

Well-informed people may sometimes call material that does not have left-hand headings "informal tables," but they do not use the word "table" before the titles nor refer to them as tables in the text. These "informal tables" are not numbered and need not be referred to in the text because, unlike true tables, they are actually part of the text, in substance if not in form. What has been pointed out about this form is not intended to discourage its use, but writers who treat it and refer to it as if it were really a table make their work look both stiff and amateurish.

Forms that may be used for tables are illustrated in Figures 2 and 3.

Table 6. Performance of Self-Polishing Wax on Linoleum

Test	Nukoat	Florglo	Waxeeze	Hygrade
Ease of Application	Excellent	Excellent	Fair	Good
Original Appearance	Good	Good	Excellent	Good
Resistance to Scuffing	Good	Fair	Fair	Poor
Resistance to Scratching	Good	Good	Poor	Poor
Resistance to Water Spotting	Fair	Fair	Poor	Fair
Response to Rebuffing	Fair	Fair	Good	Good

FIGURE 3.   Specimen Table Showing Tabular Presentation of Nonstatistical Information.

FIGURES

In the following discussion, all forms of illustrations are referred to as figures. Examination of technical books and magazines will justify this terminology.

To the technical writer, figures are functional rather than ornamental. Though in other kinds of writing they are frequently used merely to arouse interest or improve the appearance of a page, in technical writing they serve primarily as a means of giving information or interpreting results. Technical writers also should strive to produce attractive looking pages, and may think of figures with such a goal in mind, but that is not the main reason for their use.

Each type of figure presents information in a particular way; the selection of the kind of figure you wish to use depends on what kind of information you have and what you want your readers to see about the information. You need to be aware, however, that figures do not necessarily clarify. Information may be obfuscated and readers confused by the wrong type of figure or by an incorrect presentation. As in all aspects of technical writing, figures require good planning and careful execution if they are to perform their proper functions. Remember particularly the limitations of the memory discussed in the chapter on organization, and try to keep the central points of a figure to a minimum. Too often to save space, writers force their information into too few figures, creating problems for their readers. As we examine each type of figure, we will point out what it does best and discuss possible problems.

Whatever types of figures one may use, the following suggestions should be helpful.

1. *Number and caption.* Every figure should have a caption, which should be preceded by a number (usually Arabic) if more than one figure is used. It is usually best to use a single sequence of numbers even when figures differ in kind. The number and caption are normally placed underneath a figure but may be placed in a convenient location on the figure itself. (This is likely to be desirable on a graph because of the lettering at the bottom.) A caption is ordinarily phrased as a topic (see Chapter 11 for definition of *topic*); but sentences are sometimes used for figures appearing in magazine articles.

2. *The legend.* In addition to a caption, a figure may also have a legend if it needs one. The term *legend* applies to the additional lines, often in smaller type if printed, that follow the caption and explain the figure or identify its parts. If a legend follows a caption that is phrased as a sentence, the two blend together so that in appearance they constitite a single unit.

3. *Spacing of the caption and legend.* When the caption is underneath the figure, there should be at least a double space between the figure and the caption. However, the caption, and the legend if one is used, should be closer to the figure than to the text that follows, so that they will be identified with the figure, not with the text.

4. *Reference to the figure.* Every figure, unless it is purely supplementary and is placed in an appendix, should be mentioned in the text. (It *may* be referred to

even if it is in an appendix.) The reference to the figure is placed in parentheses unless it is part of the sentence where it occurs. Note the following examples:

As can be seen in Fig. 4, the dimensions . . .

The arrangement of the equipment (Fig. 5) is planned so as to . . .

If a figure is related to a long stretch of text, it should not be referred to more frequently than is necessary to keep the reader reminded of its bearing on the discussion.

5. *Placement of the figure in the text.* Any figure in the text should be placed, if possible, almost immediately after the point where it is first mentioned. Certainly it should not come very far ahead of that point, for if it does, the reader may either study it too soon or else, having passed it, may neglect to turn back to it. When possible, you should place a figure on the page where you discuss it.

6. *Partial lines of type.* If a figure is narrow enough to leave room for partial lines of type, the type should be placed on the left and the figure on the right. (This suggestion applies to material typed on only one side of the sheet and therefore regarded as the right-hand page of a book. On a left-hand page, the figure is placed on the left.) However, if two or more figures are placed on one page, the layout may be designed in any manner that gives the page a good appearance and places each figure where the reader can consult it conveniently.

7. *Drawing the figure.* Full instructions for drawing figures are beyond the scope of this text. It should be mentioned, however, that figures should be drawn in India ink, that they should be no larger than is necessary for clearness, and that each figure should be placed within a border unless its own shape creates a natural border. The figure should not extend into the margins of the page.

8. *Self-sufficiency of the figure.* A figure should be self-sufficient so that the reader for whom it is intended may study it without being forced to consult the text. Every part of a figure that cannot be identified from its appearance should be labeled or else identified by use of a key. The key may appear on the figure or may be a legend under the title.

9. *Lettering on the figure.* The lettering on a figure should be placed, when possible, so that it can be read with the bottom of the figure down. Lettering that must run at a right angle to the bottom of the figure should read with the right side of the figure down.

10. *Sidewise figure.* If a figure must be turned broadside on the page, it should be drawn so that the bottom of the figure comes on what is normally the right side of the page. (This can sometimes cause lettering to appear upside down when the page is in normal position.)

11. *Acknowledgment of source.* If a figure is taken from an outside source, the source must be acknowledged. The acknowledgment may be placed in parentheses after the caption or after the legend if you use a legend from the same source. Acknowledgment may also be made by means of a footnote.

12. *Mounting a photograph.* If a photograph or clipping is used as a figure, it should be mounted neatly, preferably by use of rubber cement, unless copy is being prepared for printing.

13. *Preparation of figures for printing.* If copy is to be printed, figures must not be fastened to the pages of the manuscript. The figures and captions are separately prepared and numbered, and the manuscript is marked to show where the figures should be placed. The figures should be drawn larger than they will be on a

printed page—perhaps twice the size. Figures for offset reproduction should be drawn, usually, in the size that they will appear in the reproduction. The full planning of layouts for display printing is beyond the scope of this book.

The preceding instructions indicate that a table or figure should be referred to in the text. This does not imply that a mere mention is the only attention it receives. Often the text will discuss a table or a figure in considerable detail, so that the reader can understand it as easily as possible. The text may direct the reader's attention to the points that are most important, or may explain the table or figure in sufficient detail to make sure that its full significance and implications are not overlooked. One test of your skill as a technical writer— a test often overlooked in this kind of discussion—is your use of the text to help the reader benefit from tables and figures, as well as your use of tables and figures to assist the text.

Thus far, all that has been said about figures applies to figures of every type. Some of the common types call for individual attention. Before they are discussed, however, it should be reemphasized that no amount of discussion can eliminate the need for constant observation of the figures in technical books and magazines.

Bar Charts

One of the simplest and most useful types of figure is the bar chart, with which every reader is familiar. Bar charts best illustrate quantities at specific times, such as output per year, or quantities per item, as in Figure 4. The type of information it presents might also be presented in a table. In fact you may often first make out a table and then convert it into a bar chart to make differences in quantity instantly and unavoidably *visual*. For example, a reader might see numbers such as 841,654 and 418,543 arranged in columns, along with several other numbers, and yet fail to become aware of the relative quantities until the table was given close attention. If such numbers were presented as bars, one of which was more than twice the length of the other, relative amounts, being clearly visible, could hardly fail to be perceived.

The bars in a bar chart are sometimes vertical (see Figure 4), sometimes horizontal (see Figure 5). They may be shaded differently in different portions of their length so as to permit comparison of parts as well as of the whole. For example, different bars in the chart might represent the total expenditures of different divisions of some organizations, and each bar might be shaded differently in different sections of its length so as to show what amounts went for specific expenses such as wages, raw materials, power, and overhead.

There are variations of the bar chart, such as pictures that represent quantities. For example, the increase in the number of passengers carried by an airline might be suggested by pictures of people, or the increase in production of a metal by pictures of ingots. This pictorial treatment is used primarily

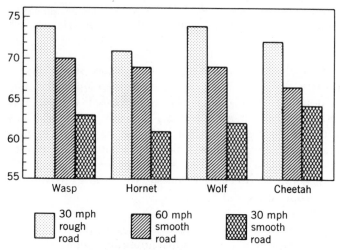

Figure 9. Interior Sound Level in Four Compact Cars (in decibels, A scale). A reading of 50 decibels is comparable to the sound level in a home; 70 decibels is comparable to the sound of a freight train at a distance of 100 feet.

FIGURE 4.   Bar Chart with Vertical Bars.

FIGURE 5.   Bar Chart with Horizontal Bars.

when an effort is being made to popularize what is written. It is intended to increase interest as well as to provide information. Properly handled, it may on suitable occasions be a useful expedient. Often, unfortunately, it is not properly handled. Instead of drawing three times as many people for one airline as for another in comparing passengers carried, writers sometimes draw one person for each, but make one three times as tall—and thus nine times as large in area. In such cases readers cannot easily form a clear, accurate impression of the facts. And since a bar chart fails to perform its function unless it uses visual methods to create a clear and accurate impression of the facts, it is best to use ordinary bars or lines, so that no dimension varies except length, or else to show the variation of quantity by varying the number, not the size, of pictures. These methods are the most honest and accurate, and are also the clearest to the reader.

Graphs and Curves

Like the bar chart, the graph or curve is a means of presenting data that might also be presented as a table. In fact it is hardly possible to construct a graph without having constructed a table first. The extra work that it takes to convert a table into a graph is often justified, however, because a graph makes it much easier for the reader to grasp the information.

Suppose you wished to show how the price of a commodity varied from month to month over a period of one or two years. To learn the low, the high,

Figure 3. Stress and Strain Plotted as a Function of Time.
Although not shown here, the initial stress is zero.

FIGURE 6.   Curve (Courtesy of General Electric Company).

and the general trends from a table, your readers would find it necessary to study the data at length. Yet all these points would be perceptible at a glance if the table were converted into a graph. If the data concerned three or four commodities, the use of curves would be even more advantageous. It would enable readers to grasp the situation in a moment, whereas they would have to read and compare dozens of numbers to get the desired information from a table.

The graph or curve differs from a bar chart in that the graph shows continuous relationships between two factors, such as strain and time, conductivity and temperature, or employment by seasons. To show the relationship, the graph has two scales, one horizontal and one vertical. The independent variable is usually plotted on the *abscissa*—the horizontal axis or scale—and the dependent variable is usually plotted on the *ordinate*—the vertical scale. The most common independent variables include time, distance, voltage, stress, and load; the most common dependent variables include temperature, money, current, and strain. Full information about the scales must appear as part of the graph. The examples shown illustrate the placement of this information. The vertical scale, ordinarily placed on the left, may be placed on the right side if the graph is so large that it is needed there.

Sometimes more than one curve appears on the same graph. In this case, if there is room, each curve should be identified by a short title. If room does not permit this lettering, a different kind of line may be used for each curve, and a key identifying the lines can be shown on some unused space on the face of the graph. Other means also—for example, a legend under the caption—may be used for identification.

Figure 6. Comparison of Sinusoidal and Actual Tides over Two Complete Tidal Cycles.

FIGURE 7.   Curve (Courtesy of Extension Service, College of Engineering Research Division, Washington State University).

Be sure that when you make a graph you choose the scales, both vertical and horizontal, in such a manner that the visual impression conveyed is justified by the facts. When a line goes up or down at a sharp angle, readers infer that the rapidity of change is highly significant. If it climbs or falls slowly, they interpret the change as having been gradual. Yet the angle at which the lines go up and down is entirely dependent on the scales. It is therefore a matter of simple honesty for the writer to use scales that create a visual impression in line with the actual importance of what has occurred.

This is a point that beginning writers, naive readers, and even textbooks often fail to appreciate: The selection of scale is the result of interpretation on the part of the writer and can be used to influence the reaction of the readers, as shown in the examples in Figure 8.

One graph suggests that the increase in sales is dramatic; the other that it is slow. Yet, if you examine the graphs carefully, you discover that they both convey exactly the same facts: that sales increased by two million dollars from May to July. The point to realize here is that either graph may be correct: If the company ordinarily sold ten million dollars worth of goods in a year, then the graph on the left might be an accurate interpretation of the increase, whereas if the company ordinarily sold ten million dollars worth of goods each month, the graph on the right might be a more accurate presentation. Without additional data, there is no way to know which interpretation is supported by the facts. While such an exaggerated comparison may make the point seem obvious, you should remember: There is no ideal scale; make sure the one you choose reflects your interpretation.

FIGURE 8. Visual Effect of Different Scales on a Graph.

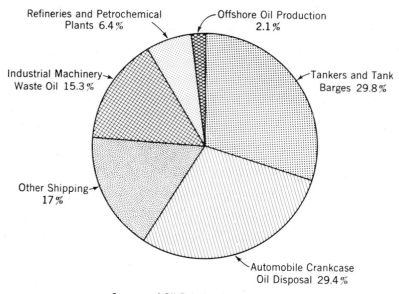

Sources of Oil Polution in the Oceans.
(Source of Information: Society of Naval Architects and Marine Engineers)

FIGURE 9.   Pie Diagram.

The Pie Diagram

Another widely used device is the "pie" diagram, consisting of a circle divided into segments. More than any other kind of figure, it permits simultaneous comparison of parts both to each other and to the whole. It might show, for example, what percentages of a corporations's receipts were allotted to meeting various expenses, to dividends, and to capital gain.

Like any other drawing, the pie diagram should state amounts as well as picture them. These amounts may be actual quantities as well as percentages.

One difficulty in a pie diagram is the problem of making the letters identifying the segments all run in the same direction. If there are too many segments, or if some of them are too small, this difficulty can become insuperable, making it inadvisable to use the form.

Figure 9 is a typical pie diagram.

Organization Charts
and Flow Sheets

Unlike the figures discussed above, the organization chart and the flow sheet are not concerned with statistical information. The very name *organization chart* shows what kind of information such a figure presents—divisions of an organization, usually represented by circles or rectangles so arranged and connected by lines that authority, relationship, and responsibility are easily seen. Such a chart is far more effective than words in conveying information of that kind. Indeed, many a body has been induced to improve its own organization when an organization chart revealed confused or illogical relationships.

The flow sheet is very similar. Its typical use would be to indicate the method by which some process is carried out. The machines used are sometimes indicated by the conventional circles or rectangles, and sometimes by simplified drawings suggesting their actual appearance. The main difference between the flow sheet and the organization chart is that the flow sheet indicates actual physical movement of materials. The direction of such movement is shown by arrows wherever necessary.

Figures 10 and 11 illustrate an organization chart and a flow sheet.

Figure 2. Quad-Cities Airport Master Planning Organization.

FIGURE 10.   Organization Chart (Courtesy of Quad-Cities Regional Airport Master Plan Study).

Figure 17. Artificial Lung (See Hill.[15])

FIGURE 11.   Flow Sheet (From an article by J. D. Hill, et al., "Prolonged Extracorporeal Oxygenation for Acute Post-Traumatic Respiratory Failure (Shock-Lung Syndrome)," *New England Medical Journal.* Courtesy of *The Trend in Engineering,* College of Engineering, University of Washington).

Left to right: Pig's Ear (Discena Perlata), Morel (Morchella Elata), Snow Mushroom (Gyromitra Gigas), and False Morel (Gyromitra Esculenta). The false morel, which is nonedible, can be distinguished from the true morel by its false pitting (mere convolutions) and its long stalk in contrast to the real pitting and short stalk of the morel. The pig's ear has neither convolutions, pitting, nor stalk. The snow mushroom has convolutions and a hidden stalk.

FIGURE 12.   Specimen Photograph (Courtesy of Edmund E. Tylutki, Department of Biological Sciences, University of Idaho).

Photographs

In technical writing you may need to use photographs for either of two main purposes: to assist verbal description, and to prove the truth of assertions. Photographs would be helpful, for example, in showing characteristic fractures of metal that might result from different causes, or in identifying insects or plant diseases. They might also illustrate the condition of equipment that had not been maintained properly, the breaks that had developed in a pavement or the lining of a canal, or the crowded condition of a factory. Sometimes, as mentioned, the purpose of the photograph might be merely to give information; but often a photograph can be used to prove the truth of an assertion by letting readers see for themselves. For this latter purpose it has no equal.

The photograph is limited however, by the fact that it can show only the surface and that it may sometimes unavoidably present both significant and nonsignificant facts of appearance with equal emphasis.

Figure 12 is a photograph used in a typical manner.

Section at
"Jamb" Stud

Section at
"Mullion" Stud

Figure 1. Horizontal Section Through a Frameless Window in Which the Glass Is Adhered Directly to the Building Frame Members.

FIGURE 13. Diagram (Courtesy of Extension Service, College of Engineering Research Division, Washington State University).

Diagrams and Drawings

There are many occasions when diagrams and drawings can help to illustrate the text. For example, electrical wiring plans, schematic diagrams showing how any type of equipment operates, or drawings intended to illustrate actual appearance are constantly used by those who write on technical subjects. The shape and relative location of objects as well as the manner in which equipment functions can often be shown better by a diagram than by any other method.

Also, unlike photographs, drawings can picture the interior rather than just the surface of an object. They make it possible to omit what is not significant and to emphasize what is important. As you read the technical writing of others, you should make it a point to observe the manner in which drawings are used; and as you plan your own writing, you should constantly be on the alert for opportunities to make your work more effective by means of drawings.

A characteristic use of a drawing for a suitable purpose is seen in Figure 13.

ASSIGNMENTS

Papers in which tables and figures are used can be based on Assignments 3, 4, and 5 at the end of Chapter 6.

II

**REPORTS,
PROPOSALS,
and ORAL
PRESENTATION**

8 | Reports— Their Importance and Qualities

Anyone who has seen reports from varied sources has noticed that they differ widely in form, length, and many other respects. This gives rise to the question, "What is it that causes reports to be considered a distinct species of writing?" The answer can be found in the following: A report usually is a communication in which the writer (or speaker, if it is an oral report) gives information to some person or organization because it is the writer's responsibility to do so.

As this description indicates, the common element in all reports, which sets them apart as a distinct species, is the element of responsibility. A report may be made by a committee to its parent organization. It may be made by a consultant to a client. It may be made by a person on the job to a superior. It may be short, simple, and intended for a single reader or it may be long, elaborate, and printed at great expense—for example, the report of a corporation to its stockholders. But regardless of all these differences, the term *report*, at least in the context of technical writing, should not be applied to a piece of writing unless the writer is responsible for reporting to the reader or readers.

There are, of course, many elements of reports that could be included in a definition, as can be seen by examining the many definitions of reports, technical writing, and technical reporting that can be found in the field. Those elements are included in the discussion of reports throughout this text so that you may consider the textbook itself to be an expanded definition of the term. Once you have read and understood the book, you will have no trouble recognizing reports when you see them.

IMPORTANCE

In any large organization, reports perform an indispensable service. They enable those in positions of authority to keep track of normal operations, to learn about unexpected developments, and to judge whether progress is satisfactory on new projects. They enable the executive to base decisions on the advice of specialists. They are the link between research and the practical utilization of discoveries. They often provide the basis for the writing of proposals and specifications. They sometimes are the only tangible result of long and expensive work. In general, they comprise one of the most important parts of the process of communication that is vital to efficient operation in government and industry.

DIFFERENCE BETWEEN PROFESSIONAL AND SCHOOL REPORTS

Chapter 2 discussed some differences between written work performed on the job and written work done in school, but further distinctions might be made. Before you begin to write reports in your professional career, you are almost certain to have written reports in the classroom. Many of the qualities that are desirable in classroom reports are also desirable in reports written on the job, but there are important differences. The value of a report written on the job depends mainly on the answer to one question: How well does it meet the needs of those who receive it? Whatever good qualities it may possess are good only because they help it to meet this one test of excellence.

In the classroom, this test does not ordinarily exist. Instructors assign reports not because they need or will use the information, but because students will profit by acquiring the information and perhaps by reporting it. As a result, students do not form the habit of constantly asking, "Does the person I am reporting to need these particular facts?" When at some later date these students write reports on the job, they are likely to include information that is not needed just because it is available and they were required to include such information in school.

A case in point is the laboratory report. Many instructors insist that a laboratory report contain information on theory and procedure because one purpose of the report is to show whether the student understands these matters and can explain them. But in industry, it is usually taken for granted that the writer understands the theory on which the work is based and has used an accepted procedure. And since readers will usually know about these matters already or else will not need to know, they will be just as well satisfied if theory and procedure are omitted, *unless they involve something new and significant*.

A second important difference between school reports and reports written on the job lies in the fact that no one expects recipients of professional reports to read all of what is written, or, in some cases, to read the report at all. Many students, when told of this fact, are shocked and immediately question why they are repeatedly told they need to learn to write well if what they write will not be read. In school, if they learned that their teachers were not reading their reports, they would be angry, and justifiably so, since in school, one of the reasons for writing is to demonstrate the ability to do so, and the primary way for the teacher to determine ability is by reading the reports. On the job, however, it is assumed that the writer can write; what is important is the information conveyed, and often, readers of professional reports can obtain the information they need by reading only parts of the reports.

What about reports that are not read at all? Do they need to be written well? Many reports are produced "for the record." Many jobs involve the construction of some object, a calculator, a building, or a bridge, for instance. The goal is the product, and when that is in hand, the temptation is great to slight the report of the activities that produced the object, since the report will simply be filed and probably not read. Giving in to such temptation is always a mistake, however. These reports may be referred to years later for any number of reasons. The passage of time increases the need for clear writing, since memories fade and individuals may not be available to answer questions. The manufacturer may wish to develop the product further, for instance, and will turn to the report as a starting point. If the report does not serve its purpose then, the writer will hear about it. Or someone may wish to modify the building or change its use and will need information about the construction. Or, as sometimes happens, the bridge may collapse, and courts may need to determine liability. A poorly written report will cause people to assume that other work performed by the writer was equally sloppy. You would not want work to return to haunt you.

But beyond such reasons, as a professional in your field, you want always to work at a level that you can be proud of, a level that will reflect credit on you, your organization, and your profession. In short, you want to be a person who can be counted on, and writing reports is one of the many tasks for which you will be counted.

In most situations, however, parts of the reports you write will be read. Reports often go to many different readers, as was explained in Chapter 2. Some readers will wish to know what you think about the subject, and thus will read just your conclusions, others want to find out how you did it, and will read your methodology, while still others may be interested only in your results. All of these readers have many activities and other people competing for their time. If they cannot immediately find the information they need, they are apt to put your report aside or send it back for revision. On the other hand, if they find that the report is organized and written so they can find what they want, they will use the information and commend the writer. Your job as a writer is to provide each reader with the information needed in a manner that makes it easily accessible.

If you realize, at the beginning of your career, the difference between what is desirable in reports written on the job and the reports you write in school, you will have taken an important step toward becoming a good report writer.

QUALITIES TO STRIVE
FOR IN REPORTS

Like any other kind of writing, a report must be considered good or bad mainly on its success in performing its function: to affect the readers as it is supposed to affect them. What specific qualities will help it to produce that result?

Some of these qualities have already been discussed because they should be cultivated in other technical writing as well as in reports. These include accuracy, adaptation of the contents and organization to the needs and abilities of the reader, and effective style. But a report makes additional demands. The key to these lies in a single word: *imagination*. As a report writer you will need to visualize what will happen after your report is submitted, and face such questions as: Whom will it reach? How much do the probable readers know about the subject already? How much do they want and need to know? How busy are they? What is the relative importance of the report at the level that the readers occupy? How much time will they give to it? At this time, it might be a good idea to review the discussion of reader adaptation in Chapter 2.

The following discussion will touch upon several specific qualities. Some of these reinforce each other, but there are times when an excessive effort to achieve one quality will mean the sacrifice of another. Thus, in considering the question of how vigorously to cultivate any of them on a specific occasion, you will need to foresee the circumstances that will probably prevail when your report is actually used and decide what the report must be if it is to perform its function under those circumstances.

Self-Sufficiency

A report should be self-sufficient. That is, anyone who is likely to read it should be able to do so without having to rely on memory or consult files for extra information.

To make this possible, you may need to include facts that could otherwise be omitted. For example, instead of assuming that you may omit facts you originally received from the person to whom you are reporting, you might well remind yourself that details that were fresh in the mind at an earlier date might be hard to recall after the passage of considerable time. Also, it is well to remember that the person you are reporting to may pass the report on to others who lack familiarity with the subject. In view of these facts, it is best to make sure that whoever the reader might be and whatever circumstances might exist at the time a report is read, it should provide what is needed for full understanding.

The material that makes a report self-sufficient comes mainly at the beginning. The opening should make clear the function of the report and the circumstances that called for its preparation so that the question, "What's this all about?" does not arise as the reader plunges into its contents.

Interest

It is natural, if you concentrate a great deal of attention on anything, to become so interested in it that you assume it is interesting to others. Applied to reports, this is a dangerous assumption. Those who read our reports are busy people. Many matters—probably many reports—compete for their attention. Whether your reports get the attention you hope for will depend in part on whether you make an immediate bid for interest.

Many a report, especially if it does not concern a colorful subject, will have a better chance of arousing interest if it is based on a principle that is a commonplace in sales work: People often buy something not because they want it but because they want the results of possessing it. They buy clothing in the latest style, for example, not because they especially want it, but because they want to be identified with the kind of people who wear clothing of that kind. A report writer can attempt to arouse interest in a report by quickly showing that it offers a prospect for producing results that the reader already desires: a better product, for example, or economy of operation or saving of time.

This does not mean that a report should resort to sensationalism or that its appeal should be specious. It means only that the interest in reports stems primarily from their practical value, which should therefore be pointed out without delay.

How can this result be accomplished? At least two concrete suggestions can be made. First, organize the body of the report so that the reader discovers as early as possible what benefits may result if your report receives attention. Your skill in organizing in this manner, yet at the same time making your plan logical and coherent, will be one of the decisive factors in creating successful reports. Second, in one or another of the preliminary parts—the letter of transmittal, the covering memorandum, the introduction, or the summary— provide a sentence or two deliberately calculated to arouse interest by foreshadowing the significance of the full report. These sentences should be specific and concrete. Just to say "This is interesting and important" is a waste of space.

Whether you can retain interest once you have aroused it depends to some extent on the effectiveness of your writing style. The likelihood that style in itself can *create* interest is slight, for readers read reports for the value of their contents, not for enjoyment. But readability, conciseness, and the color and vigor that can be obtained from concreteness will most assuredly help to retain or increase the interest, and a dull, wordy, obscure style will not.

Thoroughness

It goes without saying that thoroughness is desirable in a report as elsewhere; but whether this quality is achieved depends in many cases on whether the writer uses imagination. There are always some obvious questions that a report must answer. But answering the obvious questions is not always enough. For example, an increase or decrease in the number of people taking a training course in two successive years might seemingly point to a conclusion about the success of the course. But difference in weather, the existence of an influenza epidemic during one year, or a change in the makeup of the work force could also affect the number; and the possibility of such outside forces could not safely be overlooked.

There are other factors to be considered for the sake of thoroughness. Assuming that you have demonstrated that the action you recommend would definitely produce the result desired, might there be other and possibly better ways of producing that result? Also, is it possible that the actions you recommend would not only produce the desired result, but would also produce undesired side effects that you have overlooked? For example, you might prove that leaving stubble in the fields rather than burning it would prevent erosion, yet overlook the possibility that leaving it would permit harmful insects to multiply. Or again, a program to destroy predatory animals might succeed in reducing the number of predators but in doing so permit an increase in the population of crop-destroying rodents.

If a report is to be thorough, its writing must be preceded by a thorough job of gathering information; but writers who have answers to all the essential questions, the less obvious as well as the obvious, often fail to include them. They make the mistake of assuming that because they know all the necessary questions have been considered, they need not answer them all. They overlook the fact that readers will not share their confidence unless the readers are told enough to be sure nothing has been overlooked.

How much detail about secondary questions should a report include? There is no formula that provides an answer to this question, but at least we can realize its existence and use our judgment about whether we have given readers not only the information they will actually use, but also the extra information that will free them from worry about whether something has been overlooked.

Omission of Unnecessary Material

Like any other desirable quality, thoroughness can be carried too far. We need to ask ourselves not only, "Am I overlooking anything that should be included?" but also, "Am I including anything that could be left out?" If we do not consciously guard against including too much, we are almost certain to err in this direction.

It is not surprising that this should be true. It is natural for a person who devotes a great deal of time and attention to any subject to overestimate the extent to which it will impress others as important. And when we report to those above us, it is easy to forget that something that is a major element in our own area of activity may be a much smaller fraction of the area in which those farther up the line are involved.

The danger of including too much is increased because as we gather our material it is almost impossible to gather exactly what we need and no more. In fact we cannot be positive about just what will ultimately be needed and what will not. And since it is painful to see facts and ideas go to waste—facts and ideas that we have worked hard to obtain—it is a temptation to include them just because we have them.

Unfortunately, their inclusion sometimes does more harm than good. Readers are less likely to grasp and remember the facts of real importance if they must separate them from material that they do not really need.

The temptation to include unnecessary material is especially strong when a report leads to negative conclusions and recommendations. To see why this is true, consider the following situation. The question being investigated is whether to harvest a crop of some kind by a different method. The decision rests upon answers to such individual questions as the cost of buying and operating the new machinery, the quality of the product it turns out, the skills it demands of its operators, and its reliability. Suppose that the investigation of its reliability shows it to be subject to frequent breakdowns; and since a breakdown at a critical time during harvest would mean a heavy loss because of spoilage, this single defect is enough to bring about a negative decision. If you were reporting on the investigation you would be tempted to tell all you had learned on each aspect of the study. But if you were the reader, you would not wish to waste your time on details about cost of operation, quality of product, and operator skills only to learn later that the machine was not reliable enough to use. You would probably want full facts about the unreliability because those facts would be the basis for a decision, but would want only a minimum of information on the other questions.

Writers of reports or of any other kind of communication will always have to encounter the question, "How much should I say?" And there is no better way to answer this question than by using your imagination—by trying to visualize the moment when those who will read your report pick it up, pressed as they will be for a time, and start to read. With this picture in mind, scrutinize the contents of your report and ask yourself constantly, "Would the report perform its function just as well if I left some of this out?"

Freedom From Bias

A person who must investigate a subject and report findings often starts work with a strong expectation about the probable outcome. When this expectation is based upon sound previous knowledge and experience, it is entirely natural and unobjectionable. When it is based on mere prejudice, however—or even worse, on self-interest—the case is different. There is no surer way to forfeit confidence than by letting your reader gather the impression that you have been swayed by an eagerness to reach a preconceived conclusion.

To be sure, some so-called reports fail to meet the requirement of impartiality. Protagonists of conflicting interests, in preparing reports to use in a controversy, usually present matters from their own point of view. But these reports are not suitable models for a person entering a technical profession. A bona fide technical report is not begun with the purpose of proving something. Rather, its purpose is to present all sides with scrupulous fairness. In the long run, no one in a technical profession will profit by letting reports be or even appear to be prejudiced. Hard though it may be, you should plan an investigation in such a manner that you will not fail to discover evidence that might run counter to the conclusions you expect and hope to reach, and should present your findings in such a manner as to show that you have been impartial.

Objectivity

The term *objectivity* is sometimes used with practically the same meaning as *freedom from bias*. We will use it here, however, to refer to the quality that is the opposite of *subjectivity*. In urging that reports be objective, we are urging that they be based on concrete facts.

The need for objectivity becomes a matter of special concern whenever a report touches upon human actions or attitudes. For example, reasoning subjectively about what workers would do under certain circumstances might result in a radically different conclusion than would be reached by observing their actual conduct under such circumstances.

Some matters, of course, are basically subjective. The relative beauty of two makes of automobiles cannot be determined by measurement, nor can the question of whether a new process produces better-tasting instant coffee. But even when we deal with questions of taste, objectivity need not be surrendered. It can be attained by getting a substantial number of subjective opinions. Assuming a fair investigation has been made, though we might not be able to say objectively that Brand "A" tastes better than Brand "B," we might be able to make the truly objective statement that 70 percent of those who tasted both preferred the taste of "A" to the taste of "B."

Restraint

A report should be restrained both in substance and in style. Valuable though enthusiasm may be on some occasions, there is no surer way to lose a reader's confidence that to give the impression that your enthusiasm has overbalanced your judgment.

Restraint in substance is more than a mere concession to reader psychology. We realize this when we think of the extravagant hopes aroused by the development of drugs that seemed promising but were later found to have undesirable side effects, or the development of insecticides that unexpectedly created problems related to public health.

Restraint in style should lead you to avoid such terms as *astonishing*, *startling*, *flagrant*, or *disastrous*. Rather than writing *a tremendous increase* it is advisable to write *a substantial increase,* but even better to say *a 50 percent increase* and let the reader choose an adjective. Or again, it is better to tell your reader that a flood drove five hundred people from their homes than to refer to it as *calamitous, appalling,* or *catastrophic.*

Restraint is especially important in stating conclusions. This does not mean that you should insert such meaningless phrases as "It is believed that . . ." or "It may be concluded that. . . ." But you should not go beyond the evidence in making claims or predictions, and should recognize the possibility that later developments can always lead to results that no one could have anticipated.

Appropriate Degree of Impersonality and Formality

A personal tone in writing, as contrasted with an impersonal tone, is largely the result of frequent use of the personal pronouns *I* (or perhaps *we*) and *you,* which make it appear that the writer is addressing the reader personally. "I could hear the hum from a distance of 20 feet" is personal. "The hum was audible from a distance of 20 feet" is impersonal.

In some simple situations, especially in letters and sometimes in memorandums, a personal tone—which incidentally conveys the impression of informality too—is natural and desirable, whereas impersonality would seem stiff and aloof. Indeed, there are times when an effort to avoid referring to yourself may be awkward and even evasive. To write "It is believed that . . . ," when your purpose is to express your own belief, seems to hedge against accepting responsibility and is vague because it does not answer the question, often important, of who does the believing.

Much of the time, however, impersonality is better. If you write, "I weighed the samples at half-hour intervals," it is apparently your intention to tell us about your own actions. But if you write, "The samples were weighed at half-hour intervals," you are centering attention on what was done to the samples. Of course, if you wished to emphasize that you were the one who did the weighing, the use of "I" might be the best way to do so. Much of the time, however, an impersonal version is more likely to emphasize the facts that are really significant, because a technical writer's purpose usually is to tell about things—the samples, in the case above—rather than about the writer.

It is not always necessary, to be sure, that all parts of a report be personal or impersonal to the same degree. There is no reason that on occasion you should not use *I, we,* or *you* in the preliminary parts of a report, where you are explaining the circumstances and your own involvement, but you should probably stop using them when you begin to present the real substance. You might start, for example, "As you requested, I visited . . . to investigate . . . ," but make no reference to yourself in stating the facts that came to light during the investigation.

Of the three pronouns to which attention has been directed, *I* goes farthest toward imparting a personal tone, for reasons that have already been pointed out.

You is also decidedly personal. It conveys the impression that the reader is being directly addressed and thus involves him or her personally in what is being said. Sometimes, of course, the term *you* may appear in a report directly addressed to more than one reader. Obviously, it sets the writer apart from the reader, and it is therefore often suitable in a report written by someone such as an outside consultant. It is less likely to be suitable, however, when the writer and the reader are part of the same organization and the report will go beyond the immediate reader who is addressed as *you*.

The use of both *I* and *you* is affected by another consideration beyond mere style. The meaning of *I* depends on who is using it, and the meaning of *you* depends on who is being addressed. When it is likely that a report will be incorporated into a larger report by a different writer, *I* and *you* should be avoided when doing so would reduce the need for changes.

Sometimes, too, if a report will reach the attention of a good many people besides the person or group immediately addressed, *I* and *you* may be undesirable for another reason. The effect that the two words have on these other readers is the exact opposite of their effect on the person who first receives the report. That is, since *you* does not include these others, it is a constant reminder that what they are reading was addressed to someone else, not to themselves.

Indeed, at times a report will be of value to a good many others besides the original recipients. For example, in an actual case, a report produced by a university research division concerning the waste from a food-processing plant was written in behalf of one company and one community. It was printed, however, and sent by request to a large number of other companies and communities that faced the same problem. A personal, informal style would have been out of place where such a use was foreseen.

We is a term that can have either of two meanings, at least as it might be used in a report. It can refer (1) to the writer and those associated with the writer, as contrasted with the reader, or (2) to the group or organization to which both the writer and the reader belong. In the first use it is comparable to *I*, but a group such as a committee can use *we* without seeming so personal as an individual seems when using *I*. In its second use, *we* is often effective because it subtly suggests that the writer and the reader are both part of the same larger entity—that they are closely associated and have common objectives—presumably the interests of the group or organization as a whole.

In view of all these considerations, what is the best attitude to take in regard to personal and impersonal style? The following recommendations should answer that question.

Do what is natural, and what common sense dictates in view of the situation, the nature of your report, and the use that will be made of it after you write it. If its purpose is to tell about your actions and hence you cannot avoid *I* without sounding stiff and awkward, use it. If what you write is of primary concern to the immediate reader or readers, use *you* when you need to. On the other hand, when you are telling mainly about *things* and there is no real reason to refer to yourself, the natural and most effective way to do so is in the impersonal manner that you will find in so much of the technical writing actually being turned out in industry and government. Your style, when you do so, will not be painfully stiff and artificial if you develop reasonable skill as a writer.

As for formality and informality—when a writer uses a personal style throughout a report, that fact in itself conveys an impression of informality; but diction in general is also involved. Since the subject of formality in prose was covered in the chapter on style, the discussion here will be briefer than its importance would otherwise warrant, for readers find lapses in an appropriate level of formality quite disconcerting. You have probably experienced the discomfort felt by an audience when a speaker tells a story that seems either too informal for the gathering or too formal. Readers react similarly to a report that lacks the appropriate level of formality for the occasion.

Just how formal should you be in writing a report? As with the question of impersonality, no answer can be given that fits all occasions. You must judge each situation for itself, and choose your language and presentation accordingly. A major report aimed at a wide readership may require a high degree of formality, whereas a short report in memorandum format aimed at your immediate superior or colleagues probably would not. Imagine how you would react to the language of a report if you were the recipient and the writer had the relationship to you that you have to your readers. Would you think the writer was excessively stiff, and perhaps then wonder what the writer's feelings about you were? Or would you think the writer was being too familiar? If you can answer these questions for yourself before you submit your report, you are less likely to create problems by choosing an inappropriate level of formality.

Avoidance of Chronological Order

Many a report annoys its readers because it is written as a narrative with its contents arranged in chronological order. A typical example was a report actually written in a government bureau concerning three cabins that had been built illegally on government land in a remote area on the shore of a lake formed by a power dam. The report was submitted in late September.

The writer told first of discovering one cabin in June while he was on a horseback trip with another bureau member. The cabin was vacant and contained no evidence of its ownership. In August, on a boat trip, he found that two more cabins had been built nearby. A newspaper with a name and address on it led to a family who were able to identify the three owners. The owner of the first cabin was called upon and claimed that he considered it all right to build his cabin because a friend had told him that the power company had no objections. The bureau representative—the same one who wrote the report—informed the owner that he knew of no arrangement involving the power company, that the matter would be further investigated, but that the cabin would probably be removed. The owner said that in such a case he would move the cabin onto piles in the lake (which incidentally would also be illegal and not permitted). The owner was told that he would be informed of the results of the investigation before being required to move the cabin, but that removal would almost certainly be necessary. The report closed by saying that the other two owners had not been contacted because a survey was necessary to establish the fact that their cabins were definitely on government land.

Put yourself in the position of the reader—as the writer should have done—and ask yourself what, in that case, you would really be most interested in finding out. If you do so, you will probably conclude that the reader would have been better served by a report along these lines:

1. Identification of the subject and reference to any previous memoranda concerning it.
2. The fact that the ownership of the cabins had finally been established and that one owner had been interviewed.
3. The attitude of the owner and what he had been told.
4. The fact that interviews with the other two owners were being postponed until a survey made sure whether the cabins were on government land.

As it was written, the report consisted of two and one-half pages of single-spaced type. The reader was forced to wade through more than half of its length before coming to what really counted—the way matters stood when the report was written.

Charles Dickens once said that the way to write a successful novel was to "Make 'em laugh, make 'em cry, make 'em wait." This may be good advice to an aspiring novelist, but a report writer should by no means "Make 'em wait" for what they are most interested in learning.

A report that consists of a chronological narrative is easy to write because it poses no problem of organization. All that the writer needs to do is to start at the beginning and continue to the end. Once in a while, when a writer is reporting only to tell of personal activities, such an order may be suitable. Also, there are times when the format of a report takes care of the problem. In a laboratory report, for example, the headings identify each section and enable readers to peruse them in whatever order they please. A formal or semiformal report has special sections at the beginning to present the heart of its contents.

But on the many occasions when the report is not one of these special types, you should resist the temptation to do what is easiest. You should realize that the reader usually wants to begin not where you began but where you arrived at the end. You should tell first what the reader will be most interested in knowing first, and only when you have done this should you provide the other information you consider necessary.

Consideration of Emotional Factors

Reports, like other kinds of technical writing, should appeal to the mind rather than to the emotions. When it is our responsibility merely to present objective facts without making any attempt to say what should be done about them, this principle can be applied without modification. There are times, however, when the emotions cannot be disregarded. Though this point receives little attention in many discussions of technical writing, it is often emphasized within the broad field of industrial communication.

To be sure, an appeal to the emotions cannot be relied upon to do the constructive job of selling our ideas; but there are many occasions when *ignoring* people's emotions can destroy our chance of selling them. Why should this be true? The answer is that many actions recommended in reports affect not only things but also people—their egos, their pride, their status, their attachment to established ways. Suppose, for example, that an investigation shows a new method of doing something to be far more efficient than the method already in use. If the person who writes the report, eager to prove this, is too devastating in attacking the old method, people may be forced to defend it because the attack seems to reflect discredit upon them personally.

After all, those to whom we report are not always entirely independent in deciding what is to be done. They often need to gain the concurrence of other persons, whose objections could easily delay or seriously endanger the chance of favorable action. If we can present our findings in such a manner that we do not antagonize these other persons, we increase the chance that our recommendations will be followed.

The preceding warning is necessary because it is always a temptation to build the strongest possible case for what we believe in, and it is consequently a temptation to demonstrate that a contrary position is unthinkable. If we yield to this temptation, however, we may needlessly arouse resentment and opposition. And even if the opposition can be overcome, this antagonism may reduce the likelihood that our recommended measures will be successful.

This does not mean that when you write a report, you should be insincere in order to prevent people from disagreeing with you. However, you should give some thought to the question of who will be affected by your report. So far as possible, present your case without hurting these people, angering them, or humiliating them. Don't make anyone look bad when you can avoid doing so. Remember that even if your case is logical, people are less likely to see its logic if they feel hurt or resentful. In brief, don't increase the opposition to your ideas by ignoring the fact that people have feelings.

EXERCISES

Exercise 1

Rewrite each of the following sentences so that it shows the freedom from emotionalism, objectivity, and restraint that would be appropriate in a report. If necessary, you may make up facts that desirable style would call for.

1. The appalling devastation caused by the explosion and fire is frightful to look at.
2. The morale of union members will be completely wrecked if the reasonable demands in the contract negotiations are not met.
3. The useless paperwork demanded by the new regulations is driving the office help to distraction.
4. The cafeteria food is so bad alley cats bury it.
5. The tremendous increase in productivity will drown our competitors in a flood of superior goods.
6. This unbelievable labor-saving device will free you from kitchen drudgery forever.
7. The stench from the mill is nauseating throughout the town, creating a wave of protests from irate housewives, who have nothing better to do than complain.
8. Overloaded trucks have turned Main Street into a pock-marked no man's land.
9. Employees found frittering away their time in coffee breaks and by the water fountain will soon find themselves frittering away their time at the unemployment office.
10. The proposed work could never be accomplished in the ramshackle, run-down facilities the contractor has at his disposal.

Exercise 2

Rewrite each of the following examples so that the language is standard English and the style impersonal. (There is no intention to imply that every expression you eliminate would be wrong in circumstances where informality were suitable.)

1. Vandals in the neighborhood ripped off the company so often it went busted.
2. The production boss was such an airhead that he never could keep tabs on his crews.
3. The air conditioning always goes on the blink in hot weather.
4. When the brass called for input on the new system, we unloaded all our gripes.
5. The number of negative patient care outcomes is giving that treatment a bad rep.
6. As soon as we debug the new program, we'll access the info and whip it to you.

7. The field reps would do better if they spend more time hitting up customers and less time chewing the fat with buddies.
8. The forklift was in the shop, so the crew had to lug the crates out to the loading dock by hand.
9. His attitude was okay until he missed out on his promotion, when he started pulling a lot of sick leave.
10. As a result of prolonged high revs, the engine threw a rod.

9 | Procedure in Writing a Report

The procedure described in this chapter is suitable when the complexity of a situation or the probable length of a report creates uncertainty about how to tackle the problem. It is offered not as a formula to be followed rigidly, but merely as a systematic approach that may be modified as circumstances require. It would call for the following steps:

1. Analyze and plan the job as a whole, so that you have in mind a picture of the kind of report you must produce and what you must do to produce it. Included in this analysis is the determination of your primary and secondary readers.
2. Form a general, tentative plan for presenting your material. This can include a preliminary decision about the entire layout of the report (what elements to include) and a highly tentative plan or outline for the substance.
3. Gather the facts that you need and interpret their significance.
4. Revise your tentative plan so that your outline reflects your final ideas about organization.
5. Decide so far as possible what use you will make of tables and figures if it seems likely that they will be helpful.
6. Write the first draft of the body of your report, and of your conclusions and recommendations.
7. Make your final decision about what elements (introduction, summary, and so on) you will use as a framework for the body, conclusions, and recommendations, and write those that you have not yet written.
8. Rewrite, revise, and polish all parts as necessary.
9. Care for such final details as providing the title page and table of contents. Go over everything carefully to be sure that it contains no errors, and then go ahead with the mechanical business of copying and fastening it together.

The occasions when you will follow this procedure exactly will be few and far between. When a report is long and complicated, preparing it is a matter of pushing forward on more than one front as conditions permit. It may be necessary, for example, to gather and analyze facts on one point before deciding whether another should be included at all. Again, you may produce a rough draft of one section, complete with tables and figures, before the investigation for another section has gone beyond its opening stages. Nevertheless, the procedure suggested is sound whether applied to an entire report or to a single section of a longer report. It provides a systematic sequence of action so that you do not need to start work with no plan at all and improvise as you go.

One more point should be mentioned in connection with the procedure as a whole. The order in which a writer produces the various parts of a report is not the order in which a reader encounters them. Such parts as the title page, the summary, the introduction, and the conclusions and recommendations are brought into existence as the work nears completion; but most of these parts are placed early in the report. A person who expects to produce a report of any length and complexity should grasp this essential point at the outset.

ANALYZING THE JOB

Analyzing the job really consists of analyzing three problems—the problem that the report itself is intended to solve, the problem of what you must do to produce a report that will serve its purpose, and the problem of determining who the readers will be and what they will need for the report to be successful.

To analyze the first of these problems you will need to center your attention on the subject and purpose of the report. It is often the function of a report to answer some large, comprehensive question. As you think about the question that the report as a whole concerns, you will usually find that the answer depends on the answer to various other questions—many of which themselves depend on the answers to even smaller questions. For example, to answer the question of whether to produce some product by a new method it would first be necessary to answer questions about the quality of product that would result, the equipment needed, the space needed, the skills needed by the operators, the cost, and countless others. When such a list of questions to be answered has been drawn up, the scope of the original problem becomes clearly understandable.

But the second problem still remains: what you must do or have others do in order to produce your report. That is, you will need to figure out what methods of investigation will be necessary. You will need to decide when and by whom each portion of the total investigation will be performed. You will need to decide what kind of report (how detailed and what elements it should consist of) is necessary in presenting the information after you have gathered it. You will need to determine what your readers will require in order to understand the problem and the solution presented by the report. In brief, you will need to form a clear mental picture of the finished product, to anticipate the jobs that the production of such a product involves, and to lay plans that will result in orderly progress.

The third element in the analysis of the report writing situation was discussed in Chapter 2: analyzing the reader. What was said there applies especially to the current discussion since reports are usually targeted to specific readers. If you do not know who they are, you will find it almost impossible to answer the questions posed by the first problems involved in analyzing the job.

Admittedly, the three elements of analyzing the job are not sharply distinct. The second is to some extent based on the first, for it is obviously impossible to decide where and how to get information without first deciding what information must be obtained; the third problem involves both of the others. But in spite of this overlapping, the analysis of a report-writing job usually goes better if you think of it as the analysis of three separate problems.

MAKING A TENTATIVE PLAN
OR OUTLINE

It is almost impossible to analyze the subject of a report without forming a general idea about your outline, at least in its larger aspects. As you think of various questions, you will consider their relationships toward each other, and these relationships will point the way toward main points and subpoints of an outline. And as you think of all your questions in relation to the function of your report, a general plan for presenting your material will begin to take shape. It is a mistake, however, to think of these early ideas on organization as final. Only after you have gathered enough information to be dealing with answers instead of just with questions can you be sure that any plan you form will be satisfactory. Then some matters that looked important will dwindle into insignificance; points that seemed likely to be minor will assume importance; facts will emerge that you had not realized would call for consideration. All in all, though forming a tentative idea about organization is natural and useful, the principles urged in Chapter 4 are still sound. Only when you have gathered information and interpreted its significance can you decide definitely about the best way to organize it.

GATHERING INFORMATION

Gathering information is discussed briefly in Chapter 2 as part of the prewriting stage of the writing process. We indicated there that time spent well at this stage can help prevent difficulties in the writing stage. The subject of gathering information is so extensive, however, that entire books have been devoted to it, so a discussion of the subject in a book dealing mainly with the writing stage must be relatively general. A few comments about sources of information apply widely enough, however, to justify their inclusion. These comments will apply first to the reports you will write on the job and then to reports you may produce in a course such as the one based on this book.

Reports Written on the Job

When you are actually on the job, regardless of your field, your first step in gathering information will probably be to check the files of your own company in order to locate reports or other records that might answer some of your questions. Many companies have extensive indexes of reports and abstracts of their contents. Where no formal system for retrieval of information has been developed, personal inquiries as well as checking the files will often open leads that will enable you to avoid duplicating the work that others have done.

Published material may be the next source to try. A systematic survey of published literature is often the starting point when an extensive investigation of a scientific question is to be undertaken.

Another common method of obtaining information is by correspondence. Sometimes this involves the use of questionnaires, but a mere letter of inquiry is more frequently appropriate. Such letters are discussed in the section on letter writing. To a person on the job, the occasions when a letter is the best means of obtaining information are so obvious that there is no need to discuss the subject in detail.

In place of writing letters you may sometimes find it better to get information by interviews, either in person or by telephone. The amount of business conducted by telephone has increased over the years as efficiency experts have determined the actual costs of letter writing and the delays involved. Such interviews may be more expensive than letters, of course, but the flexibility of a personal conversation can bring out facts that would not come to light in an exchange of letters.

When the information needed for a report cannot be gathered from other people, personal observation and inspection is a possible method. A typical use of personal observation would be a visit to a factory to watch some process in operation, a field trip to observe the performance of a piece of farm machinery, or a trip to a mine to investigate its condition. Personal inspection might be the method used in an attempt to determine why accidents were frequent on a construction project, or what conditions were causing an epidemic.

Scientific tests and experimentation are of course the main source of information for many reports. Sometimes as a report writer you may do such work yourself and sometimes you may have it done for you by others whose special services are available. The contributions of scientific research to modern industry and government—contributions that are made available by means of reports—are so vast that it would be impossible to do more here than recognize their importance. However, we should not overlook the fact that reports based on scientific research are not always related to practical problems of the kind we have been discussing. Many a scientific investigation is made for no further purpose than to increase the fund of human knowledge, and many a report is written just because new knowledge exists and should be reported. The purpose and probable use of this book justify a focus of attention on practical rather than purely scientific problems; but many of the techniques recommended can be applied to reporting on scientific research that is not intended to solve a specific and immediate practical problem.

One final and general means of gathering information must be mentioned because, though it overlaps the others, it will often be the main method you will use on the job. It consists of merely doing the work that your profession calls for—sometimes with the help of others whose services are available. For many who use this book, such work will mainly consist of the scientific tests and experiments referred to above; but these and the other possible activities are so numerous and varied that no list could cover more than a small proportion of them. Actually, the work done in gathering information for a report can include any of the technical work done in our complicated industrial civilization.

It should also be noted that the information in a report does not necessarily come from a single source, and conversely, not all the sources are likely to be used on a single occasion. The fact is that although the information needed for a report may sometimes be far from easy to obtain, it will not necessarily be hard to decide what sources are promising and what method to use. The nature of the problem is likely to make the type of source and the method fairly obvious.

Reports in a Writing Course

Most of the sources of information described above may be utilized for a course in technical writing. But the problem of gathering information is complicated by the fact that resources available to a person holding a position in government or industry are often unavailable to a college student. On the job, the person who is asked to report on a problem has access to relevant files, opportunity to make necessary visits and inspections, and the time, facilities, assistance, and financial support needed in seeking a solution. In the classroom, finding a real or at least a realistic subject based on specific facts may in itself be a major difficulty—especially since the problem should not only be realistic but must also be practicable for a student to handle. Consequently, instead of starting with a problem and then figuring out the sources of information and method of investigation, students must often discover subjects they can obtain information about and then try to set up a problem that this information will be useful in solving.

There are four basic report writing situations you might consider when casting about for a subject.

1.　You may report on something you are currently doing.
2.　You may report on something you have done in the past.

3. You may report on something someone else is doing.
4. You may go to printed sources and report on what has been reported.

Each of these situations is actually similar to what you will encounter on the job, and of course a single report may cover all four areas. As an example, you may be part of a team working on a project. The report would cover background material from printed sources, the work the other team members are doing, the work accomplished, and the work in progress. But for a school report, you may well wish to cover only one area. For instance, you may be working on an ongoing project in your major field. You could easily write a report on the work. Or you may have had a summer job that involved various projects, one of which would provide good material for a report. If neither of these conditions applies, you might find a faculty member or graduate student who is working on a project and would allow you to help on a part of it, or allow you to observe enough to write a report. The latter is a common task of professional technical writers, who seldom perform the work they report about. And finally, there is the "state of the art" report, or SOTA, as it is called in government agencies. The SOTA calls for a search of all sources to discover what has been accomplished about a specific problem. Its purpose is to assemble information in one place in such a way as to focus on what is known about a single problem. These reports are often produced to give legislative bodies background for passing laws or for providing funds for remedial action. The USAID program, for instance, recently funded a SOTA on African swine fever, which threatens to decimate world hog production, to determine what steps might be taken to prevent or cure the disease. The investigator examined printed reports and bibliographies, flew to the major research centers in England, Africa, and the United States to interview experts, and eventually produced a SOTA about 800 pages long. As a student, you would not undertake such a broad problem, of course, but there is nothing to stop you from reporting on the state of the art of a narrower area of interest to you. For instance, you might report on information available about small-farm hog raising, about laser study on your campus, or crime in your community. Such a report might then be used as part of a background section of another report on a project you undertake as a result of your new understanding of your field.

Information available in books and magazines may unavoidably be a larger element in a report for a writing course than is usually the case in a report written on the job. To prevent a situation in which mere reference papers masquerade as reports it is well to remember that a school itself may provide suitable subjects. These would include such matters as parking, lighting, utilization of land, arrangement of laboratory equipment, condition of buildings, sanitation, equipment in kitchens, and many other questions that constantly arise in the physical operation of even a moderate-sized institution. In any event, unless a subject is assigned or suggested by an instructor, it may call for considerable ingenuity for a student to figure out where information can be obtained and to develop a subject that permits use of the facts available.

INTERPRETING THE FACTS
GATHERED

A report writer's job is not limited to gathering facts and stating them. Facts must often be interpreted for the reader's benefit. The nature of this interpretation is comparable to what happens when a patient goes to a doctor. The patient describes symptoms, the doctor administers tests, makes examinations, and discovers facts about metabolism, blood pressure, and other physical conditions. Then the doctor must interpret these facts about the patient's condition in order to diagnose the ailment.

Other examples would be as follows:

FACT: When a slide prepared by a medical technician is exposed to a certain stain, the smear upon it retains the color of the stain.

INTERPRETATION: A colony of bacteria has developed.

FACT: When the concentration of elk in a national forest is greatest the browse gets scantier each year.

INTERPRETATION: The herd is too large for the food supply.

FACT: The water above an industrial plant contains more oxygen than the water below, and the industrial wastes contain organic matter.

INTERPRETATION: The oxygen is being used up in the decomposition of the organic matter.

FACT: A Geiger counter ticks faster than usual.

INTERPRETATION: There has been an increase in radioactivity.

To be sure, the statements labeled *interpretation* in the preceding examples are themselves intended to be statements of fact, but the facts stated are a different kind. They are discovered not by observation alone but by observation plus reasoning, and they reveal the significance of the observed facts on which they are based.

Another kind of interpretation is the interpretation of the meaning of written material, as, for example, when the meaning of a statute is interpreted by a court decision. This kind is not so common in report writing as the kind previously described, but it is sometimes called for. For example, when there is doubt about whether certain work is covered by a contract, interpretation of the meaning of the phraseology of the contract would be necessary, and the results of this interpretation might affect the recommendations of a technical report. To be sure, everything should originally be written in a manner that reduces the possibility of a dispute about interpretation. Nevertheless, in your technical writing, you will sometimes have to demonstrate that a general or abstract statement covers a specific, concrete situation. And in doing so you will be interpreting the meaning of the statement.

The process of interpretation may call for any of the generally recognized processes of reasoning, such as generalizing on the basis of individual cases, applying a general rule to a specific case, ascertaining the cause of incidents or conditions, deciding what results will follow if certain actions are performed, or ascertaining the specific applications of a rule or regulation that is phrased in general terms.

Likewise, errors in interpretation resemble errors in reasoning in general. They include generalization on the basis of insufficient or unrepresentative evidence, or failure to consider alternative possibilities (that is, the possible existence of undiscovered facts or other possible interpretations of facts discovered). Also, writers sometimes fail to realize that an entire line of facts and reasoning may be based on assumptions that might themselves be subject to question. Much of what was said in Chapter 8 under the heading of *Thoroughness* deals with these errors.

Whether to explain in detail the reasoning by which you reach your interpretation of the facts is a matter of judgment. Certainly it is not desirable to prolong a report by explaining the self-evident. There are times, however, when readers will probably be unable to see why you interpret facts as you do unless you explain. And there are also times when it is desirable to explain even if readers might be able eventually to think their way through, for with your assistance, they will be able to see the connection faster than they could figure it out for themselves. Also, they are more likely to accept your interpretation if you show them how you reached it.

One more point needs to be cleared up: the distinction between interpreting the facts in your report and the process of forming your conclusions. Actually, there is no *essential* difference between the two. When you interpret a fact or a group of facts, you are drawing a conclusion, and when you settle upon the conclusion or conclusions for the report as a whole, those conclusions amount to an overall interpretation of all the facts. Nevertheless, at a practical, working level, a distinction is justified unless the report is short and uncomplicated. The difference lies in the scope. In stating the conclusions of a report you will be telling the convictions arrived at in connection with the major question or questions the report is supposed to answer. But in interpreting your facts you will be trying to find and reveal the significance of individual facts or small groups of facts. When you eventually draw your main conclusions, you base them on the interpretations that have occurred throughout the report.

MAKING THE DETAILED OUTLINE

As mentioned earlier, you are almost certain to form a general, tentative plan for a report before you gather all the information it will contain. But when you have gathered the information and analyzed it to ascertain its significance, you will usually need to revise the plan so that it will be a good organization for the facts as they actually turned out and to develop it into a detailed outline. In most respects, making an outline for a report is not essentially different from making an outline for any other purpose. True, you may sometimes be required to place your material under a standard list of major headings; but even then it will still be necessary to plan the presentation of the material within each main division; and there will be innumerable occasions when you will be free to plan the entire treatment. Therefore, you should not overlook the techniques of outlining presented in Chapter 4.

In that chapter it was pointed out that making an outline is primarily a matter of organizing your material. As you organize the material of a report, one consideration needs special emphasis. A report, more than most kinds of writing, is written to perform a specific function when it reaches the attention of specific readers who read it for specific reasons. Consequently, it is especially important in planning it to refrain from concentrating on the subject matter only. It is usually possible to organize the same subject matter in more than one way, and the best of these possible organizations is the one that makes it easiest for readers to find the answers to the particular questions they are likely to have in mind.

Consider, for example, a study made by a highway research laboratory to determine whether subgrade soil might be improved by mixing certain materials into it. Tests were made to determine, in regard to each admixture, its effect on the strength of the soil, the ease of compaction, the resistance to penetration of moisture, and the tendency of the soil to swell when moisture penetrated. Two of the possible organizations that would accommodate the facts are as follows:

1. An organization in which main points were based on the admixtures and the subpoints were based on their effects. This organization would make it easy for readers to learn all the results that would follow when a given substance was added; but in order to discover what substance to use if a certain result were desired they would have to look in as many places as there were substances.
2. An organization with main points based on effects. This would make it easier to see what material might be used if a certain result were desired, but information about all that would happen if a certain substance were used would be scattered in several places.

Both of these treatments are logical, but on a given occasion one or the other might be decidedly superior.

PLANNING THE USE
OF TABLES AND FIGURES

If a report calls for the use of tables and figures, you must decide whether to scatter them through the text, place them all at the end, or divide them between the two locations. It is usually best to make this decision before you do much writing. In fact it is better if you not only decide where figures and tables will be placed, but also decide what specific figures and tables you will use. If you cannot make this latter decision for the report as a whole, you can at least make it in connection with each section before you begin writing it. The reason for settling this question early is that your text itself will be affected by your decisions in regard to these nonverbal forms.

The technical aspects of using tables and figures in a report is no different from that of using them in other technical writing. Therefore the discussion in Chapter 7 makes a detailed explanation at this point unnecessary.

WRITING THE FIRST DRAFT

If the preliminary work has been done properly, the actual writing of a report is much the same process as writing anything else and should not be abnormally difficult. A few special suggestions concerning report writing, however, might be of value.

The first of these concerns the question of getting started. How to get under way may be a puzzle, of course, in any kind of writing. In writing a report, the preliminaries that must often be cared for present an additional complication. Consequently, if you find it hard to get a report started, the sensible thing to do is to postpone writing the opening and go to work on the earliest section that you feel ready to handle. If this is not feasible, start at the beginning in the simplest manner possible. Don't worry about smoothness. Just write the bare facts that are necessary if the report is to be self-sufficient, and then plunge into the first main section of the body. Later, you can revise your opening or write a new one; but it may surprise you by turning out to be better than you had expected.

In writing a report on an extensive project, you will probably not even consider writing the opening first. Rather, you will write up each section as soon as the information it will include becomes available. Your work will go faster and its quality will be better if you write while your information and ideas are fresh in your mind. Under these circumstances the preliminary parts of your report may be among the last to be written.

In any event, when you start writing, push ahead without worrying about details. If your flow of words is slowed down by an effort to improve your style, the flow of ideas may also slow down and it will be harder to preserve continuity.

If something worries you as you put it down, make a note in the margin and settle the question later.

As you write or revise you should be especially alert for spots where ordinary text is ineffective. When you have a list to present, for example, you may be able to help your reader by presenting it with each item on a separate line. A typical use of such a list (sometimes called an informal table) would be as follows:

Gradation of Samples

Percentage passing $3/4$-in. screen and retained on $1/2$-in. screen	16.7
Percentage passing $1/2$-in. screen and retained on $3/8$-in. screen	16.6
Percentage passing $3/8$-in. screen and retained on No. 4 screen	16.7
Percentage passing No. 4 screen	50.0
	100.0

The use of nontextual form may often be carried even further with good results. To show how this helps the reader, the same facts are presented below in two forms; the second makes them easier to grasp than the first.

(1)

The annual savings in labor cost that we can realize by manufacturing the newly designed bobbin rather than the present bobbin will be approximately $322,560. This figure is calculated as follows: We hire three crews to manufacture bobbins, each crew consisting of 45 workers. The size of each crew, if we make the change, will be cut to 35, a reduction of 10. Thus the labor force will be reduced by a total of 30. As we now operate, a laborer normally works 1,920 hours per year. If we made the change, the elimination of 30 laborers would mean a saving of 57,600 work hours. Our present wage to members of these crews averages $5.60. Thus by cutting the crews as indicated we would reduce cost of labor by a total of $322,560.

(2)

The annual savings in labor cost that we can realize by manufacturing the newly designed bobbin rather than the present bobbin will be approximately $322,560, as shown by the following calculations:

Number of crews		3
Size of each crew		
Present	45	
Proposed	35	
Reduction	10	
Total reduction of labor force		30
Hours worked per worker per year		1,920
Hours saved per year		57,600
Average wage per hour		$5.60
Savings in cost of labor		$322,560

REVISION

Unless you write with unusual skill, your work will probably call for considerable revision. If your schedule permits, you should not do your revising until you can look over your work with the feeling that you are a reader rather than a writer. It is possible, however, to let your work cool off too long. You should revise while your subject is still fresh enough in your mind that you will notice omissions and recognize errors.

The amount of time devoted to revision depends partly, of course, on what the project justifies, partly on the amount of time you have available, and partly on your degree of success in producing an acceptable version at the first attempt. If a report includes complicated ideas and deals with an important subject, there is no reason you should feel frustrated if it must be worked over three or four times. Most of us must face the fact that writing is not easy.

Since revision is tedious, one point needs to be emphasized. When you have revised a passage for what you think should be the final time, read it over once more from beginning to end to make sure that the changes you have made do not call for additional changes elsewhere. Many times an improvement at one point makes it necessary to change something that precedes or follows.

CARING FOR FINAL DETAILS

Caring for the final details of a report may include preparation of a title page, a summary, a table of contents, and a letter of transmittal or covering memorandum. Suggestions on the purpose and treatment of these elements are covered elsewhere in the text.

Also, one final task must always be performed: the close examination of the entire report for errors in manuscript, in form, or even in such simple matters as the order of pages. Everyone dislikes to make this final check, but it is so easy for errors to pass unnoticed that a final scrutiny is well worth the time it takes. Errors in figures are especially likely to slip in during the process of copying, and can easily escape detection. A typist can accidentally change *now* to *not*. *Dejected* can be miscopied as *ejected,* or *grout* can become *trout.* A writer who intends to change *not conspicuous* to *inconspicuous* may add the *in* and forget to remove the *not*. Such errors do not catch the eye, and close scrutiny is necessary if you are to succeed in actually writing what you intend to write.

In closing, it should be reemphasized that the foregoing procedure for writing a report cannot and need not be followed exactly. Projects that call for reports develop in so many ways that it would be not only undesirable but probably impossible for you to follow a rigid set of instructions. But at least the steps suggested are definite and their sequence is logical. Though deviation may be necessary from time to time, you need not feel uncertain about how to tackle the job of writing a report if you will regard the recommended procedure as normal.

ASSIGNMENTS

Assignment 1

Analyze the problem one of the following subjects would present if it were to be reported on. That is, draw up a list of questions that would lead to a decision about what facts were needed. Break large questions down into smaller questions that would need to be answered first. Make your chosen subject more specific, if possible, by limiting it to some particular local situation that you know about or can learn about.

1. The problems involved in setting up a car-pool system for your campus or business to reduce the numbers of automobiles driven daily.
2. The feasibility of rerouting a major highway around a community rather than through the main business district.
3. The feasibility of converting your campus or business from typewriters to word processing equipment.
4. The advantages and disadvantages of doing a major remodeling job on an existing building to modernize it rather than tearing it down and building another in its place.
5. The question of automating a manufacturing process.
6. The question of whether your campus or business should operate a food service or should hire an outside firm to provide the service.

Assignment 2

In the chapters "The Writing Process" and "Procedure in Writing a Report," a number of steps were described that applied to the process and procedures of writing in general. Pick a specific subject for a report from the suggestions in this chapter or in the chapter "Formal and Semiformal Reports" and list the specific steps necessary to write that report. Make a plan including actual dates for completion of each step, and include a tentative outline of the final report.

Assignment 3

As directed by your instructor, write one or more reports of unspecified form resulting from an investigation of one of the following subjects, a subject listed in Assignment 1, or a subject of your own choice. Like all other subjects suggested, any of these subjects would have to be defined and adapted by setting up a situation in which the report would be submitted to some person or organization that needs the information. The scope of your subject may also need to be broadened or narrowed so that you can use the information obtainable and have the size of the subject suitable to the assigned length of the report.

1. The steps to be taken to protect an endangered species.
2. The steps involved in converting an existing building to solar power.
3. The advantages and disadvantages of heat pumps over conventional methods of heating and cooling buildings.
4. The use of genetic engineering to improve varieties of trees for lumber.
5. The use of auger conveyors to automate a feed lot. The feed lot should be a definite one where specific conditions exist that you would describe.
6. The methods used to improve or maintain soil conditions for agriculture in your area. Include advantages and disadvantages and make recommendations for change if you feel current practices could be improved.
7. Land planning practices and regulations in your community. If there are inadequacies, make recommendations for change.
8. Causes of destructive fires in your area and what is being done to prevent them.
9. Styles of clothing sold in a local shopping area. Limit your survey to a specific type, such as sportswear, eveningwear, costume, with the goal of suggesting directions for expansion for a specific store.
10. Rerouting traffic flow to increase the livability of a neighborhood. Include considerations of one-way streets, traffic barriers, malls, and other methods you devise.
11. The lighting system for a building or recreational area, such as a sports field or park. Analyze the current system and make recommendations for improvement.
12. Accidents on your campus or in your business. Analyze causes of the most prevalent types and suggest remedies.
13. The use of television for teaching or training aids for a technical subject. Make recommendations for purchase; include a description of the system you propose for some specific use.
14. The supply of potable water in your community and the steps being taken to conserve water.
15. Alcoholism or drug abuse on your campus, in your community, or in your business and the steps being taken to combat it. Make recommendations if you see ways to improve current methods.
16. The use of recreational vehicles in your community. Include local restrictions, and make recommendations for ways to increase use, if local conditions allow, or to decrease use, if that is called for.

Reports— Their Variation in Form and Purpose

10

If you work for a large corporation or government bureau, you have probably found that it has developed certain types of reports to suit its own particular needs, and you should obviously conform to its preferences when writing. Many who use this book, however, will work for a wide range of employers and clients. Consequently there will be no attempt, in this chapter, to fit all reports into a neat set of classifications like a botanist's classification of plants. Rather, though some of the widely recognized types will be discussed, the purposes of the chapter will be:

1. To give you a better understanding of the wide variations of reports in form and purpose.
2. To make the nature of the commonest types familiar enough that if you are asked to write a report of a certain type but have no further guidance available, you will know what is probably expected and also realize how much latitude you enjoy.
3. To encourage resourcefulness and self-confidence so that if you are asked to write a report but given no instructions about what kind, you will know whether some generally recognized type is called for and how to produce it; and if no particular type seems to fit the occasion, you will realize that not all reports are of some established type and will proceed with assurance to produce what the occasion calls for.

The kinds of report to be covered are: reports made by filling in a form, reports in the form of letters, reports in the form of memoranda, progress and status reports, periodic reports, feasibility reports, environmental impact statements, laboratory reports, and nonformal reports that cannot be considered to be of any special type. You will find that these types are not mutually exclusive, for some of them owe their identity to form and others to purpose.

Formal and semiformal reports are briefly identified, but since they involve a substantial number of special elements, full discussion of them is postponed to a later chapter.

BLANK FORMS

Many reports are made by merely filling in the spaces on a blank form on which headings are already provided. This can happen when some activity in an organization becomes so routine that the information about it will always occupy approximately the same amount of space and will always be covered by the same headings.

Frequently the information filled in consists merely of figures. In this case making the report is not a problem of writing at all. Yet sometimes the blanks must be filled in with prose. On these occasions the printed forms are used to make sure that the organization of the reports will always be the same, to prevent the writer from overlooking any essential point, and to discourage any tendency to make the report too long. Such reports are good in that those who read them always know exactly where to find what they are looking for. Their main limitation is their lack of flexibility. Sometimes flexibility is increased, however, by a section for comments.

Writing a report on a blank calls for no skill except the ability to gather the necessary information and to write clearly, accurately, and concisely. The style should be reasonably formal in a report that will become part of an organization's permanent records, but may be informal in a report dealing with a matter of only temporary concern. An example of a report made on a form appears later in this chapter.

LETTERS

When there is no need for a report to include tables, figures, or any kind of stiff and bulky material, the natural form to use is often a letter. Most of what you need to know about reports that are letters can be learned when letters are discussed later in the book, for the same good practices apply for any letter, regardless of its use. The form for a report is no different from the form of other letters except that the use of a subject line is more likely to be desirable. In language, the letter serving as a report may not call for quite as much concern about tone, negative material, and good will if the reader is in your own organization. However, if a letter is used in making a report to someone on the outside—a client rather than an employer—tone and good will are as important as in any other letter.

A letter can be used to report on almost any type of subject. Letters that serve as reports often run longer than the average business letter. Long or short, a report in the form of a letter should be just as carefully organized as any other piece of writing of comparable length and complexity. Such devices as topic sentences for paragraphs should not be neglected. (Topic sentences are discussed in the Handbook under *Paragraphs*.) Indeed, in a letter report of two or three pages, it may be desirable to use internal headings, as in any other report. Their use is covered in Chapter 11.

Reports in the form of letters can vary widely in formality. When addressing a close associate, you might often be justified in making your style extremely informal. There are occasions, however, when your sense of fitness—your imagination in visualizing how the report will be used and whom it might eventually reach—should make it apparent that more dignity is indicated.

One final caution is needed regarding the use of the letter form for a report. If the subject is likely to be of permanent concern, the letter form is sometimes undesirable because it may cause the report to be filed along with ordinary correspondence rather than as part of the organization's permanent records. When this likelihood exists, it may be advisable to prepare a report in some other form and write a covering letter to accompany it.

MEMORANDA

In many ways a memorandum resembles a letter. Like a letter, it is one of the forms used most frequently. It is not usually used for a report that is long or one that calls for the use of material other than ordinary text, and it may range from informal to formal in style. But unlike a letter, a memorandum is strictly for use within the organization—for communication within a department or between departments. Consequently, a memorandum may be unceremonious in style and may sacrifice niceties for conciseness. This does not mean that bluntness and tactlessness are to be condoned. Tact and diplomacy help to get results in dealings within an organization just as they do in outside contacts.

Originally the term "memorandum" implied something of a temporary nature. This is no longer true; yet there are times when a memorandum is written to make immediately available some information that will later be included in a longer, more formal report. Some of the largest corporations in the nation make this use of memoranda.

For use in writing memoranda, printed forms are usually provided. If no form is provided, however, you should provide the necessary information at the top of the first page, using the following form.

To: Date:
From:
Subject:

PROGRESS AND STATUS REPORTS

Progress and status reports, if indeed they should be regarded as separate types rather than a single type called by different names in different places, resemble each other closely enough that it seems advisable to discuss them in the same section.

The term *progress report* has been widely used for many years. As the name indicates, a progress report is a report on the progress of some project. It may assume any form from that of a simple letter or memorandum to a full-scale formal or even a printed report. Progress reports may be made either when some stage of the work has been completed or at regular intervals. In the latter case they may also be considered periodic reports, the next type to be considered.

The very nature of progress reports affects their organization. Such a report is likely to include a description of:

1. The current status of the project;
2. Work completed during the period concerned;
3. Work currently under way; and, perhaps,
4. Plans and outlook for the future.

Progress reports may include conclusions and recommendations sections if the nature of the work calls for them. For example, work in one area may have run into such difficulties that the person in charge might recommend dropping that area of the project until conditions improved; such a decision would rightfully be placed in a recommendations section. Incorporated into each section would be information on unexpected developments if there have been any, and any other facts that the reader would wish to know.

In writing a progress report on an investigation it may be necessary to decide whether to include some or all of the information already obtained. If some aspects of the investigation have been concluded and if the information would be of interest, there is usually no reason to withhold it. If there is a chance, however, that incomplete information would lead to hasty and possibly erroneous conclusions, it may better be withheld until the findings are complete.

It has been mentioned that progress reports fall naturally into a past, present, future pattern, but this pattern need not be the main, overall division of material. If a report deals with a complicated project it is often better to base the main divisions on the various parts of the job and apply the past, present, future pattern to each part.

When several progress reports must be written on the same project, they should all be organized in the same manner so far as the subject matter permits.

In recent years more and more use has been made of the term *status reports*. If the reports so designated are to be regarded as anything except traditional progress reports under a different name, and if the use of the two terms is what one might expect, the key to the difference between the kinds of reports lies in the basic meanings of the two terms. The term *progress* suggests that the purpose is primarily to tell about change—about headway made in a desired direction. *Status*, on the other hand, suggests that the main purpose is to describe existing conditions.

Actually, the difference is primarily in emphasis. One type emphasizes what has been happening; the other, what condition exists because of those happenings. There is no reason to feel, however, that the two types—if such they are—contain entirely different information. It would be difficult if not impossible to tell about progress toward a desired end without telling what condition had come about as a result of that progress. In fact, it is natural to begin with a comment about that condition. Likewise, it would, to say the least, be unnatural to tell about existing conditions—status, that is—without telling what had happened since some earlier time to bring those conditions into existence.

In any event, you should be aware of the existence of both terms and feel at ease in using one, the other, or both as is customary in whatever organization you are connected with. If usage in that organization provides no definite guidance, you should feel free to use whichever term is more appropriate in view of what your report emphasizes. Regardless of your choice of terms, your main concern is to give readers the information that they need.

PERIODIC REPORTS

A periodic report, like a progress report, is distinctive because of its purpose. It is merely a report made as a matter of course at regular intervals, frequently but not solely for the purpose of keeping records. If it deals with a project that will eventually be completed, it is also a progress report, for the two types are not mutually exclusive. In its simplest form it may be merely a filled in blank form—as when a weekly report is made on accidents, production, or the percentage of production failing to pass inspection. It may also be a letter or a memorandum. From these simple forms it may range upward to the elaborate form used in the annual reports of the board of directors to the stockholders of a large corporation—an expensive, illustrated product, printed and distributed by the thousands.

In any given situation the periodic report tends to settle into a well-established form because the points it must cover are usually similar on successive occasions. Consequently, unless you are the first person to fill a position, you will often be able to pattern your periodic reports on those written in the past. Thus the problem of organization, even in a long periodic report, may not be difficult. It pays, however, to be always on the watch for special points that call for unusual treatment and to remember that even when a report follows an established pattern, the difference between good and bad writing affects its quality. And finally, you should not neglect to change the established pattern if you find that it can be improved and you are in a position to make changes.

FEASIBILITY REPORTS

A feasibility report usually results from an investigation to determine if some action should be taken. Feasibility reports fall into three basic types. Sometimes the action is known, so the investigation centers on discovering whether conditions exist for the action to be successful. An example is a Maritime Administration study to determine if sizable modern cargo ships could be powered by sail. The study determined that such ships were not feasible at the current price of fuel oil. When the price tripled, the Maritime Administration ordered an update of the study, including an analysis of commodities, trade routes, and ports that would be most feasible for computerized sailing ships.

A second type of feasibility study occurs when the conditions are known, but the action to be taken is not. For instance, the accident rate between bicycles and pedestrians on a university campus was determined to be high. A campus traffic committee undertook a study to determine causes and to discover what actions might reduce the numbers of accidents. The actions considered included banning bicycles, licensing, and stricter enforcement of existing laws. The study determined that the most feasible actions would be the creation of bicycle "walk zones," banning bicycles from sidewalks, and efforts to educate both riders and pedestrians about their responsibilities.

A third type involves a comparison of products or actions to determine which would be most feasible when both the conditions and the actions are known. Usually, such reports involve a comparison of the advantages and disadvantages of each option. Sometimes, this type of report is called a recommendation report, but since many kinds of reports have recommendations, that designation does not seem very useful.

In each of these types, the focus of the report is on the action to be taken, and the conditions under which the action would be successful. In commercial feasibility studies, cost is almost always a major consideration, but other factors might include capability, environmental impact, effect on other elements of the company, public opinion, and competition. A good feasibility report touches on all possible factors and emphasizes the important areas. Unfortunately, no rules can be given for determining important areas that would cover every possible situation. Thus, when writing a feasibility report, you need clearly to state the problem, the scope of your investigations, and the criteria you determined to be most important under the conditions you discovered. Finally, you need to state your conclusions and recommendations. The order of presentation, of course, may vary according to the needs of your readers, and, naturally, the sections may be expanded or contracted depending on the size and complexity of the problem. For instance, the updated version of the windship study mentioned above ran some 200 pages, studies of the feasibility of U.S. defense options have run into thousands of pages, and the feasibility report included among the examples in this chapter is three pages long (the letter concerning the study is two pages).

ENVIRONMENTAL IMPACT STATEMENTS

Since the passage of the National Environmental Policy Act, reports dealing with the environment have become common, especially in grant requests to government agencies. The act requires all agencies of the federal government to consider the environmental effects of proposed actions and to report on efforts to avoid adverse effects. Areas of concern include hydrology, geology, air quality, noise, biology, socioeconomics, energy, land use, and archeology or history. The reports fall into four general types: environmental information documents, environmental assessment documents, finding of no significant impact documents, and environmental impact statements. Each of these are important in the review process, but the environmental impact statement is the most complex, so it is the type we will discuss.

The Environmental Protection Agency stresses that the format for environmental impact statements should "encourage good analysis and clear presentation of alternatives." The following sections are recommended for inclusion unless a compelling reason suggests a different arrangement:

Cover sheet
Executive summary
Table of contents
Purpose of and need for action
Alternatives including the proposed action
Affected environment
Environmental consequences of the alternatives
Coordination
List of preparers
Index
Appendices

The executive summary includes enough information for readers to become familiar with the proposed project and its effects. It includes a statement of the problem, a description of each alternative, and any major conclusions. If the report is long, a comprehensive summary is sometimes included as well. In the discussion of alternatives, the EPA stresses that the discussion be balanced. Alternatives that were eliminated from detailed study must be listed along with the reasons for their elimination. The section on coordination discusses efforts made to include groups, such as the public, that might have an interest in the action.

Other elements of environmental reports are similar to those of reports in general, so the discussion in the chapters dealing with reports and report elements covers what would be said here. Since environmental impact statements tend to be large, we have included no complete example, but you may refer to the specimen following the chapter on formal and semiformal reports for an example of an environmental information document.

LABORATORY REPORTS

One of the commonest purposes of reports likely to be written by those who read this discussion is to tell the results of work done in a laboratory. Sometimes this work consists of merely running tests, the results of which may be stated without reference to the reason that the report was requested. On other occasions the report must not only tell the results of laboratory work but must also apply them to some specific problem and even recommend action. In the latter case the laboratory report form may be discarded in favor of some other form more suitable to the specific occasion.

Most of the time, however, a report on work done in a laboratory may be written under headings that have become fairly well standardized. Though there is some variation in different organizations, the following list covers most of the divisions that a laboratory report is likely to contain.

Title page, or merely a title
Object (often called Purpose)
Theory
Method (or Procedure)
Results
Discussion of Results (often called Comments)
Conclusions
Appendix
Original Data

It is not always necessary to include all these divisions. Much of the time there will be no need to explain the theory; and if the method or procedure has become standardized, the section dealing with it may also be dispensed with. The results, of course, would always be given, though they might be placed on data sheets at the end if they consist of figures. There may or may not be reason to discuss the results; but even when such discussion is needed, it may be placed in the "Results" section rather than in a separate section.

Sometimes a section on conclusions will be unnecessary because the results themselves may be the only conclusions arrived at. A section on conclusions is desirable, however, when consideration of the results has led to convictions that go beyond the results themselves. For example, the results of a test of samples of a type of steel might consist of no more than the facts about its tensile strength and other qualities, but consideration of those results might lead to a conviction as to whether the steel would be suitable for some prospective use.

An appendix need not be used unless it is advisable to add materials omitted from the main report because they are too detailed, too technical, or too bulky. As for "Original Data," much of the time the full data are included under "Results." If the full data seem needlessly detailed, however, the figures under "Results" may be reduced to averages or summary data and the full data may either be omitted or be placed at the end to permit verification. If placed at the end, they may be included as part or all of an appendix.

The arrangement, as well as the selection of points, in a laboratory report may be varied. The order listed above is logically sound, but sometimes changes may be desirable for the reader's convenience. Conclusions might be placed early in the report, perhaps following the statement of the purpose, for the conclusions are likely to be what the reader wants to know at once. In fact, some of those whom a laboratory report reaches may read the conclusions and nothing else.

To sum up: If given a specific form you should follow it as closely as possible, for the reader will grasp the contents better if they are placed in a familiar form. If no particular form has been requested, you may use the heads listed above to the extent that you need them. The result will be standard in form so far as any general standard exists.

FORMAL AND SEMIFORMAL
REPORTS

For the sake of completeness two more types of reports, formal and semiformal, must be identified here, although they are more fully discussed in later chapters.

Even though we discuss the formal report as a separate type, we must emphasize that any of the reports discussed in this chapter could be a formal report if the situation called for that type and a formal presentation.

A formal report is a report that is dignified and impersonal in tone and that presents its material in a pattern resembling the pattern used in books. The extent of this resemblance and the techniques that the pattern calls for will be expanded upon later. It is unfortunate that the term *formal* is used to identify these reports, for its use seems to imply that all other reports are informal. Actually, many reports not classified as formal are free from any flavor of informality. *Nonformal* rather than *informal* is a better term to apply to other types.

The term *semiformal report* has not been used widely enough to have acquired a generally accepted meaning. In this book *semiformal report* means a report that has the tone and general qualities of a formal report but that is simpler in design because it dispenses with some of the parts. Recognition of these reports as a distinct type is useful, because there are numerous occasions when the full apparatus of a formal report is needlessly elaborate but the use of some of its parts and compliance with its general conventions will produce exactly what is needed.

REPORTS THAT FIT INTO NO CLASSIFICATION

There are many varieties of reports that fit into no special classification, especially in their form. They are not made on blanks; they are not letters, memoranda, or laboratory reports; and they do not contain all the parts usually found in formal reports. This sometimes causes uneasiness in inexperienced writers, who often worry too much about form and too little about the reader's needs.

There is no reason, however, to be unduly concerned if you are asked to write a report and none of the recognized types seems to fit the occasion. You should be able to get satisfactory results by handling the job as follows.

First of all, consider the function that your report is supposed to perform. And since it cannot perform its function unless it secures the desired reaction from its readers, think also about who the readers will be—what their interests are and what they already know about the specific subject and the general area in which the subject lies.

Next, turn to the question of what material you will need to include if the report is to perform its function—what points must be covered and what information on each of these points the readers will want or need. As you decide what information to include, always ask yourself how long a report the readers will wish to read.

After you have gathered the necessary information, organize it by using the methods recommended in Chapter 4. These methods are flexible enough to let you form a plan that organizes your material logically and at the same time makes it easy for a reader to find answers to the questions that are likely to arise.

Unless the report will be short, decide whether the inclusion of any of the special elements discussed in the next chapter would make it more effective, and use any of them that seems likely to fit the occasion. In particular, use plenty of headings. If you have made a good outline, it will provide headings that cover whatever information you wish to present.

As you start the actual writing, decide whether the occasion calls for a style that is formal or informal—personal or impersonal. Having made your decision, be consistent in using the style that you have chosen as appropriate.

Start your report in such a manner that the reader knows at once that it *is* a report, can tell its specific subject, and can see what function it is supposed to perform.

Mention, near the beginning, any earlier reports or other written material that you will need to refer to or that are specially connected with the report you are writing. This material is sometimes headed *References* and listed before the text of a report begins. (Such a list of references is not to be confused, however, with a list of references placed at the end of the report for the sake of documentation.)

Always be sure that the date of your report is indicated at a point where it will be easy to see.

If you are writing the report as an individual, always identify yourself as the writer. One way to do this is to use the form you would use if you were writing an article. For example:

A Report on the Breakage of Laboratory Equipment
by
Robert A. Watts, Head of the Testing Bureau
July 15, 1982

When such a method seems needlessly elaborate, the necessary information can be placed in the first paragraph. If your report consists of a single page, you can identify yourself by telling your position at the opening and signing it at the end. (The phrase *Respectfully submitted* is often placed before the signature under these circumstances.) If the report is more than one page long, however, it is better to mention your position at the beginning and supply your name by signing it at the end. A personally written signature is often desirable, under any circumstances, to indicate that the writer accepts responsibility for the report.

If you are writing in behalf of a committee, it is customary not only to identify the committee at the beginning but also to list the members at the end. Sometimes, in fact, it is desirable to have the members personally sign the original, official copy. The names of members are ordinarily arranged alphabetically except that the name of the chairman comes last and the position of chairman is indicated.

The simplicity of the preceding suggestions demonstrates that writing a report that is not one of the special types is not particularly difficult. All you need to do is observe a few simple conventions as you give the reader information in a clear and easily understandable manner. If you do this, the question of whether your report can be classified as to type is likely to be inconsequential.

SPECIMENS OF REPORTS

The following reports are not offered as models. Moreover, though they include most of the types discussed, they are not intended to encourage you to draw sharp boundary lines between types. To draw such lines would be unrealistic.

The real purpose of including examples is in part to make the business of report writing more concrete. Further, it is probable that among the specimens you will find some devices that you can utilize in your own report writing, even though the reports in which they occur may not serve as patterns to follow in full. Perhaps the most important function of the examples, however, is to show how little regimentation exists in report writing. The more experience you have with reports, the more you will realize that good report writing is a matter of intelligent analysis of the reader's needs, effective organization, and clear, concise style rather than a matter of blind compliance with rigid specifications. Accordingly, you should feel encouraged to use your own judgment when you write reports.

As you study the examples, remember that no report writer ever achieves perfection. Consequently, be on the alert for alternative methods of presentation by which effectiveness might have been increased. (Illustrative specimens follow.)

Specimen No. 1*—Report
on a Blank Form

The report on the opposite page performs a dual function. As the print at its top shows, it serves as the title page of a longer report, but it is also a self-sufficient report in its own right, to be distributed more widely than the pages it precedes.

The form serves several purposes. The most obvious is uniformity; every report in this series has the same material in the same location so that readers familiar with the series know immediately where to look to find required information. By limiting space, the form forces writers to be brief so that busy individuals in other departments are more likely to read the form, and then are more likely to keep up on developments within the company.

In providing the facts necessary for self-sufficiency, which are not repeated on later pages, it makes them available both to those who receive the main report and to those who receive the single page.

It also provides the gist of the full report—giving those who depend on it what is most essential and serving as an abstract for those who receive everything. It is less technical than the full report.

It was followed by twelve additional pages—seven of discussion, one listing references, one explaining nomenclature, and three bearing a table and four figures.

Such a report as this serves the needs of those who only want to keep informed about company activities, those who are concerned with its contents but do not need full details, and those who need the full story. The tendency to include a page on nomenclature has increased in recent years. Likewise, a list of key words, used for purposes of filing and information retrieval, reflects a relatively new development in report writing.

*Courtesy of the General Electric Company.

GENERAL ✸ ELECTRIC

MAJOR APPLIANCE BUSINESS GROUP

TECHNICAL INFORMATION SERIES

Title Page

AUTHOR C. D. Denson/ D. L. Crady	SUBJECT Polymer Rheology	NO. 72–MAL–23 DATE 11/20/72
TITLE Measurements On The Planar Extensional Viscosity Of Bulk Polymers: The Inflation Of A Thin		GE CLASS I
		NO. PAGES 12
ORIGINATING COMPONENT Rectangular Polymer Sheet MAL–Plastics Laboratory		

ABSTRACT

   An experimental study of the inflation of a thin rectangular polymer sheet has been conducted to determine whether this technique can be used to measure the planar extensional viscosity of bulk polymers. Viscosities were determined at various extensional strain rates using an undiluted poly (isobutylene) of high molecular weight. Measurements were also made on the complex shear viscosity and these values were found to coincide with a "normalized" planar extensional viscosity when plotted as a function of the square root of the second invariant of the strain rate tensor. Industrial fabrication operations where the planar extensional viscosity is expected to play a role include vacuum forming of large rectangular sheets whose length to width ratio is large and in injection molds having a center gated cavity.

KEY WORDS

Polymers, Plastics Processing, Extensional Flows, Rheology

INFORMATION PREPARED FOR __Major Appliance Business Group–Plastics Lab.__
　　　　　　　　　　　　　　(name)　　　　　　　　　　　　(dept. or lab.)

REPORT APPROVED BY__R.S. Hagan__　　　　　　　　Plastics Lab.
　　　　　　　　　　(name)　　　　　　　　　　　(dept. or lab.)

Additional Hard Copies Available From __Ashley Library__

Microfiche Copies Available From - - Technical Information Exchange
P.O. Box 43 Bldg. 5, Schenectady, N.Y., 12301

MA-775 (Rev. 72)

Specimen No. 2*—Report
in the Form of a Memorandum

The report on the opposite page is included for two reasons. First, it has an irresistible appeal because of its human interest. Second, it demonstrates that no point in the world is so remote that it is immune from being the concern of report writers.

This particular specimen was chosen from a substantial number that also tell of field trips to remote Alaskan villages. Its style, like that of the others, is personal and informal—which is entirely appropriate under the circumstances. Though it sounds at the start like a personal narrative, after a few lines the writer begins to present the facts. And actually, the information about the difficulty—insurmountable as it turned out—of reaching the island is highly relevant because of its possible effect on future trips and because it contributes to the reader's understanding of the island and its population of 87 self-reliant human beings.

*Courtesy of the State of Alaska Department of Highways.

MEMORANDUM   State of Alaska

TO:
   Donald D. Friend, P.E.   DATE: August 8, 1972
   Pre-Construction Engineer
   Western District   FILE NO: 92-2519
   Department of Highways
                            SUBJECT: Local Service Road
FROM: Little Diomede Island
   Thomas A. Young
   Design, Pre-Construction Section
   Western District

On August 7, 1972 I attempted a field trip to Little Diomede Island
to determine the extent of road repairs required as per letter of
June 13, 1972 from Davis Menadelook, Village Council Secretary.

Upon arriving in Wales, I was informed that the water was too rough
to take a skin boat to Little Diomede at that time. However, I
learned that the Little Diomede Village Council President, Robert
Soolook, and Davis Menadelook were stranded at Wales. I spoke with
them and acquired the following information:

Little Diomede has a population of 87 and the only vehicle in town
is one snow machine. They have approximately 2400 feet of 10 foot
wide cobblestone type walkway through the village and one gravel
footpath, approximately one half mile long and three feet wide to
their docking area. The entire cobblestone walkway needs repairs
and about half of the footpath requires repairs. All repairs are
to be done by hand. They plan to carry rocks from the beach for the
walkway. The beach is about 200 feet away at the closest point.
They also plan to use hand shovels to repair the footpath utilizing
suitable material located alongside the footpath for almost its
entire length.

They informed me that they only need funding so that the Village
Council can hire approximately 10 to 15 laborers at a rate of
$2.10 per hour. They figure it will take three to four weeks to
complete all repairs. I told them that using 15 people at $2.10 per
hour for eight hours per day for three weeks would amount to $3,780.
However, they did not think it would take more than $1,500 to do the
required work as they anticipate a shortage of laborers and considered
this aspect in their estimate. Robert Soolook and Davis Menadelook
both agreed that $1,500 would be sufficient funds. They are not
aware of any formal Townsite Plan for the village but said that right-
of-way is considered to be no problem.

Existing utilities will not be interfered with during the repair work.

The water was still very rough by late afternoon so I was unable to
reach Little Diomede to visually inspect the proposed repair work.

I suggest that $1,500 is not an adequate amount and that $2,000 would
be more suitable to meet the required expenses of repair work to the
walkway and footpath.

TAY/vbh

Specimen No. 3*—Report
in the Form of a Memorandum

The memorandum on the opposite page is essentially a literature review. An alert employee read an article in a journal and realized that the contents had implications for her company. In the first paragraph she identifies the publication and gives the complete bibliographic information in case the recipients of the memo wish to read the article for themselves. She also indicates the reason the company might be concerned.

In the next three paragraphs, the writer summarizes the article and indicates in each case that the authors' findings differ from the findings of the Atlantic Richfield consultants who had dealt with the same subject.

The writer suggests, in the final paragraph, action that could be taken in an effort to resolve the contradictions between the article and the consultants' findings. The tone of the report is informal, as fits a memo of internal circulation and limited interest. At the time the memo was written, the matter under consideration did not require immediate action so the recommendations are presented as suggestions for consideration rather than statements. Despite the informal nature of the correspondence, the memorandum includes information like that of a more formal report—introduction, findings, recommendations—and is complete within itself.

*Courtesy of the Atlantic Richfield Company.

Atlantic Richfield Company   Internal Correspondence

Date: March 14, 1978

To: Les Ludlam

From: Patricia J. Rand

Subject: <u>PLANT SPECIES ON COLORADO OIL SHALE LAND</u>

A note by Jim Irvine, Neil West and Art Holmgren appearing in a current issue of Southwestern Naturalist (23:156–157, 1978) may be of interest to us should we begin oil shale operations in western Colorado.

These authors report finding a grass, <u>Festuca dasyclada</u>, which is listed as "possibly extinct" on the Endangered and Threatened Plant Species list of the Smithsonian Institution. They found it at the contact of the Upper Parachute Member of the Green River Formation and Uinta Sandstone Formation throughout Garfield County on south-facing slopes. This species was not collected on either the Colony or the C-b properties by our consultants.

They also found an <u>Astragalus lutosus</u> which is classified as endangered in Colorado on the eastern end of the Uinta Basin in Uintah County, Utah, when it was previously thought to occur only in Rio Blanco County. This species was reported in the Colony EIS but not found on the C-b tract.

They also substantiate the occurrence of <u>Ceanothus martinii</u> which occurs throughout the Great Basin and was presumed to be present in Colorado but uncollected. Neither of our species lists for C-b or Colony listed this plant.

We might consider examining the voucher specimens in the Intermountain Herbarium at Utah State University, Logan, before we do any more oil shale environmental work. It sounds as if the Utah botanists have been active in Colorado collecting. I'm not sure our previous consultants were aware of this.

Patricia J. Rand

PJR/lr
cc F.D. Gealy
    M.C. Roman
    D.K. McSparran

Specimens No. 4 and No. 5*—
Feasibility Report and Report
in the Form of a Letter

These two reports are taken up together because both of them grow out of the same situation—a traffic study made as the result of a citizen's request for changes in the existing conditions.

The memorandum, No. 4, progresses logically by first listing the studies made as a result of the request to determine if the changes would be feasible and then by summing up the facts that these studies revealed. The result, under the heading *Recommendations,* is a recommendation that the requested changes should not be made. The studies showed, however, that making certain other changes would improve the troublesome conditions, and the writer recommended that they be made. These were explained under *Improvements.*

The letter, No. 5, was sent—and is properly regarded as a report—to the citizen who had made the request that led to the studies. Interestingly, Report No. 5 was written by the official who received rather than wrote Report No. 4.

The writer might, of course, have merely sent a copy of the report he had received, along with a brief covering letter, but he wisely decided that something less terse and impersonal would be better when sent to someone outside the department, a fine example of reader adaptation.

The letter is excellent. First, it shows that the citizen's request was taken seriously enough to lead to a thorough investigation. Then it presents the facts that the investigation revealed. Next, it tells the decision that the facts led to. Finally, it ends on an upbeat by letting the reader know that his effort has resulted in positive action.

*Courtesy of Oakland, California, Department of Traffic and Parking Engineering.

CITY OF OAKLAND
Inter-Office Letter

To:_____ Arnold Johnson __ Attention:_____ Date__ February 5, 1971____

From:_____ Michael Pickering_____

Subject:_____ Staff Report on Kennedy
Street and 23rd Avenue _____

As the result of requests for additional openings in the median island on Kennedy Street between East 7th Street and 23rd Avenue, the following studies were undertaken:

1. Speed studies on Kennedy Street and 23rd Avenue.

2. Twenty-four hour volumes on 23rd Avenue, 29th Avenue, East 7th Street and Ford Street.

3. Truck and turning movement counts were taken at all intersections, including weaving movements on Ford Street between 29th and 23rd Avenues and at the merge of 23rd with 29th.

4. Accident studies of the entire area -- intersection and non-intersection.

5. Signal need study for 23rd Avenue and East 7th Street.

A summation of the pertinent findings is:

1. Average and 85th percentile speeds for southbound 23rd Avenue are 32 m.p.h. and 37 m.p.h., respectively; for northbound Kennedy are 30 m.p.h. and 39 m.p.h., respectively.

2. The percent of trucks for the entire study area was about 15% for the hours of 11:00 a.m. to 3:00 p.m.

3. An average of two vehicles per hour made U-turns from southbound Kennedy Street through the Rhodes and Jamieson driveway and through the existing median opening to northbound Kennedy Street.

4. Our studies showed little truck activity for the existing driveways on the east side of Kennedy Street with only nine trucks using these four driveways during the six hours we observed.

5. The accident incidences at the intersections of East 7th Street at Kennedy Street and East 7th Street and 23rd Avenue are normal for the intersection volumes at these locations.

-2-

6. Forty percent of the accidents at East 7th Street and Kennedy
    Street are caused by left-turn conflicts.

7. Five percent of the vehicles making left turns from 23rd Ave-
    nue to 29th made complete loops via Ford Street back to 23rd.
    None of these vehicles were trucks.

Recommendations:

1. No median openings be made for existing driveways on Kennedy
    Street. Usage of these driveways is light. The intersection
    of East 7th Street and Kennedy Street has demonstrated a left-
    turn accident pattern with good sight distance. The proposed
    openings in the median would be located such that sight dis-
    tances would be reduced for northbound Kennedy Street. In
    addition, a rear-end accident hazard would be created.

2. Under no circumstances should median openings be provided for
    the Campanella property on the corner of 23rd Avenue and Kennedy
    Street. Sight distances from both directions on Kennedy Street
    will be restricted, especially with any improvements made on the
    property in the vicinity of the curve on Kennedy Street. The
    median openings required to service the driveways for this
    property cannot be located far enough from the curve on Kennedy
    Street to provide adequate sight distance for vehicles traveling
    in either direction on Kennedy Street.

Improvements:

1. Left-turn pockets would not appreciably reduce the accidents at
    Kennedy Street and East 7th Street and may induce more sideswipe
    type accidents. Lane lines should be installed on Kennedy Street
    with Stimsonite reflective pavement markers at 48' and 24' centers
    on the curve. This will help reduce speed and delineate the curve
    on Kennedy Street.

2. A guide line should be painted to guide vehicles turning left from
    northbound 23rd Avenue into the proper lane on westbound East 7th
    Street.

-3-

3. New revised advance street name signing should be installed at
   East 7th and Kennedy Streets and at East 7th and 23rd Avenue.
   The existing freeway directional signing at 23rd Avenue is inadequate and misleading. In addition, signing in the vicinity
   of 29th Avenue and Ford Street should be revised.

Work orders for items No. 1 and No. 2 are attached. Item No. 3 involves both State of California and City of Oakland responsibility.
Work orders for the City's portion will be prepared and the State
will be contacted regarding the revision of their freeway guide signs.

Michael F. Pickering
Associate Traffic Engineer

MFP:pc:br

Attachments

CITY OF OAKLAND
Department of Traffic and Engineering
City Hall   1421 Washington Street   Oakland, California 94612

February 18, 1971

Mr. Charles L. Campanella
2712 East 7th Street
Oakland, California 94601

Dear Mr. Campanella:

In response to your request for additional openings in the median along
Kennedy Street between East 7th Street and 23rd Avenue, subsequent
meetings with Mr. Hurlbut of this department and recommendations suggested by the Traffic Committee, our staff initiated a comprehensive study
of the region bounded by Kennedy Street, 29th Avenue, 23rd Avenue, Ford
Street and East 7th Street.

The investigation conducted involved speed studies on Kennedy Street,
23rd Avenue and 29th Avenue along with vehicular volume counts on
23rd Avenue, 29th Avenue, East 7th Street, Ford Street and Kennedy
Street. In addition, truck and turning movement counts were taken at all
intersections including weaving movements on Ford Street between 29th
and 23rd Avenues, the merge of 23rd Avenue with 29th Avenue and the
left turn movement from 23rd to 29th Avenues. Accident studies for the
entire area were analyzed and a signal need study for 23rd Avenue and
East 7th Street was conducted.

It was found that the average speed for southbound 23rd Avenue was 32
m.p.h. and for Kennedy Street the average speed was 30 m.p.h. northbound. Truck counts taken over a two-day period showed that the percentage of trucks for the entire study area was approximately 15% for
the hours of 11:00 a.m. to 3:00 p.m. An average of two vehicles per hour
made U-turns from southbound Kennedy Street through the Rhodes and
Jamieson driveway and through the existing median opening to northbound Kennedy Street.

Our study showed little truck activity for the existing driveways on the
east side of Kennedy Street with only 9 trucks using this driveway during
the 6 hours we observed. Five percent of the vehicles making left turns
from 23rd to 29th Avenues made complete loops via Ford Street back to
23rd Avenue. None of these vehicles were trucks. The accident incidence
for the intersections of East 7th Street at Kennedy Street and East 7th
Street at 23rd Avenue are normal for the intersection volumes recorded.
Forty percent of the accidents at East 7th and Kennedy Streets were
caused by left turning conflicts.

On February 1, 1971, the Traffic Committee met and the results of our staff study and recommendations were presented. It was the decision of the Traffic Committee that no median openings be made for the existing driveways on Kennedy Street. Usage of these driveways is light, and the intersection of East 7th Street and Kennedy Street has demonstrated a left turn accident pattern even though the existing sight distance is adequate. The proposed openings in the median would be located such that sight distance would be reduced for northbound Kennedy Street and a potential rear end accident hazard would be created. In addition, no median openings will be provided for your property on the corner of 23rd Avenue and Kennedy Street. Sight distances from both directions on Kennedy Street would be restricted especially with any improvements made on the property in the vicinity of the curve on Kennedy Street. The median openings required to serve the driveways for this property could not be located far enough from the curve to provide adequate sight distance for vehicles traveling in either direction on Kennedy Street.

As a result of our study, improvements for better circulation and increased safety will be initiated. Lane lines will be installed on Kennedy Street with reflective pavement markers on the curve. This will help reduce speed and delineate the curve on Kennedy Street. Also, a guide line will be painted to guide vehicles turning left from northbound 23rd Street into the proper lane for westbound East 7th Street. New revised advance street name signs will be installed at East 7th and Kennedy Streets. The existing freeway directional signing at 23rd Avenue and East 7th Street and the signing in the vicinity of 29th Avenue and Ford Street will be revised. Some of this work will be coordinated with the State of California Division of Highways, and it is anticipated that these improvements will be completed within three to four months.

If you have any questions regarding the study conducted by our staff, please contact Mr. Pickering of this office.

Very truly yours,

Arnold A. Johnson

ARNOLD A. JOHNSON
Traffic Engineer and
Director of Parking

AAJ:MFP:pc

cc: Traffic Committee

Specimen No. 6*—Periodic Report

The report on in-service activities on the opposite page is the type sent regularly to the companies that buy Boeing airplanes to alert them to recent developments in safety, maintenance, repair, or other procedures necessary to keep the planes flying.

The format for the reports remains the same from week to week, so recipients know what to expect. Each item has a heading that includes an identification number, a few words locating the system being discussed, and a phrase indicating the problem. Although the example here omits subheadings, others in the same series used subheads when the matter being discussed was extensive.

The first problem included in the report concerns an investigation into an anti-icing device that failed during a test. The description starts with information about the conditions during the test and the damage that was discovered. The discussion continues with an account of the findings to date, and concludes with the fact that the problem is still under investigation. Later reports recapitulated the material in the original report and, under the heading, "Action Taken," detailed the modifications needed to prevent other planes from suffering similar failures.

The second item is a discussion of an oil leak. The discussion is brief because the problem had been detected before and modifications to prevent damage had been reported. The plane in question had not been modified, however, and the item is a gentle reminder to recipients of the consequences of ignoring these reports. The point does not need to be made strongly; the operators know the expenses involved in unscheduled engine replacement.

*Courtesy of Boeing Commercial Airplane Company.

August 13, 1979
Report No. 484

BOEING■747 IN SERVICE ACTIVITIES

This report provides a review of in-service activities reported during the period August 4, 1979 through August 10, 1979 and is for Boeing customer information only.

ATA 3000   ICE/RAIN PROTECTION - WING TAI - TAI VALVE - STUCK OPEN

While the airplane was being taxied from the hangar to the terminal (leading edge flaps extended), the RH wing anti-ice system was tested. When the LE flaps were retracted, the RH wing overheat light illuminated. The right-wing isolation valve and both right wing pylon valves were closed, however; inspection at the terminal revealed the bottom surface of the wing leading edge at VC flap position 23 to be buckled. The D duct web was buckled at three places, with some rivets pulled through. In addition, there was a six inch chordwise crack and buckling in the upper leading edge panel. Repairs were made for temporary service and the leading edge was scheduled for replacement. Subsequent investigation revealed the RH wing anti-ice valve P/N 67-2905-001 (Item 70 IPC 30-11-03-01) to be in the open position due to a pushed-back pin in the electrical connector. In the shop, the RH wing isolation valve would drive only 3/4 closed under air load. The actuator was found to have low torque and the operator suspected that the actuator bearings were binding, causing the clutch to slip. Similar problems with the isolation valve have been reported previously and proposals from the valve supplier to correct this problem are expected in the near future. Under investigation.

ATA 7200   ENGINE(JT9D-7) - ACCESSORY SECTION - N2 TACH DRIVE - OIL LEAK

An operator reported an unscheduled landing and engine change due to an oil leak on No.4 engine. The leak was caused by a worn N2 tachometer drive seal. Another engine had been replaced earlier in this operator's fleet and a third engine change is imminent for the same reason. P&WA modifications SB 2850 and SB 3450 to the tachometer drive bearings are designed to alleviate this condition and are not known to have been accomplished on the above engines.

Prepared by
747 Customer Support

M. J. Nancarrow
747 Customer Support Engineering Director

Specimen No. 7*—Field Report

The field report on the opposite page is like a laboratory report except that the work takes place in the field instead of in a laboratory. The report recounts the procedure undertaken to determine the direction of ground-water flow before establishing a sewage system drain field for a proposed housing development on an Indian reservation in the state of Washington. The letter uses both an "Attention" line and the salutation "Gentlemen," because the Indian Health Service contracted for the work, while the individual addressed was in charge of coordinating the various projects required before the development could be completed.

The first three paragraphs of the report constitute the introduction, in which the writer identifies the project, gives the purpose, and states specific limitations. The next five paragraphs under the heading "Field Explorations and Subsurface Conditions" give a detailed explanation of the procedures used for locating the sample sites and for obtaining the samples. Had the readers of the report been engineers, the writer might have simply identified the methods, as they are well known and standard, but since the recipients could not be expected to know the methods nor have ready access to a description of them, the writer wisely included an easily comprehended description. The next three paragraphs report the results obtained both from this field trip and from previous studies that seemed relevant to this trip. (The figures referred to have been omitted to save space; they were included in the original report, of course.)

Under the heading "Conclusions," the report lists the conclusions and recommendations drawn from the information gathered. Because this report is also a letter, the concluding paragraph provides a more personal note than the rather formal material preceding it. The report ends with the signatures of the individuals responsible for its contents.

*Courtesy of Hart Crowser & Associates, Inc., and the Indian Health Service.

J-789

**HART
CROWSER &
associates inc.**
GEOTECHNICAL ENGINEERING

May 1, 1979

Indian Health Service
1212 S. Judkins
Seattle, Washington 98144

Attn: Mr. Sam Bradshaw

RE: Subsurface Conditions and Flow Direction
      Beneath a Proposed Drainfield
      Squaxin Island Indian Reservation
      Mason County, Washington

Gentlemen:

This report presents the results of our subsurface soil explorations related to the probable effluent flow direction beneath the proposed drainfield. The project site consists of 1.0 to 1.5 acres in the southeast portion
of the Squaxin Island Indian Housing Site. The site is located on a knoll-
like topographic high situated on the generally southerly to southeas-
terly sloping land surface. A swampy-marsh area is located on the southerly project boundary (Figure 1).

The purpose of the project was to assess the probable subsurface flow direction of effluent introduced into the ground. Our scope of work included
field explorations, laboratory classification tests, and a review of the subsurface stratigraphy. The field explorations were accomplished by the
drilling of four borings using a truck-mounted, hollow-stem auger. The
borings ranged in depth from approximately 12.9 to 28.1 feet.

This report has been prepared for the exclusive use of the Indian Health
Service for specific application to the project site in accordance with generally accepted geotechnical engineering practices. No other warranty,
expressed or implied, is made.

FIELD EXPLORATIONS AND SUBSURFACE CONDITIONS

The field portion of the project consisted of the drilling of four borings on
April 14 and 16, 1979. Each of the borings was visually logged by Bruce
McDonald of Hart-Crowser & Associates, Inc. These logs are based on our

interpretations of the field and laboratory data, and indicate the various
types of soils encountered in the borings. They also indicate the depths
where these soils, or characteristics of these soils changed. These
changes are often gradual.

The borings were located in the field by hand-taping from existing physical features. The approximate ground surface elevations at the boring locations as indicated on the boring logs were interpreted from the Topographic Map, dated January, 1979, prepared by Horton Dennis &
Associates, Inc. The location and elevation of the borings should be considered accurate only to the degree implied by the method used.

Samples were obtained using the Standard Penetration Test procedure in
which a standard split-spoon sampler is driven into the soil a distance of
18 inches using a 140-pound hammer free-falling 30 inches. The number
of blows required to drive the sampler the last 12 inches, or to refusal
(generally defined as 50 blows for less than 6 inches) is known as the
Standard Penetration Resistance. This resistance, or blow count, provides
a measure of the relative density of coarse granular soils, such as sands
and gravels, and the consistency of cohesive soils such as silts and clays.

The soil samples obtained from the borings were classified in the field
and representative portions saved in air-tight jars. All samples were then
transported to our Seattle laboratory, where they were visually reexam-
ined to refine the classification made in the field. The degree of moisture
in the sample (whether it was dry, moist, or wet), the approximate percentage of fines in the sample and the relative plasticity of the fines were
particularly noted.

The natural water content was determined on all samples for the purpose of: (1) identification and correlation of the soils and (2) providing
basic engineering properties. Natural water contents are presented in the
logs of borings. Grain size analyses were performed on selected samples
utilizing the mechanical test methods in general accordance with ASTM
D442-63.

Previously completed test pits by the Indian Health Service in the area
surrounding the knoll (Figure 1) encountered GLACIAL TILL at depths of
2.3 to 3.0 feet. Test pit 10 encountered a dense gravel at 6 feet with seep-
age being observed at 5.0 feet. Test pits 8 and 9 excavated near the top of
the knoll penetrated up to 7.0 feet of sand and gravel. Borings completed
by Hart-Crowser & Associates, Inc. generally encountered a variable
thickness of loose to dense, moist, brown, relatively clean to silty, slightly
gravelly to gravelly, fine to coarse SAND (probable RECESSIONAL OUT-
WASH) overlying very dense, moist to saturated, gray-brown to brown,
silty, gravelly fine to coarse SAND (probable weathered to unweathered
GLACIAL TILL). This sequence was not, however, encountered in Boring
B-4 where clayey SILT was found to overlie silty, gravelly fine to coarse
SAND.

Indian Health Service   J-789
May 1, 1979   Page 3

Groundwater was encountered in all the borings more or less at the
RECESSIONAL OUTWASH-GLACIAL TILL contact. It is our opinion that
the observed groundwater is "perched" by the underlying relatively im-
permeable GLACIAL TILL.

The boring locations along with several of the Indian Health Service's
test pit locations are shown in Figure 1. Descriptive soil logs are presented on Figures 2 through 5. Three grain size classification tests on
representative samples are shown on Figure 6.

CONCLUSIONS

Based on the boring, and test pit logs, and general topography, we believe
the general near-surface flow direction in the vicinity of the knoll will be
towards the south-southeast. Effluent introduced into the subsurface will
move predominantly vertically until the very dense GLACIAL TILL soils
are reached. The effluent will then flow down gradient along the GLA-
CIAL TILL surface.

Prior to the siting of a drainfield over the proposed site, we recommend
that the extent of the silty CLAY encountered in Boring B-4 be determined.

We have appreciated the opportunity of providing these engineering services to you. Any questions you may have concerning the material presented can be answered at your earliest convenience.

Sincerely,

HART-CROWSER & ASSOCIATES, INC.

Matthew G. Dalton

MATTHEW G. DALTON
Hydrogeologist

Dennis R. Stettler

DENNIS R. STETTLER
Project Manager/Associate

John C. Crowser

JOHN C. CROWSER, P.E.
Vice President

MGD:vf

Attached: Figure 1, Site and
Exploration Plan
Figure 2–4 Boring Logs
Figure 5, Grain Size
Analyses

Specimen No. 8*—Laboratory Report

Laboratory reports are perhaps one of the most common types written by students during their college careers, for much technical course work involves study in laboratories or under closely controlled conditions like the situation found in laboratories. In industry too, when products or systems are studied, individuals attempt to create conditions that are controlled to the degree that the results can be trusted, and the reports written often take the form of a laboratory report or a report of physical research.

The report that starts on the opposite page tells of a study to determine if a helmet and earphone system for pilots would be as good as the system currently in use. Rather than test the system in planes in the air, which might create danger for pilots, conditions of flight were simulated on the ground at the Pacific Missile Test Center. The original report included a cover page, a table of contents, a list of references, one figure, and three tables of results. Since these elements conformed to descriptions given elsewhere in this text, they have been omitted here to save space.

The report proper begins with the summary, which identifies the problem, gives the results of the tests, and indicates the work needed to incorporate the new system into existing aircraft. A busy executive or officer would be able, from the summary alone, to make decisions based on the study.

Individuals with a need to learn more about the study could turn to the other sections that follow in standard order. The "Introduction" gives the background and purpose of the test, the "System Description" identifies the helmets and the earphones, and the "Test Description" includes detailed information about the testing procedures. Although specific data from individual tests were included only in the tables, enough information about the results is included in the "Results and Discussion" section to make it self-sufficient. It is followed by the "Conclusions" and "Recommendations" sections that give the researchers' opinions about the findings.

Although the identifying numbers might present problems in readability for anyone unfamiliar with military nomenclature, they provide exact identifications and are necessary. The report itself is straightforward and worth the time required to overcome whatever difficulties the military terminology might present.

*Courtesy of the Pacific Missile Test Center.

PACIFIC MISSILE TEST CENTER
Point Mugu, California

LABORATORY EVALUATION OF THE SPEECH INTELLIGIBILITY
CHARACTERISTICS OF THE EV-993 EARPHONES IN THE
HGU-33/P AND HGU-34/P FLIGHT HELMETS

SUMMARY

The Pacific Missile Test Center (PACMISTESTCEN) was requested by
the Naval Air Development Center (NAVAIRDEVCEN) to test and evaluate the speech intelligibility characteristics of the HGU-33/P and HGU-34/
P flight helmets configured with Electrovoice EV-993 1000 ohm impedance earphones.

The EV-993 earphones in the HGU-33/P and HGU-34/P flight helmets
met the intelligibility criterion for a military communication system and
were equal in intelligibility to the standard APH-6D flight helmet configured with H-87B/U EV-992 earphones.

Appropriate mating connectors and communications procedures analyses are required to integrate the HGU-33/P and HGU-34/P flight helmets
in Naval aircraft.

I. INTRODUCTION

A. Background

The PACMISTESTCEN was tasked by the NAVAIRDEVCEN (reference 1) to conduct speech intelligibility tests of the HGU-33/P and HGU-
34/P flight helmets with high impedance earphones.

B. Purpose of Test

The purpose of the evaluation of the HGU-33/P and HGU-34/P
flight helmets was to determine adequacy of speech intelligibility characteristics of the earphone receivers against applicable accepted criteria
and military standards (references 2, 3 and 4).

II. SYSTEM DESCRIPTION

A. Description of the Equipment to be Tested

The two evaluation flight helmets consisted of an HGU-33/P size
large and an HGU-34/P size medium configured with Electrovoice model
EV-993 1000 ohm earphones, parallel connected to a CX4708A/AIC elec-

1

tric cord assembly. The earphones were mounted in foam inserts in Gentex part number 71C2375 earcups equipped with Gentex part number 618883 replaceable earseals.

The two comparison flight helmets consisted of a large size APH-6D and a medium size APH-6D configured with H-87B/U Electrovoice model EV-992 15 ohm earphones, parallel connected to a P/O H-87A/U cord. The earphones were mounted in Sonex earcups equipped with Gentex part number 70C2196 replaceable earseals.

III.   TEST DESCRIPTION

A.   Scope of Tests

1.   Speech Intelligibility

The two flight helmet configurations were evaluated for speech intelligibility characteristics of test vocabularies recorded in an F-4J aircraft in accordance with the requirements listed in reference 2.

B.   Method of Tests

1.   Speech Intelligibility

a.   Talkers

The speech intelligibility of the two flight helmet configurations was evaluated by using experienced listeners who heard recorded Modified Rhyme Test (MRT) word lists (reference 5) mixed with a simulated jet aircraft noise spectrum. The word lists were recorded by two talkers using A-13A oxygen masks in an F-4J aircraft. The talkers were Michael Westervelt (MW) and Douglas Robertson (DR); see reference 6 for talker descriptions.

b.   Test Materials

The MRT word lists have been performance correlated with the pilot's Brevity Code by Webster and Allen (reference 4). They offer a reliable and valid measure of speech intelligibility using an easily administered procedure that forces a choice on the listener from a closed response set of six. The relationship between performance on the MRT and military aircraft communicating languages is known and therefore performance on one can predict performance on the other when physical measures of speech levels and noise levels at the listeners' earphones are known.

2

The talkers recorded the word lists shown in Table 1. The selection of the MRT word lists was accomplished using a table of random digits to assign lists by talker conditions.

c. Recording Procedures

The talkers read randomized lists of the MRT words at a two second interval in synchrony with a flashing light. The item number of the word was read aloud prior to each test word. Before each list was recorded, the talkers identified themselves, the equipment they were using and wearing, and the exact word list. This allowed them time to adopt a comfortable talking level and gave the recording technician time to establish a proper recording level. All recordings were stored on 1/4 inch magnetic tape at 7-1/2 inches per second with a Kudelski Nagra III tape recorder across the talker's helmet earphone sidetone circuit, thus recording through the intercommunications circuit of the F-4J aircraft. The oxygen system was turned on in the aircraft during the recording sessions; hence 100 percent oxygen was breathed via the miniature regulator system. Occasionally two or more pronunciations were given on a single word and in the editing and rerecording process prior to the listening tests, a selection was made of the correct or most typically American pronunciation. The talkers monitored their voice level aurally from the sidetone circuit and did not view any type of meter. This informal monitoring method kept the recordings realistic in terms of how aircrewmen actually use the system and resulted in some variation in word level on the lists.

d. Test Subjects

Listeners for the tests were nine volunteer Navy enlisted men and one Naval Officer. All met Service Group I and II aircrewman hearing standards (reference 7). The test subjects ranged in age from 19 to 37 years, with a mean age of 27.6 years. Six of the subjects were experienced aircrew members. Seven were highly experienced listeners.

e. Listening Test Procedures

The playback equipment and listening network is shown in Figure 1. The noise channel duplicated the shaped spectrum of F-4 aircraft cockpit noise in normal cruise attenuated by the mean acoustic attenuation of an APH-6 flight helmet equipped with Sonex earcups (reference 6). No attempt was made to establish exact speech-signal to noise differentials in the helmets because of the very real problem in specifying coupler levels in an irregular mating surface earcup and because of the individual differences in fit and ear structure within the helmet earcups.

3

Therefore, the noise level was approximated and the test subjects were given the opportunity to adjust the speech listening level to suit their needs in the course of the listening trials. The subject speech level attenuator was adjustable over a 45 decibel (dB) range in one dB steps. This procedure was akin to the operational environment wherein the aircrewman has an adjustable volume control over the intercommunications system audio level in the aircraft. Each test subject heard two MRT word lists through either the HGU-33/P or the HGU-34/P flight helmet earphones and two lists through the APH-6D helmet earphones. Half of the subjects heard two word lists first through the HGU-33/P or HGU-34/P earphones; the other half heard two word lists first through the APH-6D earphones. The audiometric screening, helmet fitting procedures, attenuator terminations to the helmet configurations and word list administration took place in a single session of about 30 minutes duration. The experimenter noted the subject attenuator setting used for each word list. There were 200 total items (words). Table 2 shows the distribution of helmets by size (large or medium) and type fitted to the subjects.

IV. RESULTS AND DISCUSSION

A. Speech Intelligibility

The speech intelligibility scores for the subjects wearing either the HGU-33/P or the HGU-34/P flight helmet and the APH-6D flight helmet are shown in Table 3. The HGU-33/P and HGU-34/P intelligibility scores were combined as there were no significant differences between them. The table indicates that the HGU-33/P and HGU-34/P flight helmet earphone communications systems met the intelligibility criterion for a military communications system when the subjects were afforded the opportunity to determine their own listening levels in a simulated aircraft noise spectrum. The intelligibility of the APH-6D earphone communications system was equivalent. The APH-6D earphone communications system is considered satisfactory in operational Naval aircraft (reference 4). The range of attenuator settings by the subjects was equivalent between helmet and talker conditions and the mean attenuator settings varied by only 0.8 dB between conditions. The product moment correlation between intelligibility score and listening level setting over all helmet/talker conditions was +0.37, significant at the 0.05 level of confidence. All tested helmet earphone communications configurations yielded acceptable intelligibility for a military communications system. Eight of the test subjects preferred the HGU-33/P or the HGU-34/P flight helmet communications system speech quality over that of the APH-6D flight helmet. Two subjects exhibited no preference. Appropriate mating connectors and com-

4

munications procedures analyses are required for integration of the HGU-33/P and HGU-34/P flight helmets in Naval aircraft.

V. CONCLUSIONS

The HGU-33/P and HGU-34/P flight helmets configured with EV-993 1000 ohm impedance earphones met the intelligibility criterion for a military communications system and were equal in intelligibility to the APH-6D flight helmet configured with H-87B/U EV-992 15 ohm earphones. Use of the HGU-33/P and HGU-34/P flight helmets with EV-993 earphones in a military communications system in lieu of the APH-6D flight helmet with H-87B/U earphones should entail low risk when appropriate mating connectors to the oxygen mask and/or aircraft for all flight missions are provided.

VI. RECOMMENDATIONS

It is recommended that the EV-993 high impedance earphones be approved for service use with the HGU-33/P and HGU-34/P flight helmets. Integration of these flight helmets will require appropriate mating and termination in Naval aircraft based upon communications procedures analyses.

5

Specimen No. 9*—Progress
(or Status) Report

The report on the opposite page lacks the material normally needed for an opening because, as indicated by the page number, 20, it was part of a longer report covering the progress or status of various departments at Douglas United Nuclear Industries. The departments reported progress in the areas of concern for each and the resulting reports were combined to form a full-scale formal report. One of the points of including this section is to demonstrate that there is nothing unusual about different parts of a long report being written by different people. This possibility accounts for the emphasis in this book on compliance with instructions about form and on the ability to handle an impersonal style. If all the writers whose work appeared in the same report wrote in their own personal manner, the report would be a hodgepodge of different styles that would confuse readers.

In each of the sections of this report, the writers comply with the general suggestions for progress reports even though they do it quite briefly. That is, the report indicates past, present, and future action on the points discussed, by implication or by direct statement, depending on the importance of each action. In the first section, for instance, the writer identifies the subject and gives the pounds of waste material released—and where there was a distinct change from the previous period, the writer gives that information.

In the second section, the report responds to a change in regulations, indicating that the changes do not apply. Another change in regulations is dealt with under the heading, "Radiation Control Signs." Here, action has been taken and is described. The final section deals with a problem for which a solution has not yet been discovered; the writer indicates that work is continuing and describes the major difficulty.

Each section of the report deals briefly with a separate subject, but taken together they portray the work done during the report period by the environmental engineering staff: In one case, work continues on an unsolved problem, in another, work has concluded; and in yet another, the matter is one that will probably continue as long as the plant is in operation.

*Courtesy of Douglas United Nuclear Industries, Inc.

ENVIRONMENTAL ENGINEERING

1972 Environmental Release Report

The 1972 Environmental Release Report[12] was documented and issued to the AEC-RL
on March 13. During Calendar Year 1972, 4,000,000 pounds of waste materials
including 1 gram of radionuclides were discharged to the environs in liquid effluent. This is a reduction of 50 percent from 1971 wastes released in liquid
effluent streams. During the same period, 2,000,000 pounds of wastes, including
a small fraction of 1 gram of radionuclide, were discharged to the atmosphere in
gaseous effluent streams.

Transuranium Waste Procedures

On March 1, 1973 AEC-RL requested that changes be made in procedures for handling
transuranium wastes. Evaluation[13] of DUN operations showed that no transuranium
wastes are now or will be generated in the near future. Should any program that
could generate such wastes be planned in the future, the new procedures would be
developed and submitted to AEC-RL.

Radiation Control Signs

The revision to RC-2, "Radiation Source Control" of DUN-M-1, Radiation Control
Manual requires that a sign reading "CAUTION HIGH DOSE RATES" be affixed to sources
within a Radiation Zone that are a factor of two higher than ambient dose rates.
The purpose of the requirement is to assist radiation zone workers to control their
exposure by avoiding locations so posted.

One hundred 5 X 9-inch aluminum signs with the above warning have been fabricated
and issued for use. Each sign has an alnico magnet fastened to the reverse side
so that it can be easily attached to valves or piping as required. Two holes
have also been provided in each sign so that it can be suspended where the magnets
are ineffective.

N Reactor Liquid Effluent Cleanup

Treatment of N Reactor liquid effluents is currently accomplished by soil. The
effluents are routed to a rock-filled crib for percolation through the soil before
reaching the Columbia River. Efforts are being made to find a technically more
acceptable method for liquid effluent treatment.

The major difficulty in providing an acceptable ion-exchange system for N Reactor
liquid effluents is the large volume of such wastes. The volume of N Reactor
liquid effluents requiring treatment is more than 100 times greater than that
required for other water-cooled reactors (i.e., BWRs and PWRs). And the cost for
a complete waste treatment facility for BWRs and PWRs ranged from 5 to 10 million
dollars.

---

[12] DUN-8133, "Douglas United Nuclear, Inc., Reactor and Fuel Production Facilities
1972 Environmental Release Report," TE Dabrowski, March 1, 1973. Letter, NR
Miller to PG Holsted, "1972 Environmental Releases" dated 3/12/73.

[13] Letter, NR Miller to TA Nemzek, "Additional Guidance, Immediate Action Directive
No. 0511-21, Policy Statement Regarding Solid Waste Burial (Ref. Ltr. 4/21/70
DG Williams to Contractors)," dated 3/23/73

Specimen No. 10*—A Report
that Is Not Classified as a Special Type

The following report was one of a series made to The Asphalt Institute concerning paving at airports. This report cannot be classified into any of the readily recognized forms, bearing out the statement that although this book must recognize the existence of reports of various types, there is no reason to expect that every report can be identified by name.

The purpose of the individual report is to swell the total amount of information available on a problem rather than to point the road to specific action; but this information plus the information in other reports could well lead to recommendations for dealing with the problem wherever and whenever it might arise.

The first paragraph identifies the problem and the purpose of the report. The second paragraph describes the methods used to gather information: interviews and visual inspection. The rest of the report gives the findings and the conclusions of the writer.

*Courtesy of The Asphalt Institute.

INSPECTION OF ASPHALT RAMP AREAS
at the
SAN FRANCISCO INTERNATIONAL AIRPORT

The purpose of this report is to tell what was learned when ramp areas at the San Francisco Airport were inspected to see whether their asphalt surfaces were damaged by detergents used for cleaning.

Airline representatives who were interviewed stated that they have experienced no trouble or damage to the asphalt surfaces as a result of the detergents they used to clean ramps. Visual inspection bore out their statements; the pavements appear to be structurally sound.

The difficulties being encountered at the San Francisco International Airport are similar to the problems being experienced at other airports. Oil is spilled or drips from parked aircraft and is gradually spread over a rather wide area by service vehicles which run through it. Maintenance crews sprinkle a powder on the surface to blot up the oil, and later remove the powdered material with the oil. Once a week the ramp area is scrubbed with a detergent solution.

The only problem is apparently one of house-keeping. Using virtually the same type of aircraft that sometimes cause trouble, some airlines have little difficulty in keeping the area clean, because a crew of men cleans the ramp area very shortly after a spillage of oil occurs. Other airlines, for various reasons, are not so prompt, and the spilled oil and blotter material accumulate, causing the unsatisfactory conditions.

One airline shop has fabricated a trailer on which is placed a rather large tank with a spray-bar. At the back of the trailer is a straight bristle broom which can be raised or lowered. The tank is provided with a valve through which approximately 150 to 180 pounds of air pressure is applied prior to use. The detergent used consists of one third Turco No. 2844 and two thirds deodorized spray-base solvent made by Shell Oil Company. This deodorized spray-base solvent is a highly refined kerosene with aromatics removed, and is the lowest in solvent power of naphthas and solvents produced by Shell Oil Company.

The normal procedure is to try to blot up the worst of the oil spillage as it occurs and remove the blotter material, and then scrub the area with the detergent solution once a week. During the scrubbing operations, it is attempted to hose the area clean of the solution within thirty minutes of the time it is applied. This procedure, apparently, has been satisfactory and is considered superior to procedures described in a previous report.

In comparison with the problems at Midway Airport in Chicago, the problems at San Francisco are minor. If an analysis of the pavement at San Francisco Airport were made, it would show that the pavement is structurally sound throughout and that no deleterious effects are being experienced from the detergent solution being used for cleaning.

ASSIGNMENTS

A Note on Selecting Report Subjects in the Classroom: There is so much variation in the conditions in different schools, and in the courses, training, and opportunities to gather information, that it will probably be necessary to take many liberties with the suggested subjects listed. Each subject is likely to need adaptation and sharper definition.

In adapting a subject you may need to broaden or narrow its scope. You may be forced to assume that you are writing in a different season of the year, or that some recent project in your school or community has not yet been undertaken. Or if a subject proves too large to cover in full, you may need to assume that you are conducting only a preliminary survey intended to show whether a subject should be investigated further or abandoned.

In defining a subject you will probably be assuming that you are writing the report for an actual employer or client. Thus you will need to decide whom your report is being written for and why it is needed—that is, to set up a hypothetical set of circumstances. In actual practice, of course, the information you would include would depend on the needs of the occasion. In the classroom, the hypothetical occasion must be shaped to enable you to utilize whatever information it is possible to obtain.

Many of the subjects suggested could plausibly be used for more than one type of report. There is therefore no reason that in writing any assignment you should not feel free to use a subject from the list suggested for some other assignment, broadening it or narrowing it as the occasion demands.

You may prefer to write about subjects not included in the lists provided. To develop a subject of your own, you should start out by canvassing your own resources. Review all that you have learned on any job you have held; all that you know about any technical line of activity some friend or relative might pursue; all that you have learned in any practical courses you have taken. Then try to create a hypothetical situation in which you can utilize that knowledge.

Regardless of the source of your subject, you should guard against making your reports too general, especially if some of your material must be drawn from printed references. In actual practice, a report may use general information but usually applies it to a specific case. For example, general information about killing weeds by use of chemicals is so easily obtainable in print that no one would ever be asked to write a report to present it. But the weed conditions on a specific tract of land could not be found in print, so it is quite plausible that one might be asked to write a report on the conditions on that tract and on the measures necessary to control the weeds there. Or again, anyone could find general information about fluorescent lighting systems, but it is quite possible that a report might be requested on the advisability of using fluorescent lighting in some specific building that was to be modernized or remodeled. In this report, the general, printed information would be applied to the specific facts about the building.

All this should be taken into consideration as you choose your subjects, for the wisdom of your choice will strongly affect the merit of any report you write—an unfortunate condition that will not exist when you are on the job, but cannot be avoided in the classroom. It is possible, of course, to write a poor report on a good subject; but it is impossible to write a good report on a poor subject, for a poorly chosen, unrealistic subject simply does not give you an opportunity to exercise many of the skills that make a report good.

Assignment 1

Write a report in the form of a letter using a subject from the list below or, with the instructor's permission, a subject of your own choice. Subjects 1 and 2 in Assignment 2 might also be suitable if properly adapted. The specimen letters in Chapter 15 should be sufficient guidance as to letter form.

Since this report is a letter, you would need to assume not only that circumstances make it natural for you to report to the reader, but also, if you and the reader are supposedly employed by the same organization, that circumstances make it natural to write a letter rather than a memorandum. For example, you might assume that you are writing the report while on a trip to a different town; or you might, in using some of the subjects, assume that you are reporting to an out-of-town property owner.

It is suggested that this report be 300 to 600 words long.

1. Report on the condition of a specific building that you have inspected at the request of a prospective buyer from another community who wished to know whether a trip for personal inspection would be advisable. Your report should cover such areas as exterior (including roofing, siding, windows, approaches, landscaping), interior (walls, floors, ceiling, carpeting and furnishings if any), plumbing, wiring, square footage of the building and the land, and anything else that might affect the value.

2. Report on the compliance of a public building with a specific code or law in your community, at the request of a legislative body charged with investigating problems of enforcement. The code might involve fire safety, access for the handicapped, parking facilities, or some other regulation. The building might be a local courthouse, nursing home, school, apartment, theater, or store.

3. Report on the desirability of a particular facility for a three-day meeting of an organization to which you belong. Assume you are visiting the facility and are writing to the organization's leader because you cannot make the report in person. Include in your report the cost of rooms and meals, availability of meeting rooms of the size required, and opportunities for entertainment for participants when they are not attending meetings.

4. Report on a research project being conducted at your school or at a nearby research institute or government experiment station, at the request of a corporation that is considering pursuing similar work for

commercial purposes. Include in your report a general summary of the work, its prospects for success, and its current status in relation to its long-term goals.

5. Report on the general level of satisfaction of students in your major with their studies, at the request of your high school guidance counselor. Interview enough majors to give you material for a report, but unless you reach an actual random sample or talk to all the majors, do not make claims to represent more than you have actually done. Include points of satisfaction as well as dissatisfaction.

Assignment 2

Write a report in the form of a memorandum, using a subject from the list below or, with your instructor's permission, a subject of your own choice. Since your report should be made to a reader to whom you might logically report, you will need to assume that you and the reader are connected with the same organization.

It is suggested that this report should be from 300 to 600 words long.

1. Report on the general condition and availability of parking facilities for students and faculty at your school or for employees and customers at your place of business. Indicate need for improvement of existing facilities and need for expansion. If too much space is available, indicate which areas might be converted to other use.
2. Report on the condition of equipment, the need for new equipment, or the need for additional space in any shop, laboratory, farm, or business with which you are connected.
3. Report on the kinds of information available about a subject in your field of study for use as the background section of a long report. Include the kinds and amounts of material available in the library (books, magazines, reports), the availability of expert help, and the possibility of your actually becoming involved in work concerning the subject you have chosen.
4. Report on the kinds and availability of mental health facilities for students or some other group in your school or community.
5. Report on the methods of handling complaints in your organization. Make recommendations for change if you feel another method would be more effective.

Assignment 3

Write a progress report of a length specified by your instructor. It may be in any form you desire unless you receive instructions to the contrary.

Any project that calls for a final report, unless it is small enough to be completed in a very short time, may also call for progress reports. Consequently, the subject on which your progress report is based may be any of the subjects provided for other report assignments. Another possibility is that if you have been assigned a long report in a technical writing class or a project that will last for several weeks in some other class, you might write a report telling your instructor about your progress.

Assignment 4

Assume that you are an officer or committee chairman of some organization, and that it is your responsibility to write quarterly, semiannual, or annual reports. Write a report of this nature.

The organization may be connected with your school—a social, professional, or honorary club or fraternity, for example—or a ski club, flying club, religious group, political group, or any other group that is active on the campus. As an alternative you could write a report of similar nature for a school enterprise such as a newspaper, magazine, or yearbook, or for some nonschool organization.

Assignment 5

If your course of study has included a laboratory course to supply you with subject matter, write a laboratory report on some test or experiment you have performed. The subject should call for enough writing to make the report a writing assignment. It may include tables, figures, or filled-in blanks but should not be limited to such material.

11 | Special Elements in Reports

To produce some kinds of writing a writer may need only the ability to select material, organize it, and express thoughts effectively. To produce good reports, however, these general skills are not enough. A report writer must also be able to plan the layout of a report—that is, to decide which of certain special elements will be useful in each report and to handle them so that they form an effective framework for the material.

To be sure, some reports are so short and simple that no special elements are needed; and laboratory reports are so standardized that their layout is often in accordance with the pattern already described. Nevertheless, certain special elements are needed so frequently that they call for considerable attention. They are most conspicuous in formal reports, but most writers are not called upon to write full-scale formal reports until long after they have had occasion to use these elements in reports that are less elaborate. Therefore, we will discuss them in connection with reports in general, and supplement the discussion as necessary in the chapter on formal reports.

The elements in question are the covering memorandum or letter, the title page, the table of contents, the headings, the summary, the introduction, the conclusions and recommendations, and the appendix. They have been listed in the order in which they are likely to reach the reader's attention, but they should be used only to the extent that they are needed. The body of the report is not listed because it is not a special element and performs a general rather than a special function; but it would normally comprise a substantial portion of the report and might either precede or follow the conclusions and recommendations.

THE COVERING MEMORANDUM OR LETTER

When a report that is not extremely short and simple is submitted, it is often accompanied by a covering memorandum (or a covering letter if it is sent through the mail). Such a memorandum is especially useful when a report will be of lasting rather than only immediate interest or when it will go beyond the person who first receives it.

A covering memorandum gives the writer an opportunity to make any extra remarks the occasion may call for. These remarks may add up to no more than an identification of the report, a reference to the subject that the report deals with, and a statement of the reason this subject is reported on. Remarks of this limited nature grow out of the same impulse that probably causes you, in a personal contact, to say, "Here is that book (or list, or drawing, or estimate) that you asked for." In making such remarks in a covering memorandum, however, you usually do more than just gratify your natural impulse to say something. You also make it easier for the person who first receives your report to grasp or recall important facts that concern it.

Sometimes a covering memorandum does a good deal more than this. Since the first recipient of your report is likely to be the person to whom you are immediately responsible, you may wish to report things that perhaps should go no further. For example, you may want to explain why you have emphasized some points and played down others. You may want to refer to obstacles you encountered or events that occurred while you were preparing the report. You may want to mention additional information that can be provided if it is needed. In general, you may want to give facts that would not be desirable or appropriate in the report itself; and the covering memorandum gives you an opportunity to do so.

A typical covering memorandum follows.

Date: May 3, 1982

To: Creston County Board of Commissioners
From: Rosemary Wicks
Subject: Rose Lake Recreation Area

Enclosed is my report on what work would need to be done to bring the Rose Lake area back into condition for recreational use, as it was used before it was allowed to deteriorate. This report was requested during your meeting on March 23.

I have taken into consideration the need of removing the silt that has almost filled the lake, improving the water supply, providing picnic tables and cooking facilities, and developing off-the-road parking space. I have also included a rough estimate of the cost of each of these improvements and the approximate time it would take to complete the work.

There is one complication to carrying out this project that I believe I should bring to your attention though it does not fall within the scope of the report and its inclusion would be inappropriate. Some of the nearby property owners maintain that when the area was formerly in use, it was used very late at night by noisy, rowdy groups whose parties amounted to a real disturbance. They insist, too, that their property was repeatedly subjected to vandalism.

I believe that when you discuss the project at a meeting, you should antic-ipate that several of these property owners will attend and raise strong objections.

Assuming that the project is not held up by this complication, I believe that the information I have provided will enable you to decide whether you wish to pursue this matter further.

THE TITLE PAGE

The title page normally indicates (1) the subject of the report, (2) the person or body to whom the report is made, (3) the person or body making the report, and (4) the date of the report. Also, in large organizations, it may bear the number assigned to the report and information on the project that the report concerns.

The title itself is more than just a label. Though it should not be so long as to be cumbersome, it should give the clearest possible information about the exact subject covered. *Selection of a Fin-Tip Antenna for a KC-135 Airplane,* for example, would be a better title than just *Selection of a Fin-Tip Antenna.*

If a report is a matter of mere routine—for example, the periodic report of normal operations of a division of some company—there would be no need for the title page to indicate to whom or by whom the report is made. Identifying it as a monthly or quarterly report makes these facts clear to anyone likely to be concerned.

The material on the title page should be pleasingly arranged, with the visual emphasis on the upper half of the page. Figure 14 is a typical title page for a report in which there was no need to use a specific company form.

THE TABLE OF CONTENTS

The table of contents of a report is basically the same as that of a book in that it lists all the headings used and indicates the page on which each heading appears. Since indentation makes the rank of each heading apparent, the table of contents functions as an outline of the report as well as showing where various points are taken up. Occasionally each point is preceded by a number or a letter as it would be in an outline, but most of the time these symbols are discarded or are limited to the Roman numerals identifying main divisions. When the decimal system of numbering headings is used, the numbers appear in the table of contents as well (see Chapter 5).

Like the table of contents of a book, the table of contents of a report has headings that indicate the presence of appendixes, calculations, detailed data, or any other type of material that follows the report proper. These headings are likely to include the titles of individual tables and figures when such nonverbal devices are all placed after the body of the report. When the tables and figures are scattered through the body, however, their titles appear in a separate list.

Our discussion here has been brief for two reasons. First, most of the work for a table of contents—deciding on the substance and phraseology of the headings—is done when the outline is made. Second, some of the questions that occur in connection with a table of contents are covered later, in the discussion of formal reports, because that is where they are most likely to arise.

COLLEGE OF ENGINEERING RESEARCH DIVISION
Bulletin 324

SIMULATION OF WATER QUALITY ENHANCEMENT

IN A POLLUTED LAKE

A Case Study of Vancouver Lake, Washington

by

Cheng-Nan Lin

Surinder K. Bhagat

John F. Osborn

Research reported in this publication
was supported by
STATE OF WASHINGTON WATER RESEARCH CENTER
at Washington State University

Published by the
Engineering Extension Service
January 1982

WASHINGTON STATE UNIVERSITY   PULLMAN, WASHINGTON

FIGURE 14. Well-Designed Title Page for a Report.

It is important to realize, however, that a table of contents is a valuable part of any report that is more than a few pages long. It lets the reader see at a glance what ground is covered, shows how the material is organized, and makes it easy to turn quickly to whatever information is needed.

Following is a typical table of contents for a report. Discussion of the table of contents of a formal report may be found in Chapter 12.

CONTENTS

FIGURES AND TABLES

USE OF HEADINGS

The Value of Headings

One of the useful skills you can acquire as a report writer is skill in the use of internal headings. Such headings, regardless of whether you provide a table of contents, make it easier for your readers to grasp the nature of your material, the general plan governing its organization, and the relationship among its various parts. Headings help readers find what they are looking for when they want to find out about some one area of your subject. Headings can catch attention and arouse interest when readers first glance over your work, and thus tempt them into the heart of the report before they have even taken time to read preliminary sections. Headings reduce reluctance to start reading by making the pages of text appear less formidable. And finally, the very fact that you are using headings can improve your writing by reducing the temptation to wander away from the point you are supposed to be discussing.

To provide all these benefits, however, headings must be used systematically and must cover everything in the report—a vastly different matter than casually inserting them as they are inserted in newspaper articles or most articles in magazines. The headings used are drawn, of course, from the outline on which the report is based, and their value depends on whether you have done a good job of organizing your material. Headings cannot bring order out of confusion; they can only show that order exists.

Form for Headings
of Different Ranks

Since the headings are intended to show how your material is organized, the form of each heading should indicate its rank. Headings of the same rank should be identical in form. Headings of different rank should differ in form. Headings of higher rank should look more important than those of lower rank. And finally, a form used for topic headings should not be used for any other purpose, such as the title of a list or a set of rules.

No single set of forms has become standard, but the following treatment is widely used and would be hard to improve upon.

A heading of highest rank (used before special parts such as introduction, conclusions, etc., and for the main divisions of the body) is centered, underlined, and all capital letters. It is placed three or four spaces below whatever precedes and a double space above whatever follows. It is not followed by any punctuation.

A heading of second rank begins at the left margin. It is underlined and capital letters are used only to begin important words ("upper and lower case capitalization"). It is placed a double or triple space below whatever precedes and a double space above whatever follows. It is not followed by any punctuation.

VARIABLE-SPEED SABER SAWS

Two variable-speed saws, the Zenith X-400 and the Monument 65 A were found to be superior to the others examined. Neither of them caused much effort in feeding. Each has a full-length handle above the motor housing. Each has a slide type of on-off switch that also controls the speed and can be operated easily by the hand holding the handle. Each has a rip-and-circle guide adjustable from a minimum of 1 in. to a maximum of 6 in. Each can be adjusted easily to make bevel cuts up to 45 degrees. The differences between them are mainly as follows.

Zenith X-400

Analysis of the Zenith X-400 is based on 25 hours use of a single example. Other examples were unavailable at the time of the study but the supplier indicated that the model tested was typical.

Superior Features. The Zenith X-400 is relatively quiet in comparison with most saws examined and gives less than average trouble with vibration. It handles the sawdust without letting it either accumulate on the board and obscure the guide line or blow up into the operator's face. It is provided with a shoe insert to reduce splintering.

Undesirable Features. Though the on-off switch is conveniently located, the device for locking it into position for a selected speed is a little awkward to use. The shape of the shoe does not provide good support when a cut is started or when a thin strip is cut off the edge of the board. It is somewhat heavier than average. Inspection and replacement of the brushes in the motor requires considerable disassembly and reassembly. (Such inspection and possible replacement is important because running the saw with brushes shorter than 1/4 in. can damage the motor.)

Monument 65 A

Analysis of the Monument 65 A is based on 20 hours use of five examples. There are negligible differences among the five.

Superior Features. The Monument 65 A is a relatively quiet saw that is easier than average to use. It is light enough that it can be used to cut vertical surfaces without causing more than a minimum of fatigue. It has an auxiliary handle on the left side of the housing. The design of the shoe makes it easier than average for the operator to start a cut and also provides support when a thin strip is cut off the side of the board. It can accommodate an oversize blade that permits cutting lumber up to 6 in. thick.

Undesirable Features. The lightness of the saw results in higher than average vibration. The sawdust has a tendency to collect on the board so that the guide line is sometimes hard to see, and also sometimes blows up into the operator's face. One of the least desirable features is the fact that when the saw must be disassembled, improper positioning of the wire during reassembly causes a danger of electric shock.

FIGURE 15.   Specimen Page from a Report Showing Headings of Three Ranks.

A *heading of third rank* is identical in form to a heading of the second rank, except that it does not appear on a line by itself and it is followed by a period. It is indented because it comes at the beginning of a paragraph.

Underlines are strongly recommended for typewritten work because they make the headings stand out more distinctly. Figure 15 illustrates the form described.

One feature that commends this system is that it does not include a centered upper and lower case heading. Consequently, that form can be used for the titles of lists and rules without being confused with topic headings.

Phraseology of Headings

With rare exceptions, headings that are used systematically to show the organization of a report or any other kind of writing are worded as topics. That is, each heading is a noun (or nouns) plus any modifiers needed. It is sometimes assumed that any form that is not a sentence is a topic, but this is not the case.

Though it is doubtful that most writers analyze the grammatical structure of their headings, the fact remains that their headings almost never turn out to be prepositional phrases, adjectives, infinitive phrases, or sentences. Occasionally, in a brief piece of semipopular work, headings are phrased as questions or declarative sentences; but for the most part, discriminating writers use bona fide topics because they instinctively feel that other forms do not sound right. And most readers feel a sense of strangeness when they encounter headings that are not topics, even though they may not analyze the reason for their feelings.

The following examples should further clarify the difference between topics and nontopics.

**SUITABLE AND UNSUITABLE PHRASEOLOGY
FOR HEADINGS**

(1)

Projected Use of Water (suitable)
    Industrial (an adjective and hence unsuitable)
    Domestic (an adjective and hence unsuitable)
Projected Use of Water (suitable)
    Industrial Use[1] (suitable, for *Use* is a noun)
    Domestic Use (suitable, for *Use* is a noun)

(2)

Undesirability of Kilzol Insecticide (suitable)
    For Use near Beehives (a prepositional phrase and hence unsuitable)
    For Use near Streams and Ponds (a prepositional phrase and hence
        unsuitable)

---

[1]If subpoints are likely to be far removed from covering main points, adding a few words might be desirable in some cases for the sake of self-sufficiency. For example, in (1), "Projected Industrial Use of Water."

Undesirable Uses of Kilzol Insecticide (suitable)
  Undesirability near Beehives (suitable because *Undesirability* is a noun)
  Undesirability near Streams and Ponds (suitable for same reason)

(3)

Reasons for Recycling (suitable)
  To Dispose of Solid Wastes (an infinitive phrase and hence unsuitable)
  To Salvage Valuable Materials (unsuitable for same reason)
Reasons for Recylcing (suitable)
  Disposal of Solid Wastes (suitable because *Disposal* is a noun)
  Salvage of Valuable Materials (suitable because *Salvage* is a noun)

(4)

Difficulties with Kiley Trucks (suitable)
  Repairs Expensive (unsuitable; an elliptical sentence—no verb)[2]
  Have Insufficient Power (unsuitable; an elliptical sentence—no subject)[2]
Difficulties with Kiley Trucks (suitable)
  Expense of Repairs (suitable because *Expense* is a noun)
  Lack of Power (suitable because *Lack* is a noun)

If you doubt your own ability to phrase your headings suitably by applying grammatical principles, you can probably get good results by asking yourself as you look at each heading, "Does this sound like a name? Would it seem natural to say, 'The title of this section of my report is . . . '?" If the answer is no, a change in phraseology is probably indicated.

Another point to remember about the phraseology of headings is that each heading should be as explicit as possible. When a subpoint appears in the text of a report, its main point may be on an earlier page. Consequently, a heading should be as nearly self-sufficient as you can make it without letting its length become cumbersome. For example, readers seeing *Power Lines* in an outline as a point under *Removal of Obstacles* would know at once exactly what material it covered. But if they encountered *Power Lines* as a heading on page 12 of a report when *Removal of Obstacles* was on page 6, they would have only a vague idea of what to expect. On the other hand, if they encountered the heading *Removal of Power Lines*, they would instantly know the exact nature of the material it covered. What might appear to be needless repetition in an outline can be highly desirable when the points of that outline appear in the report itself.

A report writer should never forget that, except when extremely short, reports are rarely read from beginning to end as one reads a story or an editorial. On the contrary, they are repeatedly consulted by readers who at the moment want information only on one particular point. Thus the headings should be

[2]Complete sentences would also be unsuitable but are less likely where most heads are topics.

phrased so that these readers can find what they are looking for with a minimum of effort.

General Suggestions

The question of how many headings to use cannot be answered by rigid rules; but if you observe the frequency of headings in this book and in other material where headings are used systematically, you will be able to form a good idea of what is normal. Your aim should be to place headings close enough together to keep the reader constantly aware of what is being discussed, but not so close together as to reduce continuity.

You will also notice that the presence of headings somewhat reduces the need for transitional sentences and paragraphs. And finally, you will notice that the text does not depend on the headings for completeness. That is, words such as *this,* and *these* are not used in a manner that forces the reader to turn back to the heading in order to discover their meaning.

All in all, headings have an effect that is hard to overestimate. They encourage readers to start reading a report by opening up its contents, they help constantly during the reading process, and they subject the writer to a discipline that improves the orderliness of the writing.

SUMMARIES
AND ABSTRACTS

The problem of discussing summaries and abstracts is complicated by the lack of anything approaching a general agreement about terminology. In a set of instructions for use within a single organization, the problem can be solved by the use of arbitrary definitions; but in a book for general use, this simple solution is unsuitable. The basic definitions that probably are most widely accepted in industry are as follows.

An abstract is a brief section—rarely more than six or eight lines long— the purpose of which is to tell the reader what points are covered in some piece of writing without attempting to tell what is said about them. It may be written for a report or various other kinds of writing. If it is written as a section of a report, it is one of the first parts that a reader encounters. It is an independent element rather than an integral part of the body. Such abstracts sometimes appear by themselves in publications devoted entirely to abstracts of material in some particular field.

A summary—at least the summary of a report—is a section that presents in a highly condensed form the most important information that the report contains. It is brief, but not so brief as the type of abstract described above. It is placed, as an abstract would be, extremely early in the report (one rarely finds both an abstract and a summary in the same report); and like an abstract, it is an independent element rather than part of the body. Its functions will be described in more detail later.

Unfortunately, the situation is not so simple as it would be if the distinction established above were universally accepted. If you examine a substantial number of reports from different sources, you will quickly discover two additional facts:

1. The term *abstract* is often used before a section that would be a summary according to the preceding definitions.
2. The term *summary* is used widely and legitimately in reference to a section at the *end* of some piece of writing—a section that is an integral part of the piece and is used to summarize what has already been said. (In a working-level report, however, a summary of this kind is not often used, and even when used it is likely to appear without a heading rather than to be headed *Summary*.)

The problem of terminology is further complicated by the fact that in an effort to facilitate discussion, at least three other terms have been coined for use in books of general circulation. The term *descriptive abstract* is sometimes used to identify what was called merely an abstract in our original definitions. The term *informative abstract* is used in reference to what our definitions called a summary. (In fact, one sometimes finds that the definition of *informative abstract* is sometimes followed by the statement that such an abstract is sometimes called a summary.) The term *introductory summary* is sometimes used in reference to what we called merely a summary, and sometimes in reference to a "summary" at the beginning that emphasizes such matters as an explanation of the problem or a review of the background. All these terms are used primarily in discussions and rarely appear as headings in reports.

The discussion of terminology might be further expanded to cover the occasional use of at least two other terms—*foreword* and *digest*—but we are concerned with other matters of more importance than mere terminology. The first major purpose of the present discussion is to emphasize the importance of the section at the beginning of a report that presents its most important contents in a condensed form, which for the sake of simplicity we shall henceforth call a summary. The second is to tell what a writer can do to make such a section, whatever it is called, serve its purpose effectively. It is unnecessary to deal with descriptive abstracts in the same manner, not because they are unimportant but because they are relatively easy to write.

If a report contains a summary, no other section with the possible exception of conclusions and recommendations is more important. This is the case because readers farther up the line than the writer's immediate superior may read the summary and little else. Also, readers who read the entire report on a single occasion may later refer to the summary to refresh their minds. Still others will use the summary to decide whether to read the full report. In many large corporations, summaries of reports are widely distributed so that those whom they interest may request copies of the full version if they so desire. Also, by looking at summaries that are kept in carefully indexed files, people can tell what information is available from departments other than their own and avoid needless duplication of effort. In brief, when you prepare a summary of your report you may expect that it will be read more frequently, read by more people, and read by more important people than will the full report.

The condensation that is characteristic of a summary is achieved by leaving out preliminaries, details, illustrative examples—everything except the main facts and ideas. The relationship of the summary to the full report is roughly similar to the relationship of a topic sentence to a fully developed paragraph. (There is a difference, however. The topic sentence is an integral part of the paragraph, whereas the report is independent of its summary.)

To write a good summary you should not approach the job too casually. If you try to read a few lines and then condense them, read a few more and condense them, and continue this process to the end of the report, the result will be ineffective. Instead, you should go over the entire report—or an entire section if the report is extremely long—and make notes of the matters you consider essential for inclusion. Next, write the rough draft of your summary by using these notes and by referring to the full version whenever necessary. Then polish what you have written until it reads as smoothly as it can be made to read without the waste of space on introductory and transitional material.

Make a special effort to avoid wordiness. (A summary should rarely exceed 10 percent of the length of the full version, and it is unfortunate to waste words when you are constantly being forced to cut out facts and ideas you would like to include.) If you can preserve grace and variety of style in a summary, well and good; but if you must sacrifice those qualities for the sake of brevity, there is no reason for undue concern. In a summary, conciseness outweighs all other qualities of style except precision, clarity, and readability.

In writing a summary you will need to take special pains in order to make your work accurate. Figures must be double-checked to avoid error. Facts and ideas must not be distorted by the cutting out of words. Care is needed in the selection of material for inclusion, for if its contents are unrepresentative, a summary can be misleading even though it contains no statement that was not in the original. In matters such as emphasis—in the implications as to what was important and what was less important, a summary should give the same impression that readers would receive if they read the full report.

Still another consideration in writing a summary is the question of whether the reader or readers will be the same people who read the full report. When a report contains a good deal of technical material, it is very likely that some of those who read the summary will not have the technical background of the writer or the original receiver. Consequently, a summary will often be more useful if technical information and technical vocabulary are held to a minimum. You might do well to keep in mind three types of reader: the original reader, who wants a full picture but may not check details (this person reads the introduction, summary, body, conclusions, and recommendations); the person whom the original reader depends upon to check everything (who studies the appendix, with all its data, calculations, tables, and figures to make sure that there are no errors or discrepancies); and the executive in a higher position, who may not be a specialist in the area that the report concerns, but who wants to know what is going on and who may say "Yes" or "No" to the entire project (this individual reads the summary and whatever is offered in the way of conclusions and recommendations).

To indicate how a summary compares with a full treatment of the same material, the following summary is presented. It covers all that was said in Chapter 8 about what reports are, their importance, and the difference between professional and school reports.

Example of a Summary

Though reports vary widely in form, length, and other respects, they are a distinct species of writing because they possess one characteristic in common: the responsibility of the writer of the report to the reader.

Reports are extremely important in government and industry because they are a major source of information for those who need it. Their importance to those who receive them makes skill in producing them important to the writers.

The merit of a report written on the job depends on the extent to which it meets the needs of the reader. This test of merit does not exist in the classroom. It is important for students to realize this difference and form the habit of constantly asking "Does the reader need this?" in deciding what information to include in a report and what to omit.

THE INTRODUCTION

Naturally, a report that is extremely short and simple needs little or nothing by way of an introduction. All that is necessary is to say what is required for the sake of self-sufficiency and to start smoothly enough that your readers feel at ease as they get into the subject.

When a report is longer, it may be necessary to choose between an ordinary introduction and what we will call a formal introduction. The ordinary introduction in a report is no different from the introduction in any other kind of writing. It may consist of general comment on the subject as a whole. It may give background information on whatever the report concerns. It may explain the general plan that the report is to follow. It may serve, without being so labeled, as a summary of the contents. It may attempt to arouse interest. (The presence or absence of a table of contents and a summary will of course affect the need for some of these types of material.) In general, the normal introduction may say anything that needs to be said about the subject of the report before the first main division of the material is taken up.

Example of a Normal Introduction

The Klamath National Forest in northern California contains one of the most landslide-prone regions of the world. High rainfall, steep slopes, and weak foundation materials have combined to produce hundreds of landslides. The greatest instability is found in the western part of the forest in a zone 20 miles wide centered along the Klamath River. Graphitic slates and serpentinized-ultra-basic rocks are especially unstable in this zone.

Every year on the Klamath Forest about 70 miles of new roads are built for timber harvesting and recreational access. One of the jobs of the engineering geologists is to make foundation studies for road route planning, location, and design. The foundation studies provide data for building roads which will have the least impact in terms of triggering new landslides or reactivating dormant landslides. The following deals with one aspect of these foundation studies: Making a preliminary estimate of slip-surface geometry based on the field expression of head scarps and lateral scarps.

In contrast, a formal introduction to a report is much like the introduction to a thesis or dissertation. That is, it deals not so much with the subject of the report as with the report itself. This type of introduction is likely to be used in a long formal or semiformal report that is written for some nonroutine purpose—often for the purpose of presenting the results of a special investigation. It will be discussed in more detail in Chapter 12, which deals with the formal report. Suffice it to say, at present, that in nonformal reports a formal introduction may be used to tell readers whatever they need to know about the reason the report is written, the function it is intended to perform, the scope of the problem it concerns, and the name of the person or organization that authorized it. A specimen of a formal introduction is included in Chapter 12.

Such an introduction may be followed by an ordinary introduction if one is needed; but in this case the ordinary introduction is placed under a heading such as "Background Information," "General Comment," "History of . . . ," or any other appropriate phrase.

CONCLUSIONS
AND RECOMMENDATIONS

Though some reports, because of their purpose and content, do not need to include conclusions or recommendations, such material is almost always called for in a report that presents the results of an investigation. When it is included, it should usually be drawn together in one place and headed in a manner that identifies its special nature. In this connection, the term *conclusion* or *conclusions* means "convictions arrived at on the basis of evidence" rather than merely "the section that consists of an ending."

Though conclusions and recommendations sometimes merge naturally into a single section, their nature should not be confused. *Conclusions* is a term used to identify convictions. *Recommendations* refers to actions that are called for on the basis of such convictions. We might reach certain convictions without being in a position to recommend action because the wisdom of possible actions may depend on matters beyond the scope of our particular report. It might be possible, for example, to conclude that a certain site would be physically suitable for a factory, without being in a position to recommend that a factory be built there.

Logically, conclusions and recommendations should follow the body of the report because they depend on its contents. Often, however, they are placed before the body on the grounds that the reader wants to read them immediately. Most large organizations will have a standard practice regarding placement; but a writer who is not bound by a prescribed format may place them in either location with the assurance such placement is not unusual.

In a section labeled *Conclusions,* regardless of where it is placed, every statement should grow out of facts given elsewhere. A reader has a right to assume that you are not stating conclusions except on the basis of evidence. Overstatement should be carefully avoided. When some of the evidence seems opposed to the conclusions, that fact should be conceded; and in explaining away such evidence, the report should be scrupulously fair. If the evidence points toward some conclusion but does not definitely establish it, you should frankly say that the evidence is not conclusive. There is nothing wrong with offering an opinion if it is openly identified as such.

This advice to use restraint does not mean you should resort to evasion. In a conclusion, as elsewhere, such statements as "It is believed that . . ." are obviously meaningless unless we know who does the believing.

When conclusions are placed at the end of a report, the facts and reasoning that they are based upon may be pointed out briefly; but when they appear before the body, this support is not usually offered. Wherever they are placed, conclusions should be numbered if there are more than one, but the use of numbers is not a substitute for careful grouping of material into unified elements and careful arrangement in a logical sequence. Use of numbers does not achieve organization, but merely reveals it.

Even though the nature of a report sometimes makes it natural to offer a conclusion at the end of each major division, it is advisable to draw all the conclusions together in a single section unless the report is short and simple. It would be inconvenient for a reader who wished to study the conclusions and nothing else to be forced to hunt for them on several pages.

When recommendations are called for and do not merge naturally with conclusions, they should be keyed to the conclusions (that is, given similar numbers) when such treatment is possible. Like the conclusions, they grow out of the facts and reasoning offered in the body of the report. Sometimes the recommendations grow out of the conclusions as the conclusions grow out of the body. The occasional tendency to treat *Recommendations* as a section consisting of a few helpful suggestions on incidental questions is unsound in any report based on an investigation; rather than being incidental, the recommendations are the focal point of everything else in the report—the reasons for its existence.

One final point should be heeded. Like so many other tasks in producing a report, writing conclusions and recommendations is a matter of judgment, not merely compliance with rigid rules. In the preceding discussion, we have attempted to demonstrate that these materials should be presented in a manner that helps readers to find what they are looking for as easily as possible. The organization and the choice of headings should be whatever is necessary to accomplish that result.

Following are the conclusions and recommendations arrived at in a report by the Technical Extension Service of a state university to a legislative fact-finding committee.

CONCLUSIONS AND RECOMMENDATIONS

Conclusions

The survey of mobile home courts in Carmine County indicates that though the proposed regulations may be desirable and should be applied to any court opened in the future, they should be applied to existing courts with some restraint.

Fourteen courts are operating in the county at the present time. Details about each of them are included in the body of the report. If the new standards were in effect at present, only six of them would fully comply or could be made to comply by means of minor adjustments. These six, the newest and best, really present no problem. We have divided the others into three groups, as follows.

A. Three courts comply with all the standards that relate to health and sanitation and also with most of the other standards. Forcing them into absolute compliance would put them to such expense that it would not seem justified. If minor variations were permitted in individual cases, these courts could continue to provide needed service without being changed.

B. Four courts, which have been operating for periods varying from six to fourteen years, fall far short of meeting the proposed standards. In some instances, full compliance would be impossible. In others, though not impossible, it would be so costly that the courts in question would almost certainly be forced out of business. Closure of these courts would cause real hardships. It would not only mean heavy financial loss to the owners but would also dislocate the occupants of a total of 65 mobile homes. Since space in other courts is scarce, it is hard to conjecture what would happen to these units—in which, incidentally, many of the occupants have substantial equities. In this connection we should not overlook the fact that if continued occupancy of these homes in their present locations is prevented, their occupants may find it impossible to find other places of residence in this area, the housing shortage being what it is.

C. One court, in which only six of the twelve spaces are occupied, falls so far short of acceptable conditions that it probably could not be salvaged.It is over twenty years old, has been allowed to run down, and has been losing occupancy each year. It would be unlikely to survive much longer even if the new standards were not a factor.

Recommendations

1. The courts referred to in A, above, should be granted the variations necessary for their continued operation. (These courts and those in other groups are identified in a list in the appendix.)
2. All mobile courts more than five years old (in B above) should be exempted from proposed regulations except for those concerned with health and sanitation.
3. The owner of the court referred to in C should be notified that he will be forced to close within one year after the regulations become effective unless he submits an acceptable plan for improvements within 90 days and begins to implement it as soon as it has been accepted. (He is entitled to this opportunity even though it is unlikely that he will utilize it.)

THE APPENDIX

An appendix is often desirable in a report, as in a book, because it permits the writer to make material available without placing it in the main text. In a report, a typical example of this material would be the full results of a series of tests. The average of these results might be significant, but the results of the individual tests would often serve no function except to make it possible to confirm the average and perhaps to see the extremes. Other material sometimes relegated to the appendix includes tables, figures, calculations, lists (for example, lists of equipment), specifications, and even supplemental reports.

Various reasons can make it desirable to put something in an appendix. The appended material might be technical information that most readers would not use, though they might want to have it available for examination by specialists. It might consist of detailed or supplementary information that would be consulted on special occasions but otherwise would only be in the way. It might be material to be studied separately rather than during normal reading. It might be material such as photographs, that would make the report physically hard to handle if scattered through the body.

Obviously, anything placed in an appendix is not so closely integrated with the rest of the report as it would be if placed in the body. Sometimes, indeed, it does not need to be closely integrated. But when material in the appendix is closely related to something in the main report, it should be referred to at the appropriate point in the text.

Ordinarily the appendix is preceded by a title page bearing the single word *Appendix;* but if the appendix is brief and if the first item is not of full-page size, the title *Appendix* may be placed at the top of the first page rather than on a page by itself.

When the material appended is of more than one kind and when its amount is great enough to justify the extra complication, two or more appendixes can be used. In this case, each is identified by a letter, and its title indicates the nature of its contents. (For example, *Appendix A—Tables, Appendix B—Photographs*.)

In brief, use of an appendix is often a handy method of presenting material that should be available in support of the main report without causing the main report to become needlessly cluttered and complicated.

FINAL COMMENT
ON SPECIAL ELEMENTS

Before the subject of special elements is dismissed, two more points should be brought out. The first is that in the report-writing world, uniformity is sadly lacking in the labels attached to these elements and in the details of their use. For example, what is called an *abstract* in one company may be a *summary* in a second and a *synopsis* in a third. The term *references* usually means a list of books and articles, but in some companies it means conversations, letters, and memoranda that are related to the subject of the report and are listed at the beginning. In short, although most of what has been said in the preceding discussion will apply in any position you may hold, it would be impossible for any book to give instructions that are entirely suitable for every occasion, because terminology and details of procedure vary.

Second, we should not lose sight of the fact that special elements of reports are merely tools. We should know how to use them, but we should decide in each case which ones are needed, and not use all of them just because they are available. In other words, planning the layout of a report is not just a matter of following hard and fast rules. On the contrary, it is a process of designing a product that will do its job effectively—and its job is simply to make sure that the reader can grasp the contents of the report as accurately and as easily as possible. Each of the elements is merely a means to that end.

EXERCISES

Following are excerpts from outlines made with the expectation of using the points of the outline as headings within the completed report. Revise the phraseology, when necessary, so that each point is worded as a topic and so that the principles of parallel form and self-sufficiency are applied.

<table>
<tr><td>

(1)

Steps in the new welding process
Vee out the area
Use of a sealing electrode
Completion of the weld

</td><td>

(2)

Suggestions for reducing accidents
Reduce speed limits
Prevention of drunk driving
Driver education program

</td></tr>
</table>

(1)	**(2)**
Steps in the new welding process Vee out the area Use of a sealing electrode Completion of the weld	Suggestions for reducing accidents Reduce speed limits Prevention of drunk driving Driver education program
(3)	**(4)**
Results of underloading Higher line losses Higher voltage drop Motors lose torque	Causes of forest fires Arson Lightning Human-caused fires
(5)	**(6)**
Disadvantages of aerial spraying Expensive Lack of selectivity Public protests	Need for new energy sources Depletion of former sources Rising cost of former sources Inefficient
(7)	**(8)**
Duties of Fire Inspector Checking wiring and motors Fire extinguishers Inspecting fire doors Removing flammable waste	Conditions that make bicycles dangerous Defective equipment Lack of experience of riders Riders break laws
(9)	**(10)**
Advantages of K-XZ engine Lower starting torque Maintenance is easier Requires smaller base Ease of inspection	Corrective measures Use of synchronous motors Install synchronous condensers Avoidance of overloading

ASSIGNMENTS

Assignment 1

Write a summary of Chapter 9, "Procedure in Writing a Report" of a length specified by your instructor.

Assignment 2

Write a summary of an article in your field. Photocopy the article and give it to your instructor along with the summary.

12 | Formal and Semiformal Reports

Though formal and semiformal reports have been mentioned briefly in earlier chapters, they give rise to so many special questions that they call for separate treatment.

DESCRIPTION OF FORMAL AND SEMIFORMAL REPORTS

The term *formal report* is somewhat misleading. Actually, the formality of reports can vary widely; a report can be quite formal in tone, and still not be classified as a formal report under most systems of nomenclature. Still, the term is widely used, and anyone concerned with reports should know what people are most likely to have in mind when they use it. The distinctive characteristics that the term implies, as gathered from observation of reports from many sources, are as follows:

1. A formal report has a title page.
2. It is usually accompanied by a letter of transmittal.
3. Unless it is unusually short it has a table of contents that includes or is followed by a list of nonverbal materials if any are used.
4. It has a summary (sometimes called an abstract) near the beginning.
5. It contains the kind of introductory information found in what we have called a formal introduction (Chapter 11).
6. If it contains conclusions or recommendations, it identifies them by suitable headings.
7. It has internal topical headings, usually of more than one rank.
8. It is formal in tone and impersonal in style.
9. It includes tables and figures if they can increase its effectiveness. These may either be placed in the text that concerns them or gathered together at one place at the end.

10.   It is documented by some conventional system if published information is used to any significant extent.

In this book, as mentioned earlier, the term *semiformal report* means a report that has the general qualities of a formal report but omits some of the elements. As you read the remainder of the chapter, remember that use of the elements of a formal report is not an all-or-nothing matter. On the contrary, if you are free to plan a report exactly as you think best, you can profit by following the suggestions about each element that you decide to use, but can dispense with any element that you feel can be spared. Though the term *semiformal report* has not been widely used, reports of that kind have been widely written and you need have no misgivings that they will seem strange to those who read them.

Formal reports are considered a distinctive type on the basis of form and tone, but they may be used for varied purposes. A formal report may be a progress report, a periodic report, a report on the completion of a project, or a report on the results of a special investigation. Obviously, however, it is suitable only when a report must be long enough to justify the somewhat complicated mechanism involved and when the occasion calls for dignity in treatment. Rarely would a formal report be written for a single reader. On the contrary, it is justified only when the contents will be of interest—in fact of more than temporary interest—to a number of persons.

The general nature of a formal report becomes clearer when we note that its pattern is much like the pattern followed in most books. Like a book, it may have a cover and always has a title page. Just as a book has a preface, a formal report has a letter of transmittal. Like a book, it has a table of contents that includes or is followed by a list of nonverbal materials. Like a book, it starts its sequence of Arabic numerals on the first page of text that follows the table of contents or the list of nonverbal materials. Like a book, it may have an appendix and a bibliography or list of references after the main text. It might be an overstatement to say that every element in a book has its exact counterpart in a formal report, but certainly writers who are puzzled by some question of form can often solve the problem by recalling or observing what is done in the books in their own field.

The resemblance of reports to books is worth remembering because it would probably be impossible to set down a rigid set of rules about form that would enable us to label each possible way of doing things *right* or *wrong*. But by following the conventions used in books we can at least use forms that our readers are accustomed to. The sensible attitude on questions of form is to settle them not so much on the basis of right and wrong as on the basis of effective versus ineffective communication. And we communicate more effectively when we do not depart from what is customary unless we have good reason to do so.

ARRANGEMENTS
OF THE PARTS

The most important variation in the arrangement of the parts of a formal report concerns the placement of the conclusions and recommendations. Sometimes these materials follow the body, and sometimes they precede it. The first of these alternatives is logical in that it presents evidence before conclusions; but some users prefer the second on the grounds that readers want to know the conclusions at once. This variation in the placement of materials is shown in the two following arrangements:

Title page	Title page
Letter of transmittal	Letter of transmittal
Table of Contents	Table of Contents
Summary (sometimes called Abstract)	Summary
Introduction	Conclusions
Body	Recommendations
Conclusions	Introduction
Recommendations	Body
Appendix	Appendix
Bibliography	Bibliography

When the conclusions and recommendations follow the body, the other parts are almost always arranged as shown above; but there is no set order of arranging the introduction, summary, and conclusions and recommendations when they precede the body. They may be arranged in whatever order best suits the situation unless the writer works for an organization that has a single, standardized system. Whatever it may be, the arrangement has some effect on the contents of the various sections. The reasoning behind the conclusions, for example, is more likely to be explained when they follow the body than when they precede it. And the summary will cover the introduction when it precedes that section but not when the reader encounters the introduction before the summary. Your judgment should enable you to adapt your treatment to the order you are following.

SPECIAL PARTS IN FORMAL
REPORTS

Since many of the parts that comprise a formal report appear in other reports also, they have already been discussed in a general manner in Chapter 11 and elsewhere in this text. But there is more to say about the treatment of three of these parts—the letter of transmittal, the table of contents, and the introduction.

Letter of Transmittal

The letter of transmittal in a formal report is more closely integrated with the report than is true of the covering letter or memorandum sent with reports of other kinds. In fact, it is bound into the report, usually following the title page. It may vary widely in length and contents. If the report is entirely routine, it may consist of only one or two sentences, as for example: "Following is the annual report of the Kinman Metallurgical Research Laboratory for the year ending December 31, 1982." At the other extreme, it may replace the summary and present the essence of the report in condensed form. Most of the time, however, its length is within the normal range of business letters and it presents at least the minimum of facts necessary to orient the reader to the report that follows. For example, it may identify the project the report concerns and also, if necessary, the specific subject of the particular report. It may indicate the scope and limitations of the contents if the title leaves any doubt about them. It may tell where and how the information was obtained and how the report came to be written. And finally, by its signature it may identify the writer. Of course, any of these facts that is obvious is omitted, and other facts can be included if they are needed.

Since the letter of transmittal is by its very form a personal communication from the writer to the immediate receiver, it provides the writer with an opportunity to acknowledge assistance, to express hope that the information will be useful, to offer to be of future service—in fact to say anything appropriate to the occasion.

There is no reason, unless it replaces a summary, that the letter of transmittal must provide a preview of the contents of the report, but if the contents are likely to be pleasing, there is no reason that they should not be indicated.

Being an integral part of the report, the letter of transmittal should not be conspicuously informal in style, but its style need not be as formal as that of the report proper. For example, it is not necessarily limited to the third person.

In general, the letter of transmittal in a formal report permits the writer to address the original reader personally and submit the report gracefully, and it enables the reader to start reading with the assurance that the occasion for the report is understood.

Its form is basically the normal letter form, but in case of doubt "Dear Sir" or "Dear Madam" rather than use of the receiver's name is desirable as a salutation. In governmental reports the salutation may even be the single word "Sir" or "Madam," which are the most formal of all salutations, and the complimentary close is usually "Respectfully yours."

Following is a typical letter of transmittal.

Mobray, Washington 98658
May 17, 1983

Mr. K. L. Keller
Head of the Transmission Design Section
Northern Empire Utilities System
Random City, Washington 98632

Dear Mr. Keller:

As you requested me to do in your letter of November 18, 1982, I have investigated one of the possible causes of unexplained outages that have occurred on the Northern Empire Utilities System.

My investigation concerned the possibility that the outages have been caused by the melting of ice on the transmission line suspension insulators. To check on this possibility we negotiated a contract with the Central University Technical Institute, arranging to have them test the insulators under icing conditions in a low-temperature laboratory and compare their wet-flashover resistance with the standards established by the A.I.E.E. Since the resistivity of water varies considerably, samples with water of varying resistivity were used. The variation was produced by adding sodium chloride to the water used in these laboratory tests.

A second line of investigation was also followed. We arranged for the collection of rainwater samples at 26 substations. The number of samples collected was 339. The resistivity of all these samples was tested in order that we might be able to judge the significance of the laboratory tests of the insulators in view of the actual qualities of the water comprising the precipitation in this area.

The data secured in all the tests mentioned above are presented and discussed in the following report. Our studies must be carried further before we can definitely conclude that the outages resulted from the conjectured cause; but the data obtained thus far make it apparent that further investigation is warranted. We are asking the A.I.E.E. Lightning and Insulator Subcommittee to review this report, and will submit a proposal for further investigation when we receive its comments.

Very truly yours,

John Jorgenson

John Jorgenson
Principal Technical Assistant
Transmission Design Section

The Formal Table of Contents

What has been said earlier about the table of contents applies with special emphasis to the formal report. That is, a table of contents is a list of the headings that appear on the pages of the report; headings except those of highest rank are indented to indicate their rank; they are phrased as topics, and are worded exactly as they are worded where they appear in the text; and as a result of all this they make it easy for the reader to see what ground the report covers, how it is organized, and where each point is taken up.

Since a formal report is longer and more complicated than other kinds, unusual care must be taken so that the table of contents will be easy for readers to use and so that it will observe the conventions that they are accustomed to. The danger of going astray is greatest when tables and figures must be listed. In this connection, two conventions should be borne in mind.

1. The titles of tables and figures are not mixed in among the headings that indicate topics discussed.
2. The items in any single list are arranged in the order of their occurrence. That is, an item with a smaller page number never comes later than an item with a larger page number.

In order to comply with both of these conventions, it is necessary to use a different system when the tables and figures are placed at the end of the report than is used when some or all of them are placed in the text. When tables and figures follow the text, they may be listed as part of the regular table of contents. But when tables and figures are scattered through the text, it becomes necessary to list them under a new heading, equal in rank and hence similar in form to the heading *Table of Contents*. The heading of the list of nonverbal materials can be anything appropriate: *Figures*, or *Tables*, or *Tables and Figures*, or *Exhibits*—anything that lets the reader see what kind of material is being listed.

The treatment can of course vary in detail; but the results will be acceptable if each kind of material is listed separately, if the items in each list follow the order in which the material referred to appears, and if the lists of nonverbal material are not treated as part of the main table of contents unless the report proper is complete before the first such material appears.

The terminology used to identify various kinds of material is not rigid, but comment on some of the terms should be useful. *Exhibits* is the broadest term used. Being practically all-inclusive, it can be used to head a list of specimens from outside sources or as a general heading under which figures and tables are separately listed. Tables and figures are of course different from each other in nature and would not be mixed together in a single list. The term *figure* includes graphs, charts, photographs—any of the kinds mentioned in Chapter 7. The term *plates* is sometimes used for full-page illustrations on paper different from the text pages. All the figures are usually listed in a single numerical sequence, but if there is a reason for doing so, each kind can be listed separately and given its own set of numbers.

In spite of the numerous points made in this discussion, the listing of nonverbal material is not primarily an exercise in obeying rules. Rather, the main consideration is to make it easy for a reader to see what material of this nature the report contains and to make the material easy to find. The best system to use on any occasion is the simplest system that will accomplish that result. In order to increase your resourcefulness in doing this task and at the same time become better aware of the conventions that prevail, you should constantly notice the methods used in books and reports that you encounter.

Chapter 11, "Special Elements in Reports," contains a specimen table of contents, but to illustrate how the listing of visual materials can vary, some additional specimens follow. Each consists of the last heading preceding the list of visual materials (in order to show what page it appeared on) plus the titles of tables and figures. Where a centered heading appears, it may be assumed that its form matches the earlier heading *Table of Contents*. The examples do not show every conceivable situation, but they are adequate to illustrate the principles upon which a treatment for almost any situation can be devised.

(1)

(2)

(3)

LIST OF ILLUSTRATIONS

(4)

TABLES AND FIGURES

Specimen 1 shows an acceptable method of listing visual material when it consists entirely of figures, all of which follow the written material and are placed under the heading *Appendix*. It would have been equally appropriate to have replaced the heading *Appendix* with *Figures* and to have dispensed with the separate page on which *Appendix* appeared.

Specimen 2 varies from specimen 1 in that two kinds of visual material were used, each being treated as a separate appendix. Also, the writer placed headings *Appendix A* and *Appendix B* at the tops of the pages where the appended material began rather than on separate pages.

Specimen 3 illustrates the form used when figures (presumably the only kind of visual material used) are placed in the body of the report. Whether the list of illustrations appears on the same page as the regular table of contents or on a separate page depends on the space available.

Specimen 4 is like specimen 3 except that both tables and figures were used and were placed in the body of the report.

The Formal Introduction

The introduction of a formal report is not the usual type, in which the writer begins a discussion of the *subject* of the report. Rather, it concerns the report itself. (The introduction to a thesis or dissertation is this same type.)

In order that the report may be self-sufficient, the introduction answers such questions as the following to the extent that the writer thinks they might come to the readers: What is the purpose or function of this report? What circumstances caused it to be written? Who asked, or ordered, or decided that it should be written? How much territory (assuming that the title does not precisely show the scope and limitations) does it cover? Where and how was the information obtained? Until the readers know the answers to these questions, they are not ready to receive the facts and discussion that the report offers.

There is no single form for the treatment of this introductory material. One system is to use the general heading *Introduction*, and then place the facts under specific subheadings such as *Purpose* (or *Objective*), *Scope* (or *Limitations*), *Source(s) of Information*, and *Authorization*. When it is not top-heavy, this system is good because it reduces the likelihood that any of the various kinds of information will be omitted. Another system is to use the heading *Introduction* without any subheadings. A third system dispenses with the heading *Introduction* and makes main headings of the above-mentioned subheadings. Yet regardless of the headings under which it is placed, the introductory information is the same. For convenience in our discussion, we will assume that a main heading *Introduction* is used, plus whatever subheadings are needed.

Under *Purpose* or *Objective* the introduction explains as fully as necessary the conditions or events that created a need for the report and tells the exact function that the report is expected to perform.

The material under *Scope* or *Limitations* settles any possible doubt about what the report covers and what it omits. When the title itself performs this function, a section on *Scope* is superfluous; but it is often impossible to indicate the exact limits in the title without making it cumbersome.

The discussion under *Source(s) of Information* tells where and how the facts were obtained. If their source is obvious, this section may be omitted. If some of the information comes from printed sources, they may be named in this section provided there are not more than three or four; but if numerous, they should be placed in a list of references at the end, and this list should be mentioned in the introduction.

The section headed *Authorization* is extremely brief, merely telling who authorized the report, when, and how (letter, memorandum, personal conversation, or whatever other means). In a large organization this section is often replaced by something such as a project number, which would let the reader know under what authority the work was done. Sometimes the actual document of authorization is included or reproduced here.

The introductory material under each of these headings should usually be held to one or two paragraphs. If a long discussion on any of these points is necessary, it can appear under an appropriate heading in the body of the report, and the coverage under *Introduction* can be reduced to a brief general statement and a reference to the full discussion that will come later.

All in all, the formal introduction orients the reader. Rather than discussing the subject of the report, it tells about the report itself. Consequently, there sometimes may be a need for what we referred to in Chapter 11 as an ordinary introduction, which actually opens the discussion of the subject. If so, this ordinary introduction material can be placed in the body of the report under a heading such as *General Discussion of . . .* , or *Background Information,* or *The General Problem of . . .* , or *Historical Review of. . . .* But whenever there is a formal introduction, the ordinary introduction is identified by some other heading than *Introduction*.

Some question may arise over why it is necessary to include parts of the introductory material mentioned. Since a report is usually submitted to the person who authorized it, it would seem that such a person would already possess most of the introductory facts. This is true; but as we indicated elsewhere, the passage of time may have caused many of the details to fade from memory, the report may reach readers who are not familiar with the problem, or it may be filed and used again after a long time has passed. Therefore the report should provide all the information any likely reader may need at any time, and thus be self-sufficient.

It is true, also, that some points in the introduction may have been touched upon in the letter of transmittal; but there, they will probably have been barely mentioned rather than treated in full. The slight repetition is justified because there will be times when the letter of transmittal will be read yet the report itself may be set aside until later.

Following is a formal introduction of the kind recommended:

INTRODUCTION

Purpose

In local areas, highway engineers have found it necessary to use deposits of basalt rock as a source of aggregate for the base course of highways. Some of these deposits yield an aggregate that is entirely satisfactory for such a purpose, but others yield an aggregate that deteriorates rapidly, causing failure of the highway base within a year or two.

To cope with this condition the Highway Materials Testing Laboratory has developed tests intended to indicate in advance whether basalt aggregate from any particular source will degrade rapidly. Also, it has carried on experiments to see whether treating aggregates with asphalt emulsion would slow down the deterioration of those known to be unsatisfactory in their natural condition. The purpose of this report is to present the results of these tests and experiments.

Sources of Information

Information on the sources of aggregates that have not held up as should be expected was obtained from the State Highway Department. The techniques of testing were based in part on work done in other states, as indicated by the citation of references. Most of the facts presented were obtained, however, by the direct testing and experimentation done in the laboratory.

Authorization

This report and the investigation on which it is based were authorized in a letter from Hugh D. Lockhart, Chairman of the State Highway Commission, dated April 3, 1982.

STYLE IN A FORMAL REPORT

Preparing a formal report is one of the occasions when the language should be formal and impersonal. The desired effect is secured, however, not so much by aiming at formality as by avoiding noticeable informality. Slang is obviously ruled out. Colloquialisms—including contractions—are avoided. Technical jargon, handy though it may be, should be used only as a last resort.

Formality and impersonality prevail to such an extent in a formal report that little if any use is made of first and second persons. "I" and "you" are avoided, and "we" is used only when it means the organization in which both the reader and the writer are included. One reason for avoiding "I" and "you" is that they simply seem out of place in any piece of writing where the occasions for their use are few and far between. (Notice how seldom you see them in technical books and articles.) A second reason is that the meaning of these two words depends on who is writing and who is addressed. Consequently, when material by several writers is included in a single report, as often happens, the need of revision is reduced if each contributor avoids "I" and "you."

To be sure, formality and impersonality can be misused. Such a sentence as "It is believed that . . ." dodges the significant question of *who* believes, and overuse of *the writer* in place of *I* sounds stilted. But the real objection to these forms goes beyond the formality and impersonality. Basically, in the first instance the defect is vagueness and the desire of the writer to promote certain ideas yet leave an escape if the information turns out to be mistaken. In the second, it is probably sheer awkwardness, for it is not usually difficult to construct a sentence in which neither *I* nor *the writer* appears. Formal style should not be condemned because it is bad when it is handled awkwardly, for informal style handled awkwardly is equally bad.

In summary we need only say that the purpose of formal, impersonal style is merely to achieve a reasonable degree of dignity when dignity is called for. Well handled, it is not stiff, pompous, indirect, or excessively ceremonious. It is a tool that a writer should learn to use, for informality is not always appropriate despite its many good qualities.

THE EFFICIENCY
OF FORMAL REPORTS

A person who has never had a chance to see how reports are actually used may wonder why some of the features of a formal report are necessary. To understand their value we must remind ourselves that a report is not always written with the expectation that all of those who receive it will read it from beginning to end. On the contrary, there are many more occasions when a long report will be read in part than when it will be read in its entirety.

A formal report meets whatever demands are placed upon it. Readers who cannot remember or have never known the circumstances that caused it to be written can learn the reasons as soon as they pick it up, and can decide how soon they must read it. When they read it in full, they will find that it unfolds smoothly and that they can study the technical details, skim them, or ignore them as they desire. The executive farther up the line who wants only a general picture, including the conclusions and recommendations, can get exactly what is wanted without wasting time on details. People who are interested in only part of the subject, or who want to refresh their minds on some particular point, can easily find what they are looking for. The person who wants to study the full technical details thoroughly finds them easy to locate and identify.

All in all, experience will show that a well-designed formal report is an extremely effective instrument that performs not one but a variety of functions smoothly and efficiently. (An illustrative specimen follows.)

Specimen Formal Report*

The following report was one of five that were eventually incorporated into a single report. Each of the five covered a different aspect of the same basic situation. All were the work of authors with special qualifications in the areas that they dealt with.

The basic situation was as follows: A corporation, American Metal Climax, Inc., had holdings in Wyoming—holdings in which it might eventually develop into an open-pit copper mining operation of moderate size. No immediate operation was contemplated when the reports were written, but an environmental inventory and impact analysis were considered desirable in case of future need. Hence the Kirwin Environmental Study was prepared for the company by the Rocky Mountain Center on Environment (Romcoe) in Denver and The Thorne Ecological Institute of Boulder, Colorado.

The five reports that were eventually combined covered wildlife, fisheries, limnology, recreation and scenic resources, and vegetation. The report used here as a specimen concerns limnology.

The cover of the report served also as a title page. Because of the close working relationship of the writer and Romcoe and because—though it can stand independently so far as it concerns its own area—it was to be incorporated into a larger report covering other aspects of the same situation, a letter or memorandum of transmittal was not included. For an example of a letter of transmittal, refer to the sample given earlier in this chapter, or to the preproposal in Chapter 13.

*Courtesy of Rocky Mountain Center on Environment.

Limnology

Kirwin Environmental Study

Robert W. Pennak, Ph.D.
Boulder, Colorado

Following the table of contents the report has several elements that formal reports often include. It has both an abstract and a summary. The former is the type that mainly indicates what points are covered (sometimes called a descriptive abstract, as mentioned in Chapter 11). The summary, as you may observe, is a condensed version of the actual substance of the report.

Other special elements that follow include sections on conclusions and recommendations, under separate headings. For ease of reference the items in each of these sections are numbered. If readers were to examine only the elements mentioned thus far, they would obtain a substantial, though not a detailed, picture of the contents of the entire report.

TABLE OF CONTENTS

LIST OF TABLES

ABSTRACT

The Wood River, in northwestern Wyoming, is an unproductive foothills and mountain stream that has a characteristic spring and summer runoff from the Absaroka Range. Chemical and physical conditions are typical and similar to those of other Front Range streams. The bottom fauna consists of a relatively small number of clean-water species except in the lower reaches near the water intake for the village of Meeteetse where the bottom fauna is negligible. Possible effects of mining developments on the chemistry, physics, and biology of the stream are discussed.

SUMMARY

This investigation of the Wood River in the Kirwin - Meeteetse area of Wyoming fixes "baseline" aquatic conditions and biological productivity during the open season of 1971. Five equally spaced stations were established between the altitude of Kirwin (9300 feet) and the Meeteetse water intake (6200 feet).

Wood River is a typical small Rocky Mountain stream with rubble substrate, 80 to 95 percent riffles, few pools, and torrential waters in the upper reaches. Temperature, dissolved oxygen, and free carbon dioxide concentrations are consistent with these features in other regional streams. According to local residents, there was a heavy spring and early summer runoff from the Absaroka Range. Nevertheless, Wood River had a surprisingly low silt load (less than 10 ppm) at stations 1 to 3 in June and July. Bound carbon dioxide readings ranged from 16.0 to 47.5 ppm, indicative of waters slightly harder and more productive than those of similar small Rocky Mountain streams. Hydrogen-ion concentrations ranged from pH 7.0 to 8.4; more typically the range is pH 6.8 to 7.4.

The bottom fauna is impoverished and typical of small Rocky Mountain streams, being dominated by clean-water species of Ephemeroptera (mayfly) nymphs and Trichoptera (caddis) larvae. At the lowest station (Station 5) near the Meeteetse water intake the fauna was negligible, possibly because of about two miles of upstream agriculture and pollution by cattle. The total bottom fauna list for this investigation consists of only about 40 species, in contrast to 80 to 100 species in streams east of the Mississippi. During August and September the bottom fauna made a rapid and remarkable recovery from the heavy load (50 to 250 ppm) of runoff silt during June and July. It is postulated that the fauna may undergo similar rapid recovery after short periods (one to five days) of accidental industrial silt pollution.

The periphyton film on the surface of rubble was typical and consisted mostly of organic materials, and especially algae. Inorganic silt formed a negligible portion of the periphyton, in contrast to the situation in other Rocky Mountain streams.

2

Possible ecological impacts of open pit mining, milling, pipelines,
tailing ponds, roads, and reservoirs are discussed.

CONCLUSIONS

Future ecological changes in the Wood River flowage, such as those
that could be triggered by road construction, forest clearing, and
mining activities may alter, more or less, the following limnological
conditions in Wood River: temperature
                           suspended organic and inorganic materials
                           bound carbon dioxide
                           pH
                           dissolved organic and inorganic materials
                           density of bottom organisms
                           species of bottom organisms present
                           (certain species may be eliminated)
                           quantity and quality of the periphyton

More specifically, we might infer the following impacts on the Wood
River when mining and associated construction projects are begun:

1. An open pit mine, mill site, waste dump or leach dump, as such,
would have no direct effect on the stream, providing surface drainage
were controlled.

2. Pipelines carrying tailings, etc. sometimes break at inopportune
times. Since they are often laid alongside or near a stream, the
danger of a gross siltation and stream kill cannot be overemphasized.

3. Tailing pond overflow can cause stream pollution, especially when
the pond is not adequate to permit proper settling of solids and/or when
it has outlived its projected efficient usefulness. A settling pond should
be adequate to provide a clear water effluent.

4. Reservoirs, if properly designed, can minimize the deleterious
effects of the annual runoff and summer "gully washers." Further,
reservoirs have a well-known but poorly understood fertilizing effect
on the streams below their dams, sometimes resulting in an increase in
bottom fauna and fish populations. It might be excellent public relations
if the reservoir(s) could be stocked and opened to public fishing.

RECOMMENDATIONS

1. Surface drainage from roads, mill site, waste dumps and leach dumps
should be controlled so precipitation runoff would not result in toxic
and excess silting of Wood River.

2. Settling ponds should be of adequate size so as to provide a clear
water effluent.

The Introduction is clearly intended for the reader who expects to read the full report. It tells why a study of the stretch of the stream concerned was desired, identifies the three fundamental kinds of information within the scope of the report, and comments on the nature of the stream itself. It does a good job of orienting the reader to the discussion that follows.

The full substance of the report comes under two headings—*Methodology* and *Data and Results*. The noteworthy feature of the first of these two parts is its thoroughness. It refers to certain exceptional circumstances, tells the dates of observations, identifies the locations where observations were made and sample materials gathered, and explains the reasons for the choice of these locations. Techniques used in gathering samples are described. The significance of the kind of information sought is pointed out so clearly as to be easily understandable to a reader who is not a specialist in the scientific field concerned. Tests performed in the laboratory are identified.

The section on Data and Results is understandably the longest in the report. It includes six pages of discussion supported by several tables. (Two of the latter are included as examples but the others are omitted, not being essential to the intended function of the specimen in this book.)

The discussions of both methodology and data and results are clearly written and easy to follow—as are the other materials that precede them. Throughout the report, the contents demonstrate a high degree of scientific objectivity.

3. It might be an excellent public relations gesture to consider if suitable reservoirs could be stocked and then opened to public fishing.

4. The stream survey information in this report should be kept available for comparison with future stream conditions in the Wood River flowage. This procedure provides these benefits: protection of the company against possible unjust accusations or litigation; and providing background data against which to evaluate any future effects of operations.

5. All water effluents from the projected mining operations should be monitored frequently for copper and other toxic and/or heavy metals.

6. A brief biological reconnaissance of Wood River conditions should be made once or twice per year after mining developments begin in the Kirwin area. Observations from future studies should be compared and assessed against the data contained in this report.

INTRODUCTION

Early in 1971 the investigator was asked to conduct a series of comparative biological and physico-chemical determinations on Wood River, Wyoming, from the area of Kirwin downstream to the vicinity of the town of Meeteetse. The American Metal Climax, Inc. (AMAX) has holdings at Kirwin and is interested in the future regional ecology of the upper Wood River flowage. The fundamental purposes of the study were to: (a) establish present "baseline" conditions in Wood River; (b) assess the biological productivity of Wood River, especially as compared with other streams in the Rocky Mountain region; and (c) assess the possible effects of mining developments on the stream. Very little fundamental information is available in the literature about the biology of small streams in Wyoming. The investigator welcomed an opportunity to study undisturbed conditions.

At the outset, it would seem appropriate to comment on the general physical nature of Wood River. In most respects it is similar to other small streams of the eastern ranges of the Rocky Mountains. It is relatively narrow, with the width averaging about 4.5 meters at Kirwin and about 18 meters west of Meeteetse. The substrate is chiefly rubble (the most productive type of bottom), but unfortunately the river bed consists of only 5 to 15 percent pools and it therefore cannot support a good fish population. Indeed, in four trips to the Wood River only two fishermen were observed.

METHODOLOGY

Four visits were made to the study area during the open season of 1971. The first visit was not made until 9 June, chiefly because of the very heavy runoff and the late spring near Kirwin. In addition, field operations in the autumn were cut off early because of exceptionally early snows and cold weather. However, the four visits were made at evenly spaced intervals; collectively they represent a good picture of open-season

4

conditions in the Wood River. Data were gathered at five stations
on each of the following days in 1971: 9 June; 8 July; 5 August; and
11 September. The five stations were distributed as follows and may
be easily located on the proper U.S.G.S. quadrangle maps:

Station 1 - 9300' (3120 m), near Kirwin, about 100 yards above the
mouth of Cascade Creek.

Station 2 - 8600' (2620 m), above midway between Kirwin and the site
of a proposed AMAX tailings pond; 1.3 miles above the log
bridge over Wood River.

Station 3 - 7600' (2315 m), 50 yards above Brown Mountain camp ground.

Station 4 - 6500' (1925 m), just above the first state-maintained bridge
across Wood River.

Station 5 - about 6200' (1890 m), just above the Meeteetse water intake.

These stations were selected with possible future AMAX operations at
Kirwin in mind and with a view toward similarity of substrate and
current so that biological comparisons might be as significant as
possible. All substrates studied intensively were chiefly (70%) rubble,
composed of particles ranging from 1 to 9 inches in diameter. An
additional 20 percent was predominantly unproductive boulders (more than
9 inches in diameter), and the remainder (10%) consisted of unproductive
gravel, sand, and debris.

Bottom fauna samples were taken with a standard square-foot Surber
sampler at each station on most sampling dates. The usual practice
for sampling stream bottom organisms involves taking only one or two
square-foot samples at each station, but the present investigator
took five random samples at each station so that results would be more
significant statistically. The data in this study were derived from
81 separate samples, each representing essentially all of the food
organisms in one square foot of rubble bottom. Quantitative and qualitative studies of these organisms were made by the investigator in
the laboratory at the University of Colorado. In the interest of time,
identification to the species level was not considered critical for an
investigation of this kind.

A second type of biological sampling involves the thin film of dead
organic matter and associated microscopic plant material which covers
each stone, pebble, and boulder in a stream. This material is the
basic food source for the bottom insects. It is a feature that is
highly sensitive to the presence of polluting materials in the water.
This thin film, or periphyton, is scraped from exposed surfaces of rubble
with a sharp scalpel. Five or ten minutes of scraping from three to
eight pieces of rubble constitutes a "standard" sample. The periphyton
material is preserved in vials of formalin and brought back to the
laboratory. Sixteen such samples were taken during this investigation.

5

Measurements of certain standard physical and chemical parameters
were made either in the field or on water samples brought back to Boulder
by the investigator. Temperature and turbidity were measured in the
field. The following were done in the laboratory: dissolved oxygen;
free carbon dioxide; bound carbon dioxide; suspended organic material;
suspended inorganic material; dissolved organic material; and dissolved
inorganic material. Data on stream flow and detailed water analyses
are being supplied by other participants in this study.

DATA AND RESULTS

Inventory

A. Physical and Chemical Features

Table I shows some general features of stream temperature, suspended
materials, turbidity, and general flowage conditions during the sampling
periods. According to local residents, the summer of 1971 was characterized by an exceptionally heavy snow melt and runoff in the Wood
River drainage basin; the water at stations 4 and 5 was turbulent and
more than two feet deep in June and July. It was therefore impossible
to take periphyton and bottom fauna samples.

Temperature conditions were typical of those found in most Rocky Mountain
Front Range streams (Pennak, unpublished), except that the spring
of 1971 came later than usual according to local residents; as a result,
the June, July, and August water temperatures at stations 1, 2, and 3
were about three degrees lower than would be expected in a "normal"
year. Construction of reservoirs and settling ponds would have the
effect of warming the stream in summer and keeping it warm later in the
autumn. Biologically, this situation should improve the productivity
of fish-food organisms.

The lower part of Table I illustrates turbidity conditions in Wood River
in terms of parts per million of suspended organic and inorganic materials.
(These were considered more accurate measurements than the usual turbidi-
metric determinations.) It was notable that the upriver stations had
sedimentary loads that were surprisingly light (less than 50 ppm) in
spite of the heavy runoff in June. At stations 4 and 5, however, in June
and July the turbidity was high (up to 240 ppm), presumably because of
the contributions of tributaries and because of the more erodible nature
of the shorelines at lower elevations. Natural runoff data as given in
Table I can be compared with data derived from pollution conditions such
as silting caused by road-building and some types of mining operations.
Efficient settling ponds and reservoirs should cut down on silting.

The waters of Wood River had no appreciable true color, except at Station
5 in September where the color was 10 (slight) on the platinum cobalt
scale. Such colorless water is normal for mountain streams that do not
have marshy or boggy areas along their courses.

Various chemical conditions are summarized in Table II. Several parameters
need only brief comment. Dissolved oxygen, for example, was always near
saturation or slightly above 100 percent saturation. This is the usual

6

Table I. Physical conditions in the Wood River near Kirwin, Wyoming, 1971.

	Date	Sta. 1	Sta. 2	Sta. 3	Sta. 4	Sta. 5
Temp., °C.	9 Jun	2.5	4.0	7.5	10.0	11.1
	8 Jul	3.9	5.0	7.0	12.0	13.5
	5 Aug	6.4	7.9	8.6	17.7	18.6
	11 Sep	4.2	5.6	10.4	14.8	16.8
Visible turbidity	9 Jun	light	moderate	high	high	high
	8 Jul	0	trace	trace	high	high
	5 Aug	0	0	trace	medium	high
	11 Sep	0	0	0	0	0
General remarks	9 Jun	water level high	water level high	water level high	full flood	full flood
	8 Jul	moderate	moderate	moderate	water level high	water level high
	5 Aug	average flow	average flow	average flow	average flow	average flow
	11 Sep	low flow	low flow	low flow	low flow	low flow
Suspended inorganic material, ppm	9 Jun	29.71	30.63	90.90	120.90	220.15
	8 Jul	3.82	5.46	5.13	44.13	54.13
	5 Aug	4.62		2.80	8.18	9.40
	11 Sep	8.18	6.41	9.73		4.59
Suspended organic material, ppm	9 Jun	2.11	2.54	5.03	13.89	20.55
	8 Jul	0.58	0.39	0.50	4.20	6.20
	5 Aug	0.68		0.48	1.75	1.96
	11 Sep	1.31	0.64	0.45		0.50

7

Table II. Chemical conditions in Wood River near Kirwin, Wyoming, 1971.

	Date	Sta. 1	Sta. 2	Sta. 3	Sta. 4	Sta. 5
Oxygen, % saturation	9 Jun	106.3	102.8	100.2	100.4	101.0
	8 Jul	106.4	106.6	106.0	104.2	103.0
	5 Aug	110.2		104.7		108.7
	11 Sep	103.2				126.4
Free CO_2, ppm	9 Jun	1.5	1.5	1.4	1.3	1.5
	8 Jul	1.1	1.5	1.5		
	5 Aug	3.5	1.5	1.0	1.0	-2.0
	11 Sep	1.0	1.0	1.0	0.2	-0.5
Bound CO_2, ppm	9 Jun	18.5	20.0	21.0	25.1	28.2
	8 Jul	16.0	18.0	24.5	30.5	33.0
	5 Aug	26.5	25.2	27.5	35.5	47.5
	11 Sep	29.0	29.0	28.0	38.0	44.4
pH	9 Jun	7.3	7.2	7.1	7.0	7.0
	8 Jul	7.3	7.3	7.3	7.2	7.2
	5 Aug	7.3	7.4	7.9	8.1	8.3
	11 Sep	7.1	7.7	8.0	8.2	8.4
Dissolved organic matter, ppm	9 Jun	54.45	27.74	25.76		51.52
	8 Jul	42.70			45.70	
	5 Aug		23.96		36.60	42.62
	11 Sep	29.02		35.34		46.57
Dissolved inorganic matter, ppm	9 Jun	154.55	52.42	47.22		67.34
	8 Jul	41.84		49.66	59.74	
	5 Aug		53.10		70.17	84.82
	11 Sep	68.61		88.21		122.45

8

situation in unpolluted mountain streams. Supersaturation readings are
possible because of entrapped bubbles in the field samples. Incidentally,
all preliminary analytical reagents were added to samples in the field,
so that the final laboratory titrations would be accurate.

Most of the free carbon dioxide readings ranged from 0.2 to 3.5 ppm.
These again are close to the equilibrium values with air where water is
unpolluted. The -0.2 and -0.5 ppm readings at Station 5 are a reflection
of a high photosynthesis rate by the periphyton algae but are of no special
ecological significance.

Most rapid mountain streams in the Rocky Mountains have bound carbon
dioxide values ranging from 10.0 to 25.0 ppm, a reflection of the fact
that most of the substrates are poor in carbonates and other soluble
salts. In general, the results for Wood River were somewhat higher with
14 readings in excess of 25.0 ppm. It would appear, therefore, that the
waters of this flowage are slightly harder than average. As a consequence,
the general productivity of the bottom fauna and fishing potential should
be slightly better than for the "average" small mountain stream, since
there is a general (but poorly understood) correlation between productivity
and water hardness.

The slightly harder waters were further shown by the fact that hydrogen-
ion values were higher than usual. Most similar mountain streams have pH
values ranging from 7.1 to 7.4, but a few readings for the lower stations
of Wood River were more alkaline than pH 7.4, and up to pH 8.4.

Similarly, both dissolved organic and dissolved inorganic materials
were about 50 percent higher than is customary for Rocky Mountain streams
(Amitage, 1958; Cordon and Kelley, 1961; Gaufin, 1959; McConnell and
Sigler, 1959; Pennak, unpublished). The former averaged 38.25 ppm and the
latter 73.86 ppm. These are of no special significance other than to
indicate a somewhat higher level of potential productivity. Indeed,
they are actually considerably lower than similar determinations made in
Michigan and New England trout streams, where general productivity levels
are notably high.

Bound carbon dioxide, pH, dissolved organic materials, and dissolved
inorganic materials are useful baseline parameters. In the event of
chemical pollution and settling pond construction, all of the values for
these features will change. On the other hand, reservoirs along the
course of Wood River should have no important effects on the water
chemistry.

With reference to physical and chemical factors, then, in summary Wood
River is a "typical" Rocky Mountain stream with somewhat higher than
average biological potential.

B. Bottom Fauna

Table III shows the detailed results of the bottom fauna study, a critical
item in assessing stream conditions. It illustrates particularly the
necessity of taking samples several times during the open season because

10

of the wide variations in the density of the various species from one
month to another, owing to variations in their life histories. Many
species, for example, were not taken by the Surber sampler early in the
season (May and June) because they were still in the egg or first larval
stage and consequently were too small to be retained by the net. Similarly, sampling in August and September often failed to take certain other
species because they had already emerged from the stream as adult, flying
insects.

Unfortunately, samples could not be taken at stations 4 and 5 in June and
July because of the extremely high runoff and water at these stations.

All of the genera listed in Table III are typical of small Rocky Mountain
streams. None of them are especially rare or unusual. Station 5, which
is well out toward the plains, had a poor insect fauna from the standpoint
of number of species. In part, this situation is due to the fact that
Station 5 is not a true mountain stream habitat, and ecological conditions
are not favorable for cold water species. Of greater significance,
however, is the fact that Station 5 had the appearance of some organic
pollution. It is possible that conditions in the several miles of farm
land and gross disturbances by cattle between stations 4 and 5 were
responsible for the restricted fauna found at Station 5. For example,
large quantities of cattle droppings were found along the shore upstream
from Station 5. Although little specific information can be found in
the published literature, stream biologists believe there are deleterious
effects on stream organisms from large amounts of cattle droppings.

The very large population of mayfly nymphs on 5 August at Station 4 is
remarkable (Table III). It is an excellent example of the ability of
Ephemeroptera to withstand and recover from heavy (up to 135 ppm)
runoff silting such as that experienced in this part of the stream in
June and July. Although mayflies from mountain streams are particularly
sensitive to chemical and sewage pollution, they appear to have developed
an effective resistance to temporary inorganic silting from runoff.

A total of about 40 species of macroscopic bottom organisms were found
during this study. This list is similar to those found by the investigator (Pennak, unpublished) in other Rocky Mountain streams. Chemical or
physical pollution would cut this species list to perhaps five or ten.

As shown in the summarized bottom fauna data in Table IV, stations 1,
2, and 3 had similar population densities of bottom insects. Station 4,
however, had an exceptionally high standing crop in August (4046 per sq m),
chiefly because of large numbers of very small mayfly nymphs and midge
larvae, as noted above. Presumably these insects emerged as adults
before the last samples were taken a month later. Station 5 had a negligible bottom fauna, perhaps for reasons already mentioned.

Total numbers of bottom organisms ranged from 0.8 to 378.0 per square
foot for various dates and stations. Omitting these two extreme figures,
the mean standing crop for all five stations was 23.5 organisms per square
foot. Thus Wood River must be considered a poor producer. The most
widely used stream productivity criteria in the United States have been

12

promulgated by fisheries biologists:
  Poor streams   - 0 to 100 organisms per square foot.
  Average streams - 100 to 200 organisms per square foot.
  Rich streams   - more than 200 organisms per square foot.
A similar, gravimetric system of grading streams classifies "poor"
streams as those having less than 1 gram of organisms per square foot.
Thus it is further obvious that Wood River is a poor producer. There is
no specific reason for this situation unless it is the generally short
Wyoming growing season, the relative softness of the water (16.0 to 29.0
ppm bound carbon dioxide at stations 1 to 3), and the small size of the
stream (4.5 m at Kirwin). Table V contains comparable data for other
Rocky Mountain streams; all of these have higher productivities than Wood
River; except for biomass, the data for these streams are similar to
those for Wood River. Additional data (Pennak, unpublished) show that
two small Idaho streams have productivities even lower than that of Wood
River, while data for a small New Mexico stream show a standing crop more
than 200 percent greater than that for Wood River.

Frequent intermittent silting will knock out most of the species listed
in Table III. Effluents from carefully constructed settling ponds should
not have serious effects on the stream fauna provided that the effluent
is less than 50 percent of the flow of the Wood River at the effluent
entry point. Indeed, there may be an actual "fertilizing effect" if the
settling pond effluent contains no toxic ions in excessive quantities.

C. Periphyton

Table VI contains data derived from the quantitative study of the periphyton.
Periphyton serves as an important food source for the bottom insects,
most of which are grazers over exposed surfaces. Comparable data are not
yet available in the scientific literature but on the basis of unpublished
work (Pennak, literature citation) the summer standing crop of periphyton in Wood River at the first four stations is similar to the variable
conditions found in other unpolluted streams in the Front Range of the
Rocky Mountains. In general, the data in Table VI indicate major rapid
productivity of periphyton following the spring runoff. Essentially
all of the periphyton consists of living algae with little inorganic
silt. The genera of algae represented are common in unpolluted swift
streams everywhere. If Wood River were to be subjected to continuing
mine tailing waste, etc., the periphyton picture would change quickly.
The rubble would lose most of its coating of algae, and surfaces would soon
become covered with inorganic particulate material.

Impact

Upper Reservoir. This body of water should have no deleterious effects
on the biology of Wood River. The altitude is such that temporary storage
of stream water behind the dam can have only a slight warming effect on
the flowage, not enough to influence the species and abundance of downstream organisms. On the other hand, proper manipulation of the water
level could improve biological conditions in the stream, especially if

15

greatest storage could be attained during spring runoff. The more constant
the flow in a stream, the better the biological conditions; even a small
mediation of annual high and low water extremes would have beneficial
biological effects. The reservoir could become an excellent fishing area.

Lower Reservoir. The same comments apply to both reservoirs. Limnologi-
cally, there is no preference between these two reservoir sites.

Tailing Pond. Much of the biological impact of the tailing pond depends
on the way it is constructed and its relation to the Wood River flowage.
If, for example, all of the water in the thalweg flows through this pond,
there might be problems with the efficiency of the settling process. It
would be likely that excessive tailing silt would wash downstream in the
spring and early summer, to the detriment of the biology of Wood River.
If, however, a large conduit were laid in the present thalweg so that much
of the stream water (especially in spring and early summer) would pass
under the tailing mass, then the situation would be much less serious.
It would mean that the tailing pond water overflow would join the stream
proper just below the dam. Runoff from the tailing might actually have
a beneficial effect on downstream biological productivity because of the
higher level of dissolved nutrients in the water (Pennak, unpublished).

Waste Dump, Leach Dump, Open Pit, Mill Site. None of these areas should
ordinarily produce an impact on Wood River. However, there is always the
danger of washouts and silting from heavy rains. Care should therefore
be used in laying out these sites, with special emphasis being placed on
local drainages, especially since the protective vegetative cover will
presumably be gone. A serious problem might develop if chemically toxic
substances are washed into the stream from the operations areas.

Services and Transportation. These facilities can ordinarily affect Wood
River only when they (1) are built too close to the stream and (2) are
not well maintained. When a gravel road is bulldozed alongside a stream,
much of the streamside herbs, shrubbery, and trees are destroyed. This
has the effects of (1) increasing erosion and silting, (2) decreasing
those stream fish-food organisms that rely on streamside vegetation for
breeding areas, and (3) destroying shade over the stream and thereby
raising stream temperatures.

Chronological Sequence of Impacts. When the Kirwin operation is under
construction, most of the stream impacts would be centered around the
silting produced by road building and the erection of plant facilities.

During the 15+ years of plant operations major potential impact areas
will be related to: settling ponds, leach dump drainage, waste dump
drainage, and roads and service areas.

When the Kirwin area is eventually abandoned, residual major impacts will
be the settling pond and the danger of a break in its dam, the leach dump,
and the waste dump. An intensive revegetation program will be necessary.

16

Sphere of Influence. Most of the items mentioned in this impact section
are of primary importance to the Wood River biology only in the Kirwin
property and perhaps two or three miles below the tailing pond. This
would apply unless, of course, there were to be a major plant breakdown
or severe precipitation, in which case there might be stream impact
extending much farther downstream. Because of the long distance and
because of the diluting effect of tributaries, however, it is doubtful
that there would be measurable stream impacts as far east as the
Meeteetse water intake.

ACKNOWLEDGEMENTS

The investigator would like to acknowledge the patience, help, and
understanding extended to him by the administrative personnel of the
Thorne Ecological Institute, the Rocky Mountain Center on Environment,
and American Metal Climax, Inc. He is especially indebted to James J.
Cooper who made special field work arrangements in the Cody and Kirwin
areas, and who extended much practical advice.

17

SELECTED REFERENCES

Armitage, K.B. 1958. Ecology of the riffle insects of the Firehole
River, Wyoming. Ecology 39:571-580.

Cordone, A.J., and D.W. Kelley, 1961. The influence of inorganic
sediment on the aquatic life of streams. Calif. Fish and Game
47:189-228.

Fisher, S.G., and G.E. Likens. 1970. The annual detritus budget of a
small stream. (Abstract). Paper submitted at 33rd annual
meeting of the Amer. Soc. Limnol. & Oceanogr.

Gaufin, A.R. 1959. Production of bottom fauna in the Provo River,
Utah. Iowa State Coll. J. Sci. 33:395-419.

Hynes, H.B.N. 1960. The Biology of Polluted Waters. 202p. Liverpool,
England.

_____. 1970. The Ecology of Running Waters. 555p. Toronto,
Canada.

Logan, S.M. 1963. Winter observations on bottom organisms and trout in
Bridger Creek, Montana. Trans. Amer. Fish Soc. 92:140-145.

McConnell, W.J., and W.F. Sigler. 1959. Chlorophyll and productivity
in a mountain river. Limnol. & Oceanogr. 4:335-351.

Mecom, J. 1969. Productivity and feeding habits of Trichoptera larvae
in a Colorado mountain stream. 148p. Ph.D. thesis, Univ. of
Colorado.

Muttkowski, R. 1929. The ecology of trout streams in Yellowstone
National Park. Roosevelt Wildlife Ann. 2:155-240.

Needham, P.R. 1940. Trout Streams. 233p. Ithaca, N.Y.

Pennak, R.W. 1953. Fresh-water Invertebrates of the United States.
769p. New York, N.Y.

_____. Unpublished data.

Pennak, R.W., and E.D. Van Gerpen. 1947. Bottom fauna production and
physical nature of the substrate in a northern Colorado trout
stream. Ecology 28:42-48.

Slack, K.V. 1955. A study of the factors affecting stream productivity
by the comparative method. Invest. Indiana Lakes and Streams
4:3-47.

ASSIGNMENTS

Write a formal or semiformal report of a length specified by your instructor on one of the following subjects or on a subject of your own choice.

1. Investigate career opportunities in your major area of interest. Include such factors as the number of job openings each year, potential for advancement, success of previous students from your school in finding jobs, and the preparation preferred by those organizations doing the hiring.

2. Determine some area in your field of specialization that you wish to learn more about. After getting your instructor's approval, investigate the area and write a state of the art report.

3. With the consent of the people involved, report on a research project currently under way at your school.

4. Write a feasibility study about the possibility of converting a specific vehicle (automobile, truck, or bus) from gasoline to propane. Include such factors as cost of conversion, availability of parts and propane, and payback period based on several different annual mileage estimates.

5. Write a market analysis of the profit potential of a new business in your community. The business could be locally owned, such as a craft shop, a restaurant, or a repair center of some kind, or it could be a franchise, such as a fast-food outlet. Include costs of acquiring a building, equipment, and supplies, salaries of employees, and a survey of potential customers. Estimate the annual gross required to return 2, 5, and 10 percent on the initial investment.

6. Report on the desirability of establishing a public recreation facility in your community. The facility could be indoors, such as a game center or roller rink, or out-of-doors, such as a bicycle motocross track or a preschool jungle gym for an existing park. Include such information as estimated costs, possible sources of funding, and a survey of local opinion about the facility.

7. Write a cost comparison of three types of heating for a building or a water system within a building. If the building is already constructed, include the cost of removing or altering the existing system. Make a recommendation for a type based on the conditions discovered for the building you investigated.

8. Determine a pest that attacks trees, crops, or lawns in your part of the country. Detail the types and amounts of destruction and describe biological, chemical, and other methods of control, giving advantages and disadvantages of each method. Recommend one method for a specific location and give your reasons.

9. Analyze the performance of some unit of your school or community, such as student senate, intramural recreation program, fire department, health care center, or buildings and roads department. Include interviews or surveys of affected clientele. Point out areas of excellence and areas that may need changing. Make suggestions for improvement, or indicate areas that may need further study.

13 | Proposals

A proposal, as the word itself indicates, is a suggestion or request that some particular action be taken. Usually a specific person, business, or agency wishes to do a job or solve a problem for another person, business, or agency, and writes a proposal in an effort to obtain the opportunity to do so; the proposal amounts to an offer to do the job for payment or at least with funds provided by the other party. For example, a corporation seeks, as a subcontractor, to develop and perhaps manufacture some part of a missile for an aeronautics firm which has itself, by means of a proposal, secured a government contract to develop the complete missile. Sometimes the person or organization submitting a proposal attempts to persuade those to whom the proposal is addressed to do some job—as, for example, a student group urging the municipal government to establish bicycle lanes as a public convenience. Sometimes, too, a proposal involves joint participation in a project—as, for example, the proposal by a school board to a municipal recreation board that the two bodies join in developing a tract of land for recreational use by both the school and the general public. But most characteristically a proposal is an effort by the proposer to be authorized to do a job for payment for the organization to whom the proposal is addressed.

In many cases, proposals are written in response to announcements of jobs to be done for government or industrial organizations. These announcements may be found in technical or trade journals or may be mailed directly to firms that wish to receive them. Such an announcement tells of the availability of funds to do a job and is called a "request for proposals," or "RFP." RFPs give the basic guidelines for the finished product but allow the organizations submitting the proposals to provide the details and expenses of the job. Most of the work contracted through government agencies, such as the National Science Foundation and the National Institutes of Health, is arranged in this manner. The aim is to provide through competing proposals the best work for the least amount of money.

Another area of proposal writing involves philanthropic organizations such as the Ford or Rockefeller foundations, or the hundreds of smaller funds that dispense millions of dollars every year to aid projects considered meritorious by their boards of directors. These organizations award most of their grants in response to proposals written by individuals or groups with an idea to pursue. The projects range in size from major research in heart disease to acquiring new costumes for a neighborhood barn theater, and the proposals may be many volumes long or letters of a single page.

Proposals differ from other technical writing in one important respect. While most technical writing deals with things that exist—events that have taken place, projects that have been completed, operations of machinery—proposals are concerned for the most part with the future—with projects that someone merely wants to have undertaken. This major difference makes writing proposals particularly difficult. Most people in industry become accustomed to describing with precision things that exist; but how does one describe accurately something that may be little more than an idea in someone's head? This chapter discusses methods to use in dealing with this difficulty. You are likely, sooner or later, to face this problem and indeed may find that some of the most important writing that you do involves proposals. In fact, the talent that major corporations assign to the preparation of proposals is just as high-caliber as the talent they assign to carrying out a job once the proposal has been accepted.

The importance of proposals is easily shown. Most of the projects that have been written about, after completion, and most of the changes in procedures started as proposals. Thus everyone in education, science, industry, and government at a decision-making level utilizes proposals as a basis for action.

CONVINCING THE READER

A proposal may come into existence either because those who receive it make known their desire or need for it or because those who submit it do so on their own initiative. In either case, the problem for the writer is the same: to convince the reader that the writer or the organization is capable of solving the problem. Until the reader is convinced that those who make the proposal have a reasonable chance of performing the task successfully, a go-ahead is unlikely. The reader's time or money is usually involved, and the reader doesn't wish either to be wasted.

Whether a proposal is long or short, simple or complicated, a writer can improve the chance of securing conviction by making sure that its contents answer the following questions:

1. What do we propose to do?
2. How do we propose to do it?

3. What evidence can we present to show that the methods we propose to use will actually get the desired results?
4. What evidence can we present to show that ours is the best way to get the desired results?
5. How can we demonstrate our ability to do what we propose to do?
6. What evidence must we present to show that the cost will be acceptable and, perhaps, that we can meet a satisfactory time schedule?

In providing the information called for it will be necessary to explain what methods you propose to use, to show that you have or will obtain the personnel and facilities necessary to use these methods, and to offer enough information about costs to show that your estimates are realistic. Each of these points will be covered in more detail later in the discussion.

KINDS OF PROPOSALS

Proposals are infinitely varied. As indicated before, they range in size from one page to many volumes; in subject matter from suggestions on such minor matters as better lighting for the sake of employee morale to suggestions for joint American-Soviet space exploration. Some of the main types can be described, however, under the following headings: Research, Research and Development, Sales, and Planning.

Research

Proposals for pure research are usually written by workers at universities, science centers, or in some cases, research departments of major industrial concerns such as General Electric or Dupont. Most teachers in the sciences at the university level are involved in projects that receive funding through proposals written to some funding agency. Such teachers often are happy to show interested students examples of the proposals they have written that have been successful in getting grants. A great deal of the work on such grants may be performed by students, usually working on graduate degrees. Sometimes, however, eager undergraduates help out as well. In fact, in psychology, a great many of the projects use undergraduates as subjects. You may already have taken part in a research study without being aware of the process by which the researcher received permission to proceed.

Examples of the kinds of projects that fall under the heading of research abound, but a few that have involved student help include an analysis of animal life in an Iowa cavern, the simulation of movement of fluid in the small intestine, and a study of the growth of bacteria in sewage. A proposal for a grander project was first unveiled at a New York Academy of Sciences conference: Three scientists from the Lawrence Livermore Laboratories, which is funded in part by the United States government through the University of California, proposed harnessing "black holes" within the solar system to provide for future world-energy needs.

Research of these types is frequently funded by philanthropic foundations or by government agenices but much of it is funded through other channels as well. Most of these organizations have exacting formats for the proposals they solicit. For example, a National Science Foundation pamphlet, "Guide for Preparation of Proposals," lists four major sections: Narrative, Abstract, Budget, and Cover Sheet, with details about what should be included in each. Under *Narrative*, which contains the bulk of the explanation, are listed *Major Objectives of Project, Previous Experience, Project Description, Staff*, and *Selection of Participants*. It would appear that writing a proposal for projects of this type would be quite simple, but the problem of what to put in each section remains. In unsolicited proposals to these same agencies, the problem of what to include is even greater.

Research and Development

Usually referred to as R & D, research and development accounts for a great many of the proposals written in industry, as well as some written in universities. Proposals in this area usually have as a goal the actual development of a product or process of some kind. Examples include the development of a more effective crabgrass herbicide, the development of better seals for the Wankel engine, or the development of a preliminary treatment of waste from a food-processing plant so that it can be discharged into a municipal sewage system.

Proposals for R & D projects are frequently solicited by government departments such as the Department of Defense or the Department of Agriculture, or by industrial firms such as drug or chemical companies. Once again, when requests are made for competitive proposals, the form is usually specified for ease of comparison, but even if the format is clearly given, the problem of content remains.

Sales

Sale of products and services forms the backbone of world commerce. Decision-making officials throughout industry agree that the formal written sales proposal is the most important factor in the total sales effort. A sales proposal is an attempt to persuade a customer that a product or a service will fill a particular need. It frequently describes a product already in existence, but it must also convince the customer that the product will outperform any other product in a task yet to be done. Although sales proposals often include a great deal of what is called "boilerplate"—printed material ready for inclusion in any report—successful proposals are generally tailored to the particular customer's needs and specifications. The care taken in the tailoring is one indication to the customer that the company really wants the business.

In addition, sales proposals frequently offer to develop new products to fill a need for the customer. They may be made in response to inquiries or bid requests, or they may be submitted on its own initiative by an aggressive firm as a way of expanding business. Proposals of this type resemble the R & D proposal more than they resemble the traditional sales proposal.

Planning

Planning proposals are frequently written for governmental bodies or organizations that are in the process of change and are trying to determine the best changes to make. Examples might include plans for new school districts, a new division within an organization, or a new procedure in a business. Frequently a person or committee is chosen to study the situation and write up the results as a proposal. The organization then decides on the basis of the information contained in the proposal whether the changes will be advantageous.

CONTENTS

It might seem foolhardy to attempt to give a list of ingredients for such varied types of proposals as those described above, but despite their differences, all contain similar information and are governed by the same general rules.

The first rule concerns the audience. Perhaps more than any other form of technical writing, proposal writing is audience-oriented. A reader who is at all confused about the content of the proposal will be reluctant to approve the project. Consequently, almost all large organizations insist that the material be organized to their own specifications and aimed at their own situation. Generally, too, they want the presentation of each section to move from major points to details so that readers interested in different aspects of the proposal may all start at the same place but stop where they wish. To make this possible, sections of the proposal must be clearly identified, usually with headings, so that readers may quickly turn as they please from one section to another.

Another consideration that cannot be overstressed is language. Many students, particularly those in science, feel that they will never need to write for anyone but other experts, and those "experts will know what I'm talking about." As a U.S. Department of Commerce pamphlet on proposal writing points out, however, "The reviewer may be assumed to be knowledgeable. . . . [and] may also be assumed to be a very busy person who has little patience with verbosity or fuzzy language." The concern for language is even more important in business, where proposal readers generally want the same proposal to serve managing, engineering, purchasing, and operating staffs, and thus want as much of the proposal as possible written in plain, clear language that can be understood by people of different backgrounds. All proposal readers are suspicious that the person who cannot write out the idea in plain language will not be able to carry out the idea either. Basically, if the idea cannot be understood, it cannot be approved.

INFORMAL PROPOSALS

If you are now a student, you might wait years before writing all or part of a major formal proposal, but you will find that you will be writing informal proposals much sooner. In fact, the informal proposal is a good way to advance in a job. Most jobs that students get following college tend to be routine in nature. Although it is no longer the general practice it once was, some firms used to stockpile engineers—hire more than they needed because having large numbers of engineers available looked good when writing proposals for government contracts. These engineers, usually right out of college, would be set to work on drafting or other nonprofessional tasks simply to keep them busy. Those who did only these jobs tended to stay longer at lower levels than those who looked around and found something more important to do. It is still true in any field, not just engineering, that the person who looks around to find more work advances faster than the person who does only what is required.

Ordinarily, when you find a particular problem that you would like to work on, you bring it up by speaking to your immediate supervisor. If the supervisor finds it to be a task that requires time away from other work, or requires extra money or equipment beyond what is normal for your position, a memorandum will be requested to be sent to the person with the authority to provide the required time or material. Such a memorandum is an informal proposal.

Sometimes an informal proposal is a letter rather than a memorandum. This is usually the case when you are not an employee of the organization you would like to perform the task for.

Since the proposal is informal, the contents tend to vary more than those of formal proposals. In some cases, sections are labeled, in some cases not. Nevertheless, the following elements need to be included.

Problem

Somewhere close to the beginning, a statement of the problem should be presented. The first reason is to identify the subject matter of the proposal in detail. But beyond this, the situation is often one that no one else sees as a problem. If people had, they might have solved it. As a writer of an informal proposal, you must indicate that the situation is in fact a problem, and one that can be solved. You might do this by describing the *background* to the situation—that is, how it developed into a problem. Or you might describe the *significance* of the problem, showing that it is important enough to justify spending the time to solve it.

Proposed Solution

Obviously it is not enough simply to point out the problem, so the proposal must next offer a solution. If a clear solution cannot be given, then the writer concentrates on a *proposed method* toward finding a solution. In this section the writer can show that real thought has been given to the problem and that a memo was not just written "off the top of the head." In informal proposals, just as in formal, the reader wants to be sure that the writer is genuinely interested in doing the work, and not just suggesting one thing to get out of doing another.

Statement of Request

Near the end of the proposal the writer requests permission to perform *specific tasks* that should solve the problem or provide information that may lead to a solution. At any rate, it is necessary to indicate in a proposal exactly what the writer wishes to do, since it may not be evident from the description of the solution. When the case is put into the terms of a specific request, the recipient can respond easily and quickly and need not puzzle over what to do with it. Whenever the writer can make the reader's job easier, a positive response is more likely.

While informal proposals are often used on the job, they are also used as forerunners for formal proposals. When a company or organization has an idea that it thinks another company or organization might be interested in, it will often write an informal proposal to determine whether there is enough possibility of acceptance to warrant producing a formal proposal. Because of the great amounts of time and money required to produce formal proposals, some government agencies, such as the National Oceanic and Atmospheric Administration, recommend using the informal proposal first as a standard procedure. When a great deal of detail is included, the informal proposal is sometimes called a "preproposal." If the recipient is interested, a request is made for a full proposal on the subject, which the writer then proceeds to produce.

FORMAL PROPOSALS

When proposals are solicited, the sections for inclusion are frequently specified, but the problem of what information to include remains. The following discussion is intended as a guide to the kinds of information formal proposals usually include. Although those who solicit proposals do not all want the information to appear in the same order, the sections are discussed here in an order that is often used. Whenever appropriate, sections are condensed or combined. It is always better to have the shape of the material dictate the pattern rather than to try to force the material into some preconceived format. Nevertheless, knowing what form to put the information into often helps the writer understand what is needed. (Many of the sections have already been covered in the discussion of reports, which should be consulted as necessary for further details; but there are some details to add in connection with proposals.)

Front Matter

Letter of transmittal. A formal proposal is almost always accompanied by a letter or memorandum of transmittal. Usually the letter identifies the proposal, gives the highlights, and expresses interest in doing the work. It also will usually refer to the circumstances that gave rise to the proposal, such as the request for proposals or the letter or meeting at which permission was granted to submit it.

Cover sheet/title page. The cover sheet always identifies the work as a proposal, usually in large type. It also gives the title of the proposal; the name, formal title, and address of the person or organization to whom the proposal is submitted; and the identifying number, if there is one, that links the proposal to an announcement or bid request. Usually the date of submission is given, along with the date by which the work will be completed, and perhaps the date after which the proposal will no longer be in effect. When money is requested, the total sum is sometimes included on the cover sheet. Finally, the formal name, title, and address of the person or organization submitting the proposal are given.

Summary or abstract. The summary is a brief statement of the total offering. In commercial proposals, the summary indicates that the proposal meets all specifications or clearly states the exceptions. It also identifies the problem and briefly states the proposed solution. It should be no longer than three percent of the total length of the proposal.

Table of contents. Any proposal (or report) of more than five pages should probably have a table of contents.

Statement of request. Although the statement of request is sometimes combined with the summary, in larger proposals it is a separate section. In it, the writer states clearly who wishes to do the work, what the work is, and what the cost will be. Frequently this section contains only a few sentences, but it is often used by evaluators to help rank competing proposals.

Body

Statement of the problem.   In both solicited and unsolicited proposals the section defining or describing the problem is important, even though the writer may feel that the recipients understand their own problem. In solicited proposals, it shows that the writer, too, understands the problem and has the reader's concerns in mind when setting forth a solution. In unsolicited proposals the writer often needs to convince the reader that the problem exists. In either case, the writer might spend a paragraph to several pages defining the problem, its ramifications, its significance, and its relation to larger problems, issues, activities, or missions. Often, in the statement of the problem, the writer includes the material covered in this discussion under the heading *Background*.

Background.   The background, history, or, in proposals for pure research, the state of the art concerning the problem is given in this section. Usually the conditions leading up to the problem are described, indicating why the problem is now being considered. If previous attempts at solution have been made, they are described along with their results and shortcomings. A review of the literature on the subject is often given as well. What the writer needs to show is an understanding of the total context of the problem and an awareness of previous work in the area.

Scope.   The section on scope sets the limits within which the proposal writer will work. When contracts are involved, this is a vitally important section, since the writer does not wish to be held accountable for more than what has been agreed upon. Even if for no other purpose than preserving good will, a clear understanding of the limits of the work to be done is important. For instance, if you proposed to sample opinion on a stated subject, you would not want the reader to think you were going to sample every person affected when you planned to sample only 5 percent. In this section, then, you state exactly what you plan to accomplish, and often, what you do not. The scope section is usually separate from other sections because of its importance in protecting both the writer and the reader from misunderstandings.

Methodology.   The section explaining how the work is to be accomplished is perhaps the most important in any proposal. The reader wants to know that the methods used will in fact produce the results promised. It is not enough simply to find a solution; it has to be a solution that will withstand all kinds of attacks. When competing proposals with similar costs are compared, the one offering the more comprehensive methodology is often the one chosen. For these reasons, the methods used to solve the problem or do the job are always given in detail. When the methods are standard and are understood by both writer and reader, they need only be identified, but any prospective deviations from them are fully described. When the methods are innovative, they are described step by step, with reasons for each step included, in enough detail to convince the reader that they will work.

Facilities.   Usually the section describing facilities follows the section on methods, for it amplifies that section. It does little good to promise work in sterile conditions, for instance, if the proper laboratories are not available. Equipment to be used is frequently described in this section, although sometimes it is listed separately. Equipment might range from normal laboratory equipment to massive roadbuilding machinery, but it must clearly be capable of doing the job. If you promise to gather certain information from published sources, for example, you would do well to indicate that the library you plan to use is likely to contain the information you will need. In major proposals, one of the further reasons for including this section is the description it contains of what the contractor will be getting for the overhead charges, which often range from 40 to 60 percent above the cost of actually doing the work.

Personnel.   The people who will be doing the work are listed in the personnel section. When they are known, their names are given, and frequently a page or two is included giving a biography of each person complete with pertinent publications. The purpose, of course, is to convince the reader that the people can do the work proposed. When the personnel are not yet hired, the grade or degree of competency of those to be hired is described. Part-time personnel are listed, and the percentage of time they will be devoting to the project is indicated. When major personnel will be working only part-time on the project, their other commitments are listed as well.

Advantages and disadvantages.   Often the section listing advantages and disadvantages of a proposal is omitted because it is thought that the first is obvious and the second ought not be pointed out. Both reasons are short-sighted. This section can sum up a proposal in an honest manner that goes a long way toward convincing the reader to grant permission to proceed. Listing all the advantages in one place lets the writer summarize the total offering. Such a list might include the products that will result, possible side benefits, improved methods of production, savings, or simply the knowledge that will be accumulated. The argument for listing the disadvantages of a proposal is even stronger. As writer, you want to have the current proposal accepted, but of more importance is the proposal after the current one. You do not want the reader to accept something that will turn out badly and result in suspicion and ill feeling. Dissatisfied customers not only refuse to give repeat business to the source of dissatisfaction, they also warn others of their experience. Nothing spreads through a field faster than a reputation for questionable practices. Therefore, in plain, clear language, the disadvantages of the proposal should be mentioned.

Duration. This section tells how long it will take to complete the project. If the project is divided into segments, then firm schedules for completing each are given. Delivery dates are also listed where appropriate.

Costs. The section on the costs of the proposed project is crucial. Sometimes the writer concludes with a statement of costs in hopes that after going through everything else, the reader will be convinced that the expense is justified; sometimes costs are detailed in a section separate from the proposal so that they will not influence other deliberations; sometimes they are presented first on a special budget sheet. In any case, all costs should be itemized under headings such as salaries, capital equipment, expendable equipment, miscellany, and overhead. Often only estimates are possible, but they obviously should be made with the greatest care. In industry, at least, expensive cost overruns are rarely tolerated.

Reports. Finally, there should be a section detailing what reports will be written, giving dates for progress and final reports. Often included, too, are the names or titles of the people responsible for the reporting, and the form that the reports will take.

EVALUATION OF PROPOSALS

It might seem that in a chapter on writing proposals, a section on evaluation would be inappropriate, but generally writers produce better products if they understand how their work will be judged. The following set of questions was devised by an agency within the Department of Health, Education, and Welfare to aid readers in preliminary review of proposals. By giving a scale of values to the questions, the reader can make a point comparison between competing proposals. Although the questions were intended for particular kinds of research and development proposals, they apply to proposals of all kinds. Writers would do well to ask themselves the questions before submitting a proposal. If any question receives an inadequate response, the section concerned should be restructured.

1. Understanding of purposes, objectives and tasks—30 points.
   a. Does the bidder demonstrate clear understanding and acceptance of the requirements presented in the RFP (request for proposal)?
   b. Are the tasks outlined in the proposal clear and well-defined?
   c. Are there important omissions in the specified tasks?
   d. On a scale of 30 points please assign a rating to this area.
2. Technical quality of methods proposed—30 points.
   a. Are sufficient time and resources specified to accomplish the quality outlined in the proposal?

b. Does the proposer emphasize quality as an important criterion when presenting methods?

c. Will the quality of the proposed methods be monitored throughout the contract period?

d. Is this monitoring sufficient to insure quality?

e. On a scale of 30 points please assign a rating to this area.

3. Quality of management plan and planning—10 points.

a. Has a management plan been designed to insure receipt of materials at certain specified times?

b. Does the proposal clearly identify working relationships within the contractor staff and with this agency's staff?

c. Is sufficient technical management assigned to the task to insure production and quality of output?

d. Will information be available to this agency sufficient to permit analysis of cost and effectiveness?

e. On a scale of 10 points please assign a rating to this area.

4. Qualification of staff—20 points.

a. Have individuals assigned to the task had prior experience in the required technical areas?

b. Have key personnel been assigned to the project for large enough percentages of time?

c. Have project directors and those assigned management roles been experienced in similar management positions?

d. Is there sufficient depth in the staff to provide backup and overload capabilities?

e. On a scale of 20 points please assign a rating to this area.

5. Corporate capability and experience—10 points.

a. Has the organization had previous experience in planning and managing efforts of this type?

b. Has the organization previously managed efforts of the size and complexity of this agency's?

c. Is the organization of sufficient size and stability to undertake the responsibility called for?

d. Is there any "track record" of performance available, indicating consistent meeting of schedules, with quality output, within fiscal limits, or the inverse?

e. On a scale of 10 points please assign a rating to this area.

Although only one of the preceding questions deals directly with expenses, costs are a primary consideration. In commercial proposals, the first consideration is cost, followed in order by performance, reliability, economy of operation, and early delivery of the product.

Beyond these considerations are many others, but one of the most important is the tone of the proposal. Those addressed are extremely concerned that the responsible officials of the proposing organization are genuinely interested in doing the work and are committed to providing complete satisfaction, even beyond normal guarantees. In other words, reviewers are likely to pass up a proposal if they feel that they would have to put up with sharp dealing—or with indifference.

Recently, the National Institutes of Health, which award close to 4 billion dollars each year in grants, released a study of proposals they had rejected. The proposals were unsuccessful most often because of: 1) questionable project design, 2) inadequate explanation of the research, and 3) lack of experience of the investigator. Other major reasons included: vague experimental purpose, poorly prepared applications, unsuitable methods, problems dealt with inadequately, poor knowledge of the literature, and "The results will be confusing, difficult to interpret, or meaningless." The important point to note is: Most of these reasons derive from the presentation of the material, that is, from how the proposals were written, rather than from the nature of the research. In other words, if the investigators had prepared their proposals more carefully, they might have been successful, and in the world of research and grantsmanship, as well as in business, a successful proposal often means the difference between working and looking for another job.

As a final word on evaluation, all proposal writers should heed the warning given by the National Science Foundation concerning a major reason for rejection of proposals:

It cannot be emphasized too strongly that the reviewing panels must base their recommendations largely on the information contained in the proposal. The most frequent fault of proposals is vagueness.

(Illustrative specimens follow.)

Specimen No. 1*—Research Proposal

The proposal shown here is the kind produced regularly by students and faculty at universities and research institutes throughout the world. Besides the material shown, the original proposal included a letter of transmittal, a cover page, a resumé, and a list of publications, all of which were omitted to save space because examples may be found elsewhere in the text. Although the proposal is quite short, it is complete and was successful in obtaining the funds requested. On the West Coast, cattle and horses succumb to the effects of eating tansy ragwort, but sheep apparently are not affected. The proposal suggests a study to determine the level of tansy needed to create a toxic effect in sheep, and to discover if the apparent immunity derives from the kind of stomach sheep have, or from some attribute of the liver. The "Summary" describes the study and makes the specific request.

Under "Purpose and Description of the Project," the background and methodology are spelled out. The first part of the study is not described in detail because the methods have been adequately related elsewhere and can be found in the cited references. The second and third parts, however, being original to this study, are given in enough detail to be convincing. Implied in the proposal, but not spelled out, is the hope that once the area of protection is determined, studies could be implemented to discover the mechanism, which perhaps could be utilized in some way to aid cattle and horses.

The language throughout is formal and technical. The recipients are expected to understand such terms as *histopathological* and *bolus*, yet an understanding of the intent of the proposal does not require such an understanding. The "Budget" is detailed, but again, technical terms are not explained. "Other payroll expenses," for instance, includes required retirement benefits. and "indirect costs" is a term for overhead.

*Courtesy of A. Morrie Craig, Corvallis, Oregon.

A PROPOSAL FOR RESEARCH

DETERMINATION OF DOSE/RESPONSE LEVELS
OF PYRROLIZIDINE ALKALOID IN SHEEP

Submitted to the Oregon Sheep Commission

by A. Morrie Craig, Ph.D.
Assistant Professor, Veterinary Medicine
School of Veterinary Medicine
Oregon State University

Institution
Oregon State University, School of Veterinary Medicine

Summary
The Oregon State University School of Veterinary Medicine proposes to conduct a series of experiments over twelve months concerning tansy ragwort toxicity in sheep. These initial studies are designed to produce a toxic effect in sheep by giving crystalline alkaloid and to study why sheep under natural conditions are seemingly unaffected by tansy ragwort. We would like to request partial funds of $5450 to enable us to accomplish a specific phase of needed research.

Purpose and Description of Project
Recent studies have shown that sheep can consume approximately 200 percent of their body weight of tansy ragwort and do not show effects of liver toxicity. Cattle or horses, however, that consume approximately 15 percent of their body weight succumb to liver cirrhosis that develops as a result of consumption of tansy ragwort. Two dominant questions resulting from these studies are as follows: (1) What is the toxic level for sheep, and (2) Why are sheep seemingly unsusceptible to the toxicity of the tansy ragwort plant?

Senecio jacobaea (tansy ragwort) is predominant in Northern California, Oregon and Washington. The pyrrolizidine alkaloids of S. jacobaea are responsible for the tissue damage to an ever increasing number of livestock. The pyrrolizidine alkaloids that are found in S. jacobaea are composed of six different molecular species. These alkaloids are structurally very similar and have similar mechanisms of action.

A. Morrie Craig
To - Oregon Sheep Commission
Proposal - Determination of Dose/Response ...
Page 2

During the coming year we propose a three-part study:

(1) A crude mixture of the six alkaloids will be isolated from S. jacobaea by Campbell's technique[1], as later improved by Deagan[2] and Segal[3,4];

(2) This crude mixture of alkaloid will be resolubilized and fed by bolus to several groups of sheep at various concentrations. These concentrations would both imitate and be greater than the levels of alkaloids in the feeding experiment recently completed at Oregon State University. Liver enzymes will be analyzed every two weeks. Gross and histopathological studies will be performed on all sheep at the conclusion of the experiments. These experiments would determine a dose/response curve for sheep under laboratory conditions. The dose to cause liver destruction would be established and could be related back to field consumption.

(3) Part 3 is scheduled for the second six months. The resolubilized alkaloid at concentrations determined in phase 2 will be infused into the jugular vein of sheep. This infusion isolates the rumen from the liver which will help separate detoxification action in these organs. If sheep are not susceptible to rumenal or intravenous innoculation of alkaloid, it is implied that the liver is the detoxifying organ in the sheep. Again, liver enzymes will be analyzed on the sheep. Gross and histopathological studies will be performed at the conclusion of the experiments.

These experiments are designed to distinguish both between effects in the rumen and effects in the liver and to determine toxic levels. Results would indicate whether protection from tansy toxicity occurs in the rumen of the liver of a sheep. This would aid investigators in determining that sheep are, in fact, not susceptible to pyrrolizidine alkaloid toxicosis.

Animals used in this experiment will be kept in accordance with the "Guide for the Care and Use of Laboratory Animals", DHEW Publication No. (NIH) 74-23.

A. Morrie Craig
To - Oregon Sheep Commission
Proposal - Determination of Dose/Response . . .
Page 3

BUDGET

Enzyme kits	2,360	
Cost of 40 sheep	1,200	
Supplies	1,890	
Subtotal		5,450 (funds requested)
Feed and housing for sheep/$0.90 per day	2,128	
Technical support	5,298	
Other payroll expenses (OPE)	1,007	
Pathology and histopathology services	1,000	
Direct cost total	14,883	
Indirect costs (34%)	5,060	
Total funds requested		5,450
TOTAL	19,943	

Note:

Funding/support provided by Oregon State University, School of Veterinary Medicine includes:

- principal investigator's time;

- routine laboratory supplies and equipment.

References

1. J. G. Campbell. Proceedings of the Royal Society Edin, 66, 103–135, 1955.
2. J. T. Deagan and M. L. Deinzer. Lloyatta, 14, 395–397, 1974.
3. H. J. Segal. Toxicology Letters, 1, 279–284, 1978.
4. C. W. Qualls and H. J. Segal. J. Chromatography. In press.

Specimen No. 2*—Research and
Development Preproposal

Full-scale formal proposals are often so long that using one as a specimen would be impractical, but the following preproposal illustrates the manner of their construction. The additional material in a full, formal proposal on the same subject would consist mainly of much more detail and a more extensive analysis of costs.

The page opposite to this comment was printed on heavy paper and served both as a cover and as a title page. It shows to whom the proposal is submitted, and it names the three bodies that would cooperate in the project—assuming that a full proposal was encouraged.

The title page is followed by the letter of transmittal, which orients the reviewers to the project, provides evidence of the care with which the material submitted had been prepared, and makes the actual request. Note that even at this early point attention is directed to the fact that the value of the project could extend far beyond the benefit to the particular location involved.

*Courtesy of William A. Smith, City Engineer, Moscow, Idaho.

A Preproposal To The
U.S. ENVIRONMENTAL PROTECTION AGENCY
For

A RESEARCH & DEMONSTRATION STUDY ON THE REMOVAL OF IRON & MANGANESE FROM POTABLE WATER SUPPLIES USING OZONE

SUBMITTED BY
CITY OF MOSCOW, IDAHO

IN COOPERATION WITH
UNIVERSITY OF IDAHO, MOSCOW, IDAHO

AND
STEVENS, THOMPSON & RUNYAN, INC. BOISE, IDAHO

CITY OF MOSCOW

MOSCOW, IDAHO
83843

122 East Fourth

Telephone: 882-5553—Area Code: 208

October 31, 1972

Dr. Gordon G. Roebeck, Director
Water Supply Research Laboratory
Taft Environmental Research Center
Environmental Protection Agency
Cincinnati, Ohio 45268

Dear Dr. Roebeck:

Enclosed for your consideration are copies of a preproposal for a research and demonstration study on the removal of iron and manganese
from potable water supplies using ozone. Professor Calvin C. Warnick,
Director of the Idaho Water Resources Research Institute, talked with
you last summer with regard to the possibility of EPA participating in
the proposed water treatment system.

Reviewers are respectfully reminded that this document is a preproposal and therefore does not contain detailed information on equipment,
budgeting, study programming, manpower requirements, etc. The intention would be to include such detail in the formal proposal and grant
application.

However, the preproposal is based upon a comprehensive literature review. Particular emphasis is directed to the reference by F. L. Evans,
"Ozone in Water and Wastewater Treatment." This recent publication
is a compendium of the published literature on ozone, and offers a thorough review and report of previous work with ozone.

As demonstrated herein, the City of Moscow is committed to improving
the water quality of the City and will commit a substantial amount of the

Dr. Roebeck   -2-   October 31, 1972

City's resources to the proposed research and demonstration plan. We
hope that the Environmental Protection Agency will view this study as
being of significance to water supply systems throughout the nation and
will assist in funding of the project.

Sincerely yours,

U OF I CHEMICAL ENGINEERING DEPT.

Robert R. Furgason, P. E.
Professor and Chairman

CITY OF MOSCOW

W. A. Smith
Director of Public Works

STEVENS, THOMPSON & RUNYAN, Inc.

Richard O. Day, P. E.
Boise Manager

RRF/WAS/ROD:sm

Encl.

The preproposal proper, as can be seen from the table of contents, follows a pattern in which, first, the Ozone Water Treatment System is explained. (Two figures are provided to make the general nature of the process easier to grasp.) Then, the development of such a system is justified. Next, the capabilities of the cooperating agencies are established.

This pattern provided answers, in a logical order, to the questions most likely to arise in the minds of the reviewers.

CONTENTS

FIGURES

PROPOSED WATER TREATMENT SYSTEM

PROJECT OBJECTIVES

The purpose of the proposed water treatment demonstration program is to investigate the feasibility of using ozone to remove iron and manganese from potable water supplies. Results from the operation of the ozone treatment demonstration will be used to establish the practicality and cost effectiveness of ozone treatment for removal of iron and manganese from domestic water supplies. Specific objectives are to:

1. Determine in a pilot plant operation the optimum operating conditions and design criteria for ozone oxidation of iron and manganese, using the water from wells No. 2 and No. 3 of the Moscow, Idaho, water system. Water from these wells typically analyzes 2 - 10 mg/l iron and 0.1 - 2.0 mg/l of manganese. Pilot plant control variables will include ozone dosage levels, residence time, pH and ozone injection method.

2. Determine the optimum filtration rates possible for satisfactory removal of the iron and manganese precipitates.

3. Utilize the design criteria and optimum operating conditions derived from the pilot operation to design, construct and operate a full-size demonstration plant of approximately 3.7 million gallons per day capacity.

4. Establish actual construction and operating cost data from the full-scale demonstration plant in order to make an accurate economic comparison of ozone treatment to other accepted iron and manganese removal methods.

PROPOSED TREATMENT SYSTEM

Features of the proposed treatment system are as follows:

1. Full-scale, pressurized filters utilizing chlorine-potassium permanganate oxidation will be initially installed on the combined discharge from City wells No. 2 and No. 3. This oxidation process is temporary and is intended to remove only the iron, not the manganese.[9][10]

2. An ozone pilot unit will be constructed simultaneously in parallel with the above full-scale filtration system. The purpose of the pilot plant is to obtain accurate scale-up design data for the full-sized demonstration plant. Wherever possible, the pilot plant equipment will be designed to be incorporated in the final plant equipment, for example, the ozonator.

1

3. Upon successful completion of the pilot studies, the full-scale ozone oxidation equipment will be installed replacing the chlorine-potassium permanganate oxidation facilities, thus providing complete removal of the iron and manganese. At this point the full-scale demonstration plant will be complete so that accurate operating and cost data can be obtained and documented.

The above operational plan facilitates iron removal from the water supply during the initial development of the demonstration system. This is desirable from the City's standpoint because it removes most of the aesthetically objectionable material as soon as possible with only a minor amount of additional cost involved. Figures 1 and 2 illustrate the proposed system operation.

FINANCIAL OBLIGATIONS

It is requested that the Environmental Protection Agency fund one-half of the cost of the total project by supporting the full cost of the ozone research-demonstration facilities. Thus the EPA grant would include: (1) a pilot plant designed for a capacity in the range of 20 - 200 gal/min consisting of an ozonator, gas-liquid contactor and a precipitate removal system; and (2) ozone demonstration equipment consisting of full-size ozonators and gas-liquid contact system.

The City of Moscow will provide the pressurized filtration equipment, the chlorine-potassium permanganate feed system, personnel, building and other In-Kind services.

Costs are estimated as follows:

City of Moscow

Pressurized filter system and the $KMnO_4$ - Cl_2 system	$200,000
In-Kind services	25,000

EPA

Ozone research-demonstration system including O_3 pilot plant ($75,000)	225,000
TOTAL COST	$450,000

2

Figure 1. Treatment Process Phase I

3

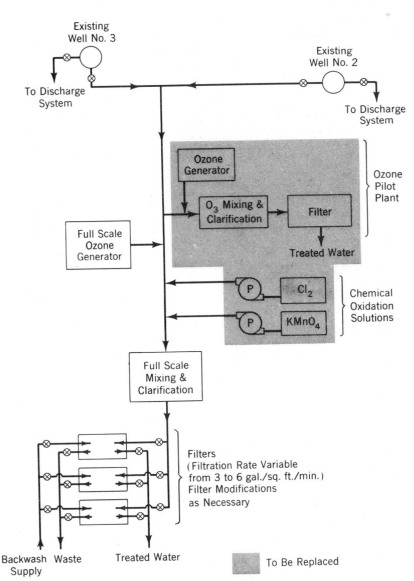

Figure 2. Treatment Process Phase II

The nature of the proposal having been explained, the reviewers are prepared to evaluate its justification.

The first part of this section covers the widespread need for such a project and then tells why Moscow, Idaho, would be a desirable location for it. Thereafter, it provides more information about the Ozone Treatment—concentrating attention, at this point, not on the technique but on what the treatment should be able to accomplish, its use elsewhere (primarily in Europe, where it is common), the reasons that it has not been extensively explored in the United States, and the reasons that experience with it elsewhere does not cover the needs for the work proposed on this occasion.

PROJECT JUSTIFICATION

PROJECT NEED

The problem of iron and manganese removal from domestic water supplies is not unique to Moscow, Idaho. A 1970 study by the Bureau of Water Hygiene, U.S. Public Health Service[8] analyzed community water supplies across the nation from Vermont to California. Nearly 1000 water supply systems were involved. This study revealed that iron and manganese exceeded the Drinking Water Standards (DWS) for the delivered water in 159 water systems. No other constituent exceeded the DWS more times than iron and manganese. Of the communities tested 32% of the groundwater supply systems had iron and manganese in excess of the DWS recommended limits for delivered water.

The Bureau of Water Hygiene report found that the smaller water systems had more difficulty in meeting water quality standards than did the systems serving larger populations. The recommendations of the report contain the following statement pertaining to iron, manganese, color, turbidity, etc.

> *"Further research is necessary to understand what causes these problems as well as how to more effectively remove objectionable constituents."*

Two other recommendations of the Community Water Supply Study are relevant to this proposal. They are:

> *"The presence of coliform organisms at a level above the DWS limit in 120 systems indicates that better source protection and/or disinfection should be instituted as soon as possible,"*

and

> *"Inasmuch as tastes and odors remain a common complaint in public water supplies, more effort should be made to employ the present technology of oxidation and adsorption to control or remove these troublesome constituents. Further research is needed to reduce cost of such treatment to make it more practical and acceptable."*

The Moscow, Idaho, water system is an ideal supply for a demonstration study of ozone technology applied to water treatment: first, the system is intermediate in size serving a current population of about 10,000; and second, the test system can be built and run without affecting the normal operation of the basic water supply.

The City of Moscow uses two main groundwater aquifers as sources for its domestic water supply. One aquifer serves as the source for two deep wells, No. 6 and No. 8 (approximately 1300 ft.), both of which are relatively free of iron and manganese. These two wells produce

5

an average of 3.2 million gallons per day and can satisfy normal water demands during fall, winter and spring periods. However, during the heavy demands of the summer months, water must be drawn from wells No. 2 and No. 3, which are shallow wells (approximately 500 ft.) and are characterized by a very high iron and manganese content. Thus during the construction period of the demonstration plant, the normal water supply will not be disturbed and any unforeseen difficulties can be dealt with while the normal system is isolated from the proposed water treatment plant since the ozone treatment is needed only for wells No. 2 and No. 3.

Although the poor quality water from wells No. 2 and No. 3 is currently being used only for peaking purposes, it is projected that the population growth by 1975 will require these wells to be used on a regular basis.[10] Thus if a research demonstration project is to be done with a minimum amount of system disruption, the time to do it is now.

The City of Moscow has investigated alternate sources of water. One apparent solution to this problem would be the development of new deep wells which would draw water from the same aquifer as wells No. 6 and No. 8, providing the same high-quality water. However, the expense of developing deep wells requires an excessive capital outlay.[10]

Previous investigations first examined the feasibility of a surface supply. In 1970, Stevens, Thompson & Runyan, Inc., completed a study[11] for the development of a joint surface supply for the City of Moscow, the City of Pullman, the University of Idaho and Washington State University under an intergovernmental, bi-state agreement. The study concluded that development of a surface supply was feasible, provided conjunctive use of all existing wells (including wells No. 2 and No. 3) was a continuing source of supply.

Since the existing wells would still be necessary even if a surface supply were developed, it became apparent that consideration should be given to the concept of water treatment in lieu of development of new deep wells. This concept would consider the feasibility of treating the water from wells No. 2 and No. 3 to provide a suitable quality of water, and compare the expense of such treatment to the cost of developing a comparable supply through construction of new deep wells.

Accordingly, the City had preliminary pilot tests performed by Keystone Engineering & Products Co., Inc., in 1971.[9] These tests were limited to chemical treatment and filtration of water from the two wells. Generally, the results showed that this process would require uneconomic chemical costs to achieve complete removal of the iron and manganese. Keystone recommended that an ion-exchange process be tested as an alternate method of treatment.

In November of 1971, the Mayor and City Council authorized Stevens, Thompson & Runyan, Inc., to investigate feasible methods of water treatment to remove the iron and manganese from wells No. 2 and No. 3.[10] This report found ozone treatment to be one of two feasible methods of mineral removal from the sources involved. The intergovernmental group that sponsored the earlier surface water supply study, the Pullman-Moscow Water

Resource Committee (PMWRC), took action in August 1972 in support of a planned program to encourage the construction of a water treatment system that would remove the iron and manganese from the groundwaters. Professor Warnick, chairman of this intergovernmental group, made contact with Dr. Roebeck of the Taft Engineering Center, EPA, to seek advice and counsel on possible support of a research and demonstration grant.

The University of Idaho, located in Moscow, Idaho, maintains its own water system. The University's water is supplied by deep and shallow wells reflecting the same problems as those of the City. The results of the proposed demonstration plant will substantially influence the type of water treatment installation that the University would install to improve its own supply.

OZONE TREATMENT

The impetus to use ozone as a treatment method originates from the ability of this material to solve simultaneously several of the problems associated with potable water supplies. Ozone is a very powerful oxidant second only to fluorine. Due to its strong oxidizing power it is known to be able to deodorize, decolorize and sterilize water supplies in addition to oxidizing the iron and manganese.[5] If properly handled, it has few adverse environmental side effects since it decomposes readily into oxygen. The proposed treatment plant can be designed so that no free ozone can escape into the atmosphere.

The current use of ozone for water supply treatment is concentrated in Europe and to a limited degree in Canada. There are a number of operating systems that use ozone as a disinfectant and as a polishing treatment for odor and color removal. The only known system which utilizes ozone principally for iron and manganese removal is in Duesseldorf, Germany, with a capacity of 38.5 million gallons per day.[7]

There are several reasons why ozone has not been widely used in the United States. With the development of large and relatively inexpensive sources of chlorine, ozone at over twice the cost could not compete economically with chlorine as a general disinfectant. Also, chlorine escapes from a system quite slowly, thus providing a disinfectant residual that is carried throughout the water system. Ozone, on the other hand, has a half-life of only 20 minutes in water and its decomposition to oxygen is catalyzed by many metals such as iron. As a result, many regulations require a chlorine residual to be present in the water system which precludes the use of ozone. However, ozone is a more powerful disinfectant than chlorine and the presence of a chlorine residual is not necessarily good when organic materials are present. Also, the cost of ozone has decreased substantially in recent years.

Nonetheless, ozone treatment remains as a very viable alternative to other methods for iron and manganese removal and as an agent for eliminating odors, color and microorganisms. The fact remains there is little published in the literature on the design and operating characteristics of a treatment plant for iron and manganese removal using ozone. The available information on the Duesseldorf plant[7] which is many times larger than the proposed system for Moscow, Idaho, does not duplicate the effort herein proposed. Thus the objective of the research-demonstration plant is to generate usable data to evaluate the economic feasibility of such a treatment system and to make this information available to small municipalities across the country who share in the problem.

7

Having explained the system and justified the request, the preproposal takes up the question of whether the three cooperating agencies involved are capable of handling the job. It explains the responsibilities each would accept, providing a figure to illustrate the organizational structure. It presents a copy of the official resolution authorizing the request to show the firm commitment of the municipality to carry the project forward. It covers the experience in similar work of the university department involved, and its available facilities. It provides evidence of the qualifications of the engineering corporation that would cooperate in the work.

This material is followed by a list of references showing familiarity with the background of previous work in the field (one such reference had been mentioned earlier as particularly significant) as well as listing literature written by personnel who would participate.

CAPABILITY

The proposed organizational structure for the demonstration process is shown on Figure 3. The City of Moscow will act as grant applicant and administrator. The actual demonstration process will be a cooperative effort between the City of Moscow, the University of Idaho's Chemical Engineering Department and Stevens, Thompson & Runyan, Inc. Performance of the individual work elements to be conducted by the City, the University and the Consultant are shown on Figure 3. Division of the work elements is based on using the individual strengths of each of the three team members.

Summaries of the capabilities of the City, University and Consultant, and of key personnel who will work on this project also are included.

CITY OF MOSCOW, IDAHO

Public Works Department

The City is managed by a Mayor-Council form of government. Departmental supervision is administered with a full-time Administrative Assistant and Director of Public Works/City Engineer. Subdivision within the Public Works Department includes a Department of Sanitation, Department of Water, Department of Streets, Department of Building Inspection and a Department of Engineering. The coordinator of the Departments of Sanitation, Water, and Streets is the Superintendent of Public Works, who was recently promoted to that position from Water Department Superintendent. Each department operates at the direction of a superintendent. Currently there are two foremen and seven laborers working under the Water Department Superintendent, who perform the daily tasks of maintaining and improving the City water and sewer collection systems. The existing crew can maintain the proposed project equipment within the scope of its present duties.

Wells No. 2 and No. 3 are located adjacent to each other and are enclosed in a masonry structure which serves as storage, office and maintenance area for the Water Department. In a preliminary review of this proposed project, the building and site were found to be more than adequate for the proposed treatment and filtration facilities. The Water Department has a full complement of maintenance tools and equipment which would be available to the project, should it be funded.

The City Council, in their meeting of October 16, 1972, passed a resolution authorizing the City Engineer to pursue EPA participation in a research and demonstration grant (see following resolution).

City Financing

Should this grant be awarded, the City is committed to funding their proportionate share of the project as previously outlined in this pre-proposal. Currently the City is retiring $309,000 in both long-term revenue and general obligation bonds. The assessed evaluation

8

for taxing purposes is approximately $11,750,000 and the current assessment is at the rate of 32 mills. Consequently, bonding for this proposed project is fully within the City's capabilities. In addition, the current budget (1972) has set aside $103,000 toward capital improvement of the water collection and distribution system. Accrued water revenue will allow an increase in this figure in the 1973 budget.

The City is in a sound posture to accept the responsibility of funding, coordinating and administrating the contracts for the proposed facility.

9

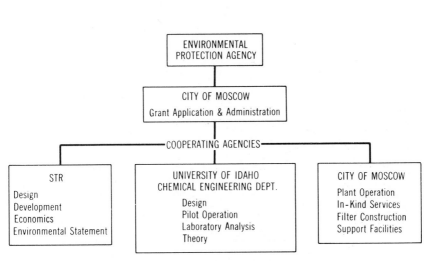

Figure 3. Organizational Structure

CITY OF MOSCOW

RESOLUTION

WHEREAS, the waters from City wells No. 2 and No. 3 contain quantities of iron and manganese which exceed current drinking water standards, and

WHEREAS, acceptable quality of water from these wells is necessary to meet future demands of the City of Moscow, and

WHEREAS, treatment of such waters to remove the iron and manganese utilizing ozone has been investigated and found to be possible, and

WHEREAS, ozone treatment for mineral removal is unique and possibly eligible for federal assistance,

NOW, THEREFORE, BE IT RESOLVED, that the City of Moscow make application to appropriate federal agencies for grant assistance for a Research and Demonstration Study on the Removal of Iron and Manganese from Domestic Water Sources using Ozone, and

BE IT FURTHER RESOLVED, that W.A. Smith, Director of Public Works, be authorized to act on behalf of the Mayor and City Council in making such application, and

BE IT FURTHER RESOLVED, that the City of Moscow participate in the demonstration process as a cooperating agency in conjunction with the University of Idaho and Stevens, Thompson & Runyan, Inc.

APPROVED this 16th day of October, 1972.

Larry Merk _____
Mayor

Councilmen:

Glenn Utzman _____
Larry Kirkland _____
Paul Mann _____
C. H. Bond _____
Cliff Lathen _____
George R. Russell _____
ATTEST:
Marvin Kimberling _____
City Clerk

11

UNIVERSITY OF IDAHO

Ozone Research

The Chemical Engineering Department at the University of Idaho has been engaging in research involving ozone for over 10 years. Projects utilizing ozone for pollution control work particularly in the pulp and paper industry are underway as well as studies that have involved the physical properties of ozone.[1][2][3][4][6] Dr. Furgason and Dr. Edwards have participated as consultants in a number of projects when ozone was utilized including odor control for a blood drying plant, purification of brewery water, waste water treatment from a chemical plant and ozone bleaching of paper pulp. Of particular significance from an equipment standpoint is the ozonation facilities currently available in the Chemical Engineering Department. A portable ozone test unit was developed and built by the Department through a grant from the Water Resources Research Institute. The portable unit contains a small ozonator, a stirred tank reaction system, pH level and temperature control, an ozone injection system, feed and recirculation pumps and other peripheral equipment. The test unit has been used at several locations throughout the West to examine the feasibility of ozone treatment of various materials. It was used on the Moscow water system and did demonstrate that ozone was capable of removing the iron and manganese (see Appendix). It is not, however, large enough to be used for design and scale-up purposes.

The Department also has a large 51-tube ozonator capable of producing 20 lbs/day of ozone which is currently used in several research projects. The ozone equipment of the Department are the only ozone generation facilities available at the University of Idaho or at Washington State University (nine miles away). Thus these facilities serve a regional need and are used by researchers on both campuses.

In the proposed research-demonstration program, the expertise of the University faculty primarily involving Dr. Furgason and Dr. Edwards would be an integral part of the project. Analyses of the plant's operation would be incorporated into the graduate program of the institution, thus providing for extensive study of the system. Publications of the results would appear when appropriate.

Chemical Engineering Department Facilities

The Chemical Engineering Department occupies extensive facilities in the relatively new $2,500,000 Buchanan Engineering Building on the University of Idaho campus. Approximately one-half of the laboratory building is available for Chemical Engineering activities, including 19 specialized laboratories (250 - 750 ft.^{2}) for research work, two large 3500 ft.^{2} and 2500 ft.^{2} general purpose laboratories, an analytical laboratory, a well-equipped shop and a full complement of study and office areas.

12

STEVENS, THOMPSON & RUNYAN, INC.

Stevens, Thompson & Runyan, Inc., was formed in 1920 as a partnership by J.C. Stevens and R.E. Koon under the name of Stevens & Koon. Upon the retirement of Mr. Koon in 1951, the name was changed to Stevens & Thompson. Mr. Stevens retired in 1954 and Mr. Runyan became a partner in 1955. To assure better continuity of service to clients, the firm was incorporated in 1961. STR is an independent consulting firm in which all the owners are active managing officers. It has no affiliation with any manufacturing, construction or other commercial interests.

The initial assignments of the firm were in the public works field, but through the years the firm has had a steady growth in personnel, responsibility and type of projects. Today the staff of approximately 200 is engaged in a variety of projects including water and sewerage planning and design, industrial waste treatment systems, urban planning, architecture, highway and airport planning and design. In addition, the firm participates in several joint ventures with other consulting firms.

STR has direct access to GE-235 and GE-635 computers as well as the IBM 360. Sending and receiving consoles are located in both the Portland and Seattle offices. This readily available tool with its numerous programs for solving engineering problems supplements other computer services utilized by Stevens, Thompson & Runyan, Inc.

Sanitary engineering laboratory facilities maintained in the Portland office are equipped for routine water and sewage analyses in conjunction with design and research projects conducted by the firm. Much of the equipment is portable to permit setting up field laboratories and pilot plant studies.

All projects undertaken by STR are under the direct management of one of the officers of the firm. Also, regardless of the size of the project, STR utilizes a team approach in which all the disciplines represented on our staff are available and responsible to the officer in charge. For a large, multi-facet project several teams may be established, each of which is responsible to the officer in charge. Since the credentials of a consulting firm are based equally on the experience and reputation of the firm and the capabilities of its personnel, STR has developed a professional staff of specialists who contribute a wide range of engineering and planning services to its clients.

STR has considerable experience in the field of municipal water supply and distribution. Recent experience includes the design of municipal water treatment facilities at:

Eugene, Oregon	100 mgd
Bellingham, Washington	24 mgd
Anacortes, Washington	24 mgd
Oregon City, Oregon	10 mgd
Anchorage, Alaska	8 mgd
Fairbanks, Alaska	4 mgd
Orofino, Idaho	2 mgd
Castle Rock, Washington	1 mgd

13

STR has provided engineering services to the City of Moscow for many years. The firm has completed the following projects for the City's water system:

Preliminary Treatment Report	1953
Distribution Analysis	1962
Reservoir, Pump Station	
Control System Design	1968
Water Supply Study[11]	1970
Water Treatment Analysis[10]	1972

Consequently, the personnel at STR are thoroughly familiar with the City's water system and its inherent problems.

REFERENCES

(1) Edwards, L.L. "Heat Transfer in Thermally Decomposing Ozone," Ph.D. Thesis, University of Idaho, 1966.

(2) Edwards, L.L. and Furgason, R.R. "Heat Transfer in the Thermally Decomposing Ozone System: Theoretical Investigation," CEP Symp. Ser. 61, 229, (1965).

(3) Edwards, L.L. and Furgason, R.R. "Heat Transfer in Thermally Decomposing Ozone: Experimental Investigation," IEC Fundamentals, 7, 440 (1968).

(4) Edwards, L.L. and Strecker, S. "Operation of Welsbach C-1-D Ozonator at Low Temperatures," Report to Welsbach Corporation, Philadelphia, Pennsylvania (1966).

(5) Evans, F.L. "Ozone in Water and Wastewater Treatment," Ann Arbor Science Ann Arbor, Michigan (1972).

(6) Furgason, R.R., Harding, H.L. and Langeland, A.W. "Use of Ozone for the Decolorization and Deodorization of Pulp Mill Wastes," submitted to TAPPI.

(7) Hopt, W., "Water Purification with Ozone and Activated Carbon (Duesseldorf Process) II: Uses of Ozone," Gas-Wasserfach, Wasser-Abwasser, 111, 156 (1970).

(8) "Community Water Supply, Analysis of National Survey Findings," U.S. Department of Health, Education and Welfare, Bureau of Water Hygiene, (July 1972).

(9) "Pilot Plant Test of Wells Nos. 2 & 3 for the City of Moscow," Keystone Engineering & Products Co. Report (April 1971).

(10) "Water Treatment - An Alternative Analysis for Moscow, Idaho," Stevens, Thompson & Runyan, Inc., Report (April 1972).

(11) "Water Supply Study," Stevens, Thompson & Runyan, Inc., April 1971.

The proposal proper was followed by an appendix consisting of two parts. The first (omitted) consisted of eleven pages describing the qualifications of the key personnel in the three agencies that would work on the project.

This was followed by a report on laboratory-scale tests already conducted—tests that demonstrated that the Ozone Treatment would actually produce the desired results. The report consisted of one page of discussion, a flow sheet illustrating the laboratory setup and process, and a table showing the results of the tests performed.

OZONE OXIDATION TESTS

ON MOSCOW WELL NO. 3

March 27, 1972

The principal objective of the tests was to determine the capability of ozone to oxidize iron and manganese from the water from Moscow Well No. 3. For three days prior to sampling, the well was intermittently operated for 2 - 3-hour periods by city personnel in order to obtain representative samples. A test sample of 60 gallons was withdrawn from the well and transported to the University of Idaho Buchanan Engineering Laboratory for testing.

Figure 1 demonstrates the flow process utilized in this test. Raw water was metered to a reaction vessel through an adjustable feed pump. Flow rate of raw water was set at one gallon per minute for all tests. From the reaction vessel the raw water was pumped through a venturi, at which point the ozone was introduced, and then returned to the reaction vessel. Retention time within the reaction vessel, after introduction of the ozone, was set at one minute by regulation of the height of an overflow weir. Ozonated water left the reaction vessel over this same weir.

Gas flow from the ozonator was regulated by a rotometer for application rates of 0.5, 0.75 and 1.0 scfm, providing ozone production of 0.1667, 0.201 and 0.206 mg O_3 per second, respectively. Ozone was produced by a Welsbach Model No. C-D-1 generator.

Samples were taken from the effluent for each different gas flow after five minutes of operation at that flow rate to achieve stabilization. Chemical analyses were run on raw water for iron and manganese. Chemical analyses for iron and manganese residual were run on treated water from the reaction vessel after filtration. All analyses were performed in accordance with the latest edition of "Standard Methods."

The test demonstrated the principal objective of the study in that significant oxidation of the soluble minerals could be accomplished using ozone without pH adjustment. For both iron and manganese the mineral concentrations were reduced over 90% but did not quite reach drinking water standards; however, no attempt was made to further reduce the concentrations or optimize the operating conditions.

27

Figure 1. Schematic Flow Diagram of Portable Test Unit

Run No.	Gas Flow scfm	Titrant Sample	ml	PPM	mg/sec

OZONE

Ozone in Effluent - 100 ml Samples

No. 1 Unfiltered	0.5	1	26.35		
		2	26.40	.016	
No. 2 Unfiltered	1.0	1	18.5		
		2	16.5	.0106	
No. 3 Unfiltered	0.75	1	21.1		
		2	20.9	.0128	

Ozone from Generator - 10 ml Samples

No. 1	0.5		82.1	5.0	0.1667
No. 2	1.0		94.6	6.2	0.206
No. 3	0.75		92.2	6.04	0.201

Run No.	Gas Flow scfm	Sample	Unknown	Trans-mission%	PPM	Avg. PPM

IRON

Raw		1	Iron	25.0	5.0	
		2		25.0	5.0	5.0
No. 12 Filtered	0.5	1	Iron	66.0	0.40	
		2		76.0	0.25	0.325
No. 23 Filtered	1.0	1	Iron	76.0	0.25	
		2		79.0	0.225	0.2375
No. 34 Filtered	0.75	1	Iron	90.0	0.1	
		2		95.0	0.05	0.075

MANGANESE

Raw		1	Mn	82.5	0.93	
		2		(95)(4)	0.90	0.91
No. 1 Filtered	0.05	1	Mn	96.0	0.265	
		2		96.5	0.21	0.2375
No. 2 Filtered	1.0	1	Mn	97.00	0.17	
		2		98.0	0.13	0.15
No. 3 Filtered	0.75	1	Mn	98.5	0.09	
		2		98.0	0.13	0.11

ASSIGNMENTS

Assignment 1

In the list of assignments following the chapters on reports, find a subject you would like to report on, and write an informal proposal suggesting that you do the research and write such a report. Write the informal proposal to your teacher, or to a third person or agency that would be interested in receiving the information.

Assignment 2

After you have received approval of your informal proposal, research the subject adequately enough to write a formal proposal. The formal proposal should be aimed at the same audience as the informal proposal.

14

Oral Presentation of Technical Information

Anyone who is involved in a technical field will be called upon from time to time to present technical material orally. While a written discussion of this subject will, of course, be no substitute for a course in speech, the presentation of technical information presents special problems that often are not covered in a college speech course. The discussion in this chapter is intended to bridge the gap between the speech courses you may take as a student in college and the time when you must rise before a group of people and speak as a professional in a technical field.

When would such a necessity arise? One occasion would be a meeting of a few people to whom you wish to explain a project you propose to undertake. Such meetings are common in all proposal negotiations. Again, after submitting a report on a finished project, you might be asked to discuss aspects or ramifications of the project that were rightfully not included in the written report. It cannot be assumed that any report, however good, can cover every aspect of a problem that might interest people.

A typical occasion when you might need to speak to a larger group would be a public meeting at which you present the attitudes or intentions of your employer or profession concerning some project of general interest. For example, a forest ranger might speak to an environmentalist group explaining logging procedures in nearby forests; a home extension agent might speak to a group about methods to save money in shopping for food and in preparing meals; a health department agent might discuss the problems involved in following certain organic diets. The list could go on indefinitely, but is long enough already to show that men and women in technical professions are constantly called upon to speak on matters in their fields as well as write about them.

Still another type of speaking occasion for those who write on technical subjects occurs in connection with technical articles. If you look over technical journals you will observe that many of the articles are papers that have been read at professional gatherings. A first reaction to the idea of reading a professional paper may be that this activity is peculiar to the academic profession, but such is not the case. For example, in a recent issue of a periodical in the field of paper manufacture, more than half of the articles were printed versions of talks at conventions and similar meetings by speakers employed in industry.

PREPARATION

Although preparing material for oral presentation is in many respects the same as preparing to write about it, there are important differences, because listening and reading are different processes. These differences are especially noteworthy when the material in question is a technical report, where a large amount of detail needs to be conveyed accurately. What is said here, therefore, is especially applicable to report situations, but not limited to them.

When people read what you have written they do not necessarily read in the same manner from beginning to end. For instance, they may read some parts carefully, pausing to consider them if they feel inclined; other parts they may skim because the topics lie outside their particular interests or because they accept what you say without needing to evaluate your supporting evidence. Sometimes they may turn back to refresh the mind on what they read earlier, or they may skip a section and turn, with the help of headings, to material that comes later. In general, readers are masters of the situation, free to decide for themselves how they wish to approach the report.

But when people listen to a speech, options such as these are not open. They must accept what you offer. Unless you repeat, they hear each fact only once. Unless they want to risk missing something that follows, they can think about what you say only while you are saying it. They can't skim; they can't turn back; the time they devote to each aspect of the subject depends on the time you spend on it.

Of course, a listener may interrupt with questions, but unless the group is extremely small, many people are reluctant to interrupt out of consideration for others who might prefer to have you proceed. Thus when you speak to even a small group—actually speak, rather than engage in a general discussion—you have a captive audience. Captive, that is, to the extent that in most cases they won't stand up and walk out—but there is no way that you can hold their minds captive. You can not force anyone to listen who doesn't want to; you must persuade people to listen willingly by the skill with which you prepare your material.

Audience Adaptation

The first step in preparing your material does not deal with the material at all, but with the audience. You need to analyze the audience to determine the best method of presentation. There have been many methods and checklists suggested to aid in analyzing an audience for public speaking, but the best method is sitting down with a paper and pencil and jotting down notes about the composition of the group that will be listening to you: What brings them together? How do they feel about the subject? How much do they already know? In addition, you need to decide what the purpose of your speech is: Is it, for example, to inform the audience of action already taken, to persuade them to approve action you wish to take, or to persuade them to engage in action themselves?

For instance, if you plan to speak before a civic group about the need for an improved sewage system to reduce pollution in the local river or lake system, there are many factors about the audience that you need to consider. Are they generally in favor of pollution controls, or do they feel that environmentalists are impractical idealists? If the latter, then your task would be to organize your material to show how an improved sewage system is in fact practical. On the other hand, the audience may well feel that a better system would be desirable, but too expensive for the town. Your task would then include showing how the system could be financed without raising taxes, or that it was more important to have a new system than to keep taxes low. The organization and emphasis of your presentation would reflect these considerations.

Another factor that would cause your presentation to differ from audience to audience would be the extent to which your audience is already informed on the subject. How much background information must you include before you can discuss the issue at hand? What terms must you define? In other words, it is not enough simply to decide that you understand your audience; your presentation must reflect that understanding if you are to be an effective speaker.

Arrangement of Material

The preceding considerations should, of course, affect the arrangement of your material. Before more is said on the subject, however, one point should be made clear: Nothing that follows is intended to advise that the demands of logic or continuity be ignored. Overall, the method for organizing information to be given orally should follow the suggestions on organization in Chapter 4. The suggestions that follow should be used after the demands of logic are met. For instance, the logic of a subject may require the consideration of a number of points, but the order of presentation may well be arbitrary. Thus you are free to plan your delivery to capture your listeners' attention so that they will more easily assimilate your information.

Listeners are likely to be fully attentive to what a speaker says at the beginning of a talk. Later, unless the speaker is unusually effective or the subject is one of absorbing interest, their concentration becomes less intense. For this reason the skillful speaker provides variety from time to time to revive interest. As the listeners sense that the speaker is nearing the end, they once again become more attentive. It is sound strategy, therefore, to utilize these natural tendencies as you arrange your material.

At first it might seem obvious to arrange your material in the order of decreasing importance, but this method creates the same effect in speaking as it does in writing. That is, if the material becomes less and less interesting or significant as time goes on, people are not only naturally inclined to become less attentive, but justified in doing so. Consequently, the following pattern is one method used to overcome the difficulties of oral presentation. It should be remembered, however, that other methods may be more appropriate to a specific situation, and that this method should be followed only to the extent that it is compatible with logical, coherent treatment.

The first step is orientation: You need to let the audience know what to expect, and how you plan to present your information. Once they have a good idea of your plan of organization, they can fit the information better, as they receive it, into the overall scheme and can better judge its relationship to the problem or situation you are talking about. In other words, prepare your audience: Tell them what you are going to say and how you plan to say it.

Then, as you get into the body of your speech, take up some part of your material that will make a strong impression. When this has been covered, turn to something less vital and build from it with stronger or more interesting parts. Such a pattern will help capture the listener's attention by the strength of the first material, and then retain it because of the increasing interest or importance of the remaining information.

At the end, summarize the material to leave the listeners with the entire picture rather than the last detail. Such an arrangement utilizes the natural tendency of listeners to pay closest attention at the beginning and also makes use of the fact that people remember best what they have heard last.

What we have suggested about arrangement is not to be confused with the adage, "Tell them what you plan to say, say it, then tell them what you said." The adage grew out of an incomplete understanding of ancient Greek rhetorical theory, and is never good advice, however often it may be repeated, for, among other problems, it implies a superiority of the speaker over the audience. We are suggesting that you give your listeners insight into your organizing principle, then give them the information they need, and finally put the information into a context that accords with the purpose of your talk.

Preparing Notes

People vary so much in how natural they feel when they use notes that it would be undesirable to urge a specific manner to prepare them, just as it would be undesirable to insist on a single method for organization. As you gain experience in public speaking you will discover kinds of notes you prefer and will prepare them accordingly. The following suggestions are intended for students who have not yet discovered their individual preferences.

1. How detailed and extensive your notes should be depends, of course, on how familiar you are with your subject. It is possible to have too many notes. When this happens, you may keep looking at them so often that you seem to be speaking to your notes rather than to the people present. Or perhaps, not needing them all, you may talk so long without looking at them that you will have difficulty finding your place when you need to refer to them. It is not the function of notes to encompass all the material to be presented. Rather, they should serve to remind you of the organization of your material. Another function of notes is to provide exact quotations or figures where these are required. No audience expects a speaker to have quotations or figures memorized, and will in fact have greater confidence in a speaker who refers to notes at such times.

   In this connection, there is no reason your notes need be equally detailed for all parts of your presentation. Where you are especially familiar with your material, they may be sketchy. Where you are less familiar, they should be more detailed.

2. If time permits, you may find it helpful to make notes in considerable detail so that you will really think your speech through, and later make a new and shorter set for actual use—a set that has been cut down to what you find you really need as you run through the others. It is inadvisable, however, to attempt to use new notes unless you have time to become thoroughly familiar with them.

3. While an outline is always useful for organizing information, it may not be the best form of notes for a speech. Using an outline to speak from sometimes produces a mechanical-sounding result—making it seem as if you are running down the points of an outline rather than thinking about what you are saying, particularly if you simply follow the points in the outline rather than use them as notes to jog your memory for more extensive discussion. All speeches are more effective if the remarks seem spontaneous. Many speakers find that they get the best results from notes consisting of key words and phrases which keep the ideas in the right sequence and launch the topic without further assistance. An example of such key words and phrases would be:

   Sources of pollution: *Mines   Farms   Homes   Boats*

   These few words might be enough to cover what would need to be said about the pollution of a lake—the facts about each source being so familiar that details would be unnecessary.

4. Although it is obvious that your notes should be legible, don't worry about their appearance. Irregularities, once you are familiar with them, may serve as landmarks. The unique appearance of each one may help you find your place faster than you could find it by looking for typed or longhand words that look much the same as other words on the page. But by all means be thoroughly familiar with the physical appearance of your notes as well as with their content, regardless of what their distinctive features may be.

5. It is harder to judge in advance how long you will need to cover the material in notes than to read from a previously written text. Most people need to feel more concern about talking too long rather than talking too briefly. You can provide for this possibility by marking your notes to show what may be omitted if time runs short. For instance, suppose you asserted that fire was not being guarded against adequately, and were supporting your assertion by citing three examples. You might include all three in your notes but put two in parentheses or otherwise mark them to indicate that they could be omitted or only referred to briefly if necessary. Remember, too many simplistic examples bore an audience, but important detail should never be omitted.

6. In addition to notes that you definitely will use, supplementary notes, to be used only if needed, are frequently helpful. For instance, you might include additional statistical data in supplementary notes. Having them available would enable you to answer possible questions without cluttering your regular notes with figures you might not need.

LANGUAGE

Assuming that you have satisfactory notes, you still need to think about putting your facts and ideas into words. It would be a glaring omission if a discussion on this point failed to take up the subject of "standard" English, or "correct" English, as it is sometimes called.

Standard American English and Dialects

Actually, except for some college professors, newscasters, and an educated minority, few people use standard American English as a matter of course. Most people in the United States speak some variety of three major dialects, usually referred to as Northern, Midland, and Southern—terms which correspond somewhat to their original distribution. The dialects derive in part from the origin of the early settlers in the area, in part from the period of settlement, and in part from naturally differing development of spoken language. For instance, settlers coming from England at an earlier period spoke a different dialect of English than those who came later. Thus, the dialect spoken in New England in some respects represents a development of the language from an older period than does the dialect spoken in Pennsylvania.

These dialects differ in grammar, vocabulary, and of course pronunciation. Thus in Southern, the construction "hadn't ought to" is not only common but acceptable, while in Northern the same thought is expressed by "ought not to." In vocabulary, although few Northern speakers would be confused by the Southern "bucket," they would tend to use "pail"; but not many Northern speakers would recognize that the Southern "pulley bone" is the Northern "wishbone."

While this is an immensely oversimplified discussion of dialects in the United States, it does suggest the problems you might have if you are speaking outside your own dialect area, or if you speak a minor variety of one of the major dialects. In addition, on more or less formal occasions such as a speech, your audience will probably expect you to use some level of standard English, which in the United States is largely Northern and Midland in composition, and will view any dialect as funny, peculiar, or incorrect. The same general principles apply to other dialects and other languages as well.

If you are unaware of the difference between your own speech and standard English, you may well have difficulties in communicating with your audience, for they may be distracted from your message by your language and miss the points you wish to make. Some people, of course, use reasonably standard English as a matter of habit because they have been exposed to it all their lives, have picked it up through extensive reading, or have had intelligent instruction in the classroom. If you are among their number, you are fortunate.

If you are not, however, you should face the fact that most of those with whom you will wish to communicate in your career know standard English reasonably well and will notice deviations that might pass unnoticed in your previous surroundings. Consequently, the sensible thing to do is become concerned about your speech and try whenever you speak to use the language and terminology you will encounter in your professional field. It will not be enough to be able to avoid deviations when you concentrate on doing so. When you talk to people about technical topics, you will need to concentrate on the substance of what you are saying without worrying about your English.

What kinds of deviations do you need to concentrate on? It would be unrealistic to maintain that everything that has ever been labeled "error" must be regarded as equally serious, especially in speech rather than in writing. Some deviations, in fact, are not very serious, and the distinctions are being dropped from American dictionaries. Many a literate person, in speaking, uses *like* when standard English calls for *as, who* at the beginning of a sentence when traditional grammar calls for *whom*, or "It was me" instead of "It was I."

At the other extreme are deviations so well known that you may be avoiding them already: *ain't*, for example, most double negatives, and changes in case such as "My colleague and *me* agree. . . ." The deviations that persist are those less well known. Typical examples are *come* for *came* in the past tense, "We *was*" for "We *were*," or *good* for *well* in adverbial constructions as in "The motor runs *good*." These deviations, if you use them, may cause some people to regard you as illiterate.

The above discussion should not be read as permissive: Deviations from standard English in either speaking or writing can gravely hinder a career. In addition, there are real errors in grammar and usage that do not arise from dialect differences. For instance, "It makes three days that I am here" is not correct usage in any American dialect. Correct language is largely a matter of convention; that is, a group of people have agreed, through custom, that certain combinations of words will mean certain things. Other combinations are confusing. If a technician is confused about a set of directions, or if an audience does not clearly understand what is said, action may be postponed, or perhaps worse, the wrong action may be taken. The Handbook sections *Standard Usage* and *Glossary of Usage* will help you identify the kind of language that calls for attention.

Formal versus Informal Style

Avoiding conspicuous deviations does not mean becoming pompous or excessively formal. Just how informal you should be depends largely on your relationship with those to whom you are speaking—your position, perhaps, in relation to theirs. It might be advisable to be less informal in addressing superiors than in speaking to subordinates, but a degree of formality that seems natural in writing may sound stiff in speech. For example, one might write, ". . . people with whom I talked," but it would be both natural and perfectly acceptable in speaking to say ". . . people I talked to." Or again, contractions would be natural and unobjectionable used orally, though it would be preferable to avoid them on some occasions in writing.

In general, unless you are frankly reading from a prepared text rather than putting your thoughts into words as you speak, a moderate degree of informality, if compatible with your personality, is well-advised.

Personal versus Impersonal Style

When your language is informal, it is likely at the same time to become more personal. This is all to the good. It is entirely appropriate that speech should be less impersonal than writing. When people read what you have written, they are in direct contact only with the printed page and are not necessarily thinking of the writer; but when you are speaking, what they hear comes directly from a person, and it seems only natural that what comes directly from a person should be expressed in an honest or candid personal manner.

Variety

Finally, there is no reason that during a talk your style need be the same at all times. Even when for the most part you use moderately formal language, an occasional injection of informal personal style may be not only unobjectionable but also highly effective. In fact, such a conversational style gives listeners a feeling of involvement similar to what they feel when you talk directly to them in personal conversation. They feel that you are speaking to them instead of at them, and they are much more inclined to listen. A good opportunity for such a change is the time when you have finished one point and are about to take up another. A shift to informal, conversational style during the transition not only conveys the impression that you are relaxed but also gives your listeners a chance to relax. In written material, such a device not only would seem out of place, but it would serve no purpose because readers can pause and relax for a moment whenever they desire, rather than depending on you, as a listener does, to provide the opportunity.

DELIVERY

Needless to say, no amount of advice about delivery can approach the value of coaching by a competent instructor, and even if you can obtain such coaching, you will derive its full benefit only as time passes and you gain experience. After all, you have been talking a long time and can hardly expect to change your abilities overnight. Still, you do not need to feel discouraged about improving your skills.

In discussing this subject we will first assume that the occasion is one on which you are speaking to a small group—eight or ten to perhaps fifteen persons. Later we will add a few ideas that would apply mainly when you speak to a larger group.

GENERAL SUGGESTIONS

1. Consult your dictionary in advance to check the pronunciation of any word you are likely to use and feel the slightest uncertainty about.

2. Talk fast enough that your remarks don't drag—which may be hard if you customarily talk unusually slowly—but not so fast that you find it difficult to enunciate clearly. If you speak clearly, don't worry about your audience's ability to comprehend the words themselves. Experts in communication estimate that the average individual can distinguish aurally 400 to 800 information bits a minute—much more than you can speak even at your fastest. If you talk too slowly, you give your listeners the opportunity to let their thoughts wander, and once you've lost them, you have difficulty getting them back because they are no longer oriented to your information. Nevertheless, you should vary your rate of speaking from time to time. When an idea is harder than average to grasp, you should consider repeating the information at some point in your speech rather than slowing down and risking losing the audience. When a statement is especially significant, consider pausing a moment at its end to give listeners time to think about it.

3. Enunciate distinctly. Avoid dialect pronunciations such as *gonna* for *going to*, *dunno* for *don't know*, *yuh* for *you*, and *em* for *them*. When a word ends in *ing*, don't pronounce it as if it ends in *n*. On the other hand, don't go to the opposite extreme and pronounce words like *cupboard* as if it were *cup board* or *often* with a *t*. With this in mind, note that no objection is raised on the grounds of pronunciation to recognized contractions such as *can't* and *doesn't*, which are pronounced as they are spelled.

4. Speak loud enough that no one must exert a special effort to hear and understand you, but not so loud that you seem to be shouting and sound like a television with the volume too high. If some members of the audience seem to be talking among themselves and making a distracting noise, a trick many seasoned speakers try is talking *softer*, rather than louder, to force the listeners to concentrate on listening. Save the extra volume for points that call for extra emphasis.

5. Avoid a monotone, which as the word implies is a monotonous tone that fails to accentuate words that need to stand out. In varying pitch, however, do not vary it in the same manner in every sentence—which is just as monotonous as no variation whatsoever.

6. By varying speed, volume, and pitch as recommended above, stress words that call for special emphasis, either for their own importance or for clarity. For example, compare the meanings between "Is *that* your only objection?" and "Is that your *only* objection?" The value of proper stress will become apparent if you listen to effective speakers, including the leading television commentators.

7. Keep most of your sentences relatively short and simple. Listeners, unlike readers, cannot look back at what came earlier if they lose the gist of what you are saying.

8. Don't indulge in oratorical flourishes like those too often heard in political speeches. Your language should sound natural—especially when you are speaking to small

groups. For example, "in the United States," or "in this country," sounds less affected than, "in this great democracy where we are fortunate enough to reside." Follow the advice of the critic who said that a major cause of bad writing (and bad speaking) is the writer's striving to be considered a person of genius rather than being satisfied to be regarded as a person of sense.

Transitions

Transitions call for a little more comment than the items in the brief recommendations above. They need a somewhat different treatment in speech than in writing—especially writing where headings are used. The presence of a heading often minimizes or eliminates the need for a transitional sentence or paragraph. A reader who sees a heading knows that you have finished one point and are about to take up another. When you are speaking, however, it is often necessary to use a sentence or two to introduce your new subject or point. If you have already told the audience in your introduction the major aspects of your talk, it is much easier to introduce a new point by reminding the listeners of that fact. Moreover, you can sometimes dramatize a transition by physical movement: If you are standing, by a change in position; if you are sitting, by an action such as setting aside a sheet or note card and picking up another. You must be careful, however, that such attempts at dramatization do not distract from what you are saying. It does no good to emphasize a transition if your audience is going to remain thinking about the transition device rather than the subject to come.

Visual Aids

Visual aids to a talk come in many shapes and sizes. There are videotapes, slide projectors, movies, drawings, tape recordings (which aren't visual, of course, but are frequently used to aid a speech and so are mentioned here), objects of any kind—the list could go on and on. The primary point to remember, however, is they are *aids*, not substitutes for your talk. Visual aids should be used whenever you need to support or clarify a point with further evidence that can be shown more easily than told, and in talks about technical material, this is often the case. Likely aids include tables, bar charts, and graphs. Before preparing any, you should review the chapter, "Tables and Figures," for many of the same rules apply. However, when illustrations are used in connection with a speech rather than with writing, further factors need to be considered.

Any visual materials must be large enough and clear enough for everyone present to see and understand easily. Visual aids should not be so numerous that your talk seems to be merely comments on exhibits—unless that is your intention, as in a slide show. They should also be somewhat simpler than you might use in writing, since the audience can not study them as long as every member might wish.

Even though an exhibit may not be complicated, you should point out its salient features. This is advisable for two reasons: First, those present can discover what is significant to your talk faster with your assistance, and second, you lose contact with them if their attention to the illustration is not integrated with attention to your remarks. With this in mind, it is easy to see why you should not introduce a visual aid until you wish your listeners to consider it, and why, if possible, you should withdraw the visual aid from view after you have completed discussing the information it illustrates.

Audience attention is also the reason handouts should be avoided during a speech, despite the frequency of their use in technical talks. Few people can listen well while passing papers around, and fewer people can resist reading the handout instead of listening. The best solution is to avoid handouts altogether, but if they are necessary, pass them out after the speech. If you plan to use them as illustrations during the speech, label them clearly and pass them out prior to beginning. Then you can call attention to the proper handout at the proper time in your talk. Unfortunately, however, you will not be able to regain attention easily after you have discussed the handout.

SPEAKING TO A LARGER AUDIENCE

Everything that has been said about speaking to a small group also applies when the number of listeners is larger, but when it is larger, some additional suggestions may be helpful. Some of them, to be sure, may well be borne in mind even when the number listening is small, but are taken up in connection with the larger groups because that is when they become especially important. Others are of concern only when you speak in front of enough people to be regarded as an actual *audience*.

The Opening Moments

For your own benefit, when you talk to a sizeable audience, you should take a few moments to adjust to the situation. That is, you need to establish eye contact by looking over the audience, thus causing them to look at you, to form an impression about how loud you must speak in view of the size of the room and the number present, or to position yourself to use the microphone comfortably if one is provided.

For the most part, however, what you need to do is based on the needs of the audience. For the first few moments people are preoccupied, at least subconsciously, in forming an impression about *you*—in reacting to your appearance and manner, and in getting used to your voice and way of speaking. Consequently, you should not be in a hurry to plunge into the substance of your talk. If you notice what is done by effective speakers, you will find that many begin by making a few preliminary remarks that do not constitute the real introduction. If you do this, the remarks should appear spontaneous rather than prepared. You might express your appreciation of the opportunity to speak. You can thank the person who introduced you. You can compliment the preceding speaker if there was one. You can talk for a moment in a general way about the occasion for the meeting and the importance of whatever it is concerned with. You must be careful, however, not to trivialize your own talk by attempting to be humorous if the event does not warrant it, or by calling attention to your lack of preparedness in a false show of humility, or by any other mannerism that will distract from your message. Remember, the audience came to hear what you had to say, and most of them want you to get on with it.

Speed

It is even more important, when the audience is large, to avoid talking so fast that you fail to enunciate clearly. If you naturally talk slowly this caution is probably unnecessary, but if words come easily to you, don't let yourself, as you pick up speed, indulge in rapid-fire delivery and thus become hard to understand. If a formula for an ideal speed could be devised, it would include several ingredients, instinctively blended in proportions suited to the occasion. Among them would be what seems natural to you, the size of the room and the audience, the quality of your voice and the distinctness of your articulation, and the nature of your material—whether at any moment it is easy or hard to grasp. And during any talk an occasional change of pace is desirable to avoid monotony.

Articulation

Clear articulation, mentioned earlier as desirable even when only a few are present, becomes immensely important when your audience is larger. Whatever problems in understanding that might exist in a small group are intensified as the size of the audience increases. With more people in a room the possibilities for distractions are more numerous, the general noise level is higher, and the distance between the speaker and much of the audience is greater. Every speaker is aware that the hardest part of an audience to hold is that part in the back of the hall. With greater distance, the possibility for eye contact becomes less, and therefore the relationship between speaker and listener becomes more impersonal. In addition, if listeners cannot easily make out the expressions on a speaker's face, they have more difficulty in understanding the intent of the speaker's words. If poor or sloppy articulation is added to noise, distractions, and distance, the problems for the listeners may make continued close attention more trouble than they feel it is worth. You can do little about the size of the audience, but careful attention to articulation can minimize its effects.

Use of a Prepared Text

The nature of technical information is such that often the only way to be certain of the accuracy of what you say is to read it. When you prepare a text with this in mind, it is doubly important to remember the points made in this chapter. A written text is of necessity different from a spoken text, but if you plan to write out what you say, it is easy to forget this fact and write as if it were to be read, not heard. The best way to make sure you have allowed for your audience's listening problems is to read your text aloud several times. If possible, read it to a friend or colleague, just as you would a draft of an article or report. If you cannot, read it aloud by yourself. Your own ear will tell you much about the presentation.

For instance, you will discover places where what you have written is unsuitable for oral presentation, although it would have been fine on the printed page. Sometimes you will find problems in making pronunciation distinct. Consider, for example, Pope's lines

When Ajax strives some rock's vast weight to throw
The words, too, labor and the line moves slow.

Pope wrote the lines to make a point about poetic style, and no adverse criticism is intended. They are quoted here to show how hard it is to pronounce some combinations of sound. You may also become aware, when you read your text aloud, that some word has been repeated so many times it becomes unpleasant, that you have accidentally written distracting rhymes, or that some sentences are so long and involved that it will be difficult for the audience to follow them easily, even though the sentences may have been easy to follow when read.

In addition, reading the text aloud will help you to become thoroughly familiar with it, so that you may look up more frequently and appear to be talking to the audience rather than to the lectern. Professional readers, to accomplish this, do not read at all, but deliver lines, apparently read, from memory. Charles Laughton told of a time he was supposedly reading Dickens when he glanced down at the page to further the illusion and his eye caught the actual words he was then supposed to be reading. They so surprised him that he forgot what he was saying, and was forced to actually read for a while until he regained his composure and could remember the lines again.

If your talk is to be printed after you have delivered it, you should edit your text before you turn it over to the printer. When you prepared it you had in mind, or should have had in mind, inflection, vocal emphasis, and the many other resources available to the speaker for making meaning clear. The same text might not be clear if the words alone had to be depended on.

There are times, however, when such editing cannot be considered. When you are reading a prepared statement or the report of a committee, written for the express purpose of becoming a matter of record, changing it after you read it, except for minor emendations, could not be justified. The general principle is: Sometimes what you read orally will be exclusively your own, written at least ostensibly to be read before an audience although it may also be published later. In this case, you should feel free to exercise your judgment in adapting it to the nature of the new medium. On other occasions what you are reading will have been prepared primarily to become part of the written record, in which case the oral version and the written should not differ.

Oratorical Style

Nothing said here is intended to recommend against being emphatic or, within reasonable limits, enthusiastic. But unless you are an accomplished speaker, you would be ill-advised to try to be oratorical—especially in any talk that a book on technical writing should concern itself with. Remember, science convinces not by rhetoric, but by content. Be sparing in the use of dramatic effects. Exercise restraint if you feel tempted to become emotional—even though your own emotions may be aroused by your message. You are more likely to be effective in a technical talk by making your facts pertinent and your ideas clear.

ATTITUDE

Perhaps most important of all is your attitude as a speaker. Any audience, of whatever size, will react quicker to your attitude than to anything specific you have to say. And your attitude is almost impossible to hide or disguise. It will show in your choice of words, in your posture, your gestures, your tone of voice. If you feel superior to your audience, or arrogant, they will respond by rejecting both you and your message. If you are bored by your material, your audience will be bored. If you show fear or lack of confidence, your audience may well interpret this as doubt about your material, and your message may lose credibility.

The subject of fear, or "stage fright," is one that every speaker should face. At some point, everyone has felt some degree of fear at facing a gathering of people. Accomplished speakers avoid fear in two ways. First, they learn their material so well that they have total confidence in their ability to remember all the pertinent information. Their knowledge is gained by practice, by reviewing the material so many times they have no doubts about it. Second, accomplished speakers concentrate on the needs of the audience. They have no doubts about the value of the information to the audience. The confidence they gain also helps them avoid the distracting mannerisms many persons have when they speak, such as putting their hands to their faces, slouching in an unnatural way, or swinging a leg back and forth.

The most important factor to remember in any speaking situation is your audience. Without an audience, you would have no speech. In a very real sense, the members of the audience have honored you by coming to hear what you have to tell them. Although they may appear to you to be a mass of people, they listen to you as individuals, and they will react to you as individuals. You are not, finally, speaking to a mass, but to John, and Jack, and Jane, and Mary. You owe them the respect to present your information honestly, in a clear fashion, so that they will feel they have spent their time well. If you do this, even though your delivery is not perfect, they will listen to what you have to say, for that is why they came.

ASSIGNMENTS

Assignment 1

Copy the following Audience Analysis Worksheet on a separate sheet of paper and fill it out, using as much space as necessary for completeness. You may use the audience you plan to address in one of the following assignments, or the audience for another speech you plan to give.

<u>Audience Analysis Worksheet</u>

<u>Situation:</u>
  1. What brings the audience together?

  2. How will the audience be seated?

  3. Will the atmosphere be formal or informal?

  4. What time of day will the talk be given—will anyone be arriving late, or feeling tired, or will everyone be eager to listen?

  5. What distractions will there be, such as note-taking, handouts, coffee, food, cigarettes, noise?

<u>Audience:</u>
  1. How many people will be listening?

  2. Will the listeners be technical, administrative, lay, or mixed?

  3. What is the range of knowledge about the subject?

  4. How much do the listeners need or want to know about the subject?

5. How do the listeners feel about the subject—for, against, neutral, or mixed?

6. Are there prejudices or misconceptions to be considered?

7. Is there any individual or group within the audience who needs to be addressed?

Speaker:
1. What is the relationship of the speaker to the audience— higher, lower, or equal, inside or outside the organization to which the audience belongs?

2. Will the listeners view the speaker as an expert, a sales representative, or a neutral person?

3. Will the speaker be able to set the tone of the meeting?

4. Will others be speaking or will there be only one speaker?

Purpose:
1. What is the main purpose of the talk?

2. Why should the audience listen?

3. What should the audience know after listening?

4. What should the audience do after listening?

5. What attitude should the audience have after listening?

6. Is the final attitude a change?

7. Are there any secondary purposes for this talk?

Assignment 2

Prepare the information you would include in a formal proposal for oral presentation. If you are in a class, confine yourself to a length specified by your instructor, but in any case, keep in mind the time limitations of your audience. The purpose of the talk, of course, is to gain permission to proceed with the project. You may find subjects in the list of assignments following the chapters on reports, or you may propose a subject of your own choosing.

Assignment 3

After you have completed a written report, prepare it for oral presentation. Although the length must in part depend on the nature of the material and the length of the original, keep in mind the needs and desires of your audience.

Assignment 4

Take an illustration that you have devised for a written report and reshape it for presentation as a visual aid. Unless you plan to actually use it with a talk, you need not make it the actual size you would need, but you should indicate the dimensions that would be necessary for the size of audience you have in mind. Keep in mind how appearances change as the size of the aid changes.

Assignment 5

Prepare a talk for two different audiences. For instance, you might prepare a talk based on a proposal to perform a task, first to the people who are in position to give approval to proceed, then to a general audience of people who may be affected by the project. Subjects might include rerouting traffic to create a mall, turning a parking lot into a park, designing a public transportation system, revising a college curriculum, or any of the suggested subjects in the list following the chapters on reports. Make sure that the audiences are different enough to require major changes in presentation. Keep the talk brief, and deliver both versions to your class, if you have one. Following the second presentation, let the class discuss the effectiveness of the two versions.

BUSINESS
CORRESPONDENCE

15 | General Principles of Business Correspondence

If we regard the term *technical writing* as broad enough to include whatever writing you will be responsible for doing on the job, there can be no doubt that letter writing is an important part of the subject. If you cannot write effective letters, you will be seriously handicapped in your career.

Study of letter writing is especially desirable because of the difference between letters and most of the other material written by those trained in technical professions. In many professions you will spend much of your time thinking not about people but about things—about steel, or oil, or concrete; about transistors, or chemicals, or geological formations; about smog, or insecticides, or machinery. And when much of your writing deals impersonally and objectively with such subjects, you can profit by a reminder that there are times when writing is done for the sake of personal communication, when different techniques are therefore desirable. A study of letter writing constitutes such a reminder.

To make it easier to visualize the finished product, attention is first centered on letter form. Next, the substance of letters in general is discussed. In the following chapters, specific types of letters are considered, in order to illustrate how the general principles of letter writing apply in a variety of characteristic situations.

CORRECT FORM IN BUSINESS LETTERS

This discussion of letter form is not exhaustive, for covering every possible question of form would occupy too much space and would make it harder to identify and remember the points that are essential. If you need additional information on some points, you may find it in a special section in the best collegiate dictionaries. The information that follows, however, will answer most questions that arise in ordinary correspondence.

Parts of the Letter

The parts that appear in *every* letter are (1) the heading, (2) the address, (3) the salutation, (4) the body, (5) the complimentary close, and (6) the signature. The heading is the section that tells the address of the writer and the date; the address is the section telling whom the letter is addressed to; the salutation is the "Dear Sir" or its equivalent; the complimentary close is the "Sincerely yours" or its equivalent; and the signature includes the longhand signature of the writer plus the typed material that accompanies the longhand signature. These and certain other parts that appear with varying frequency will be discussed later in more detail.

Mechanical Details

Stationery.   The stationery should be good, unruled bond paper, preferably white, 8½ by 11 inches large. If the first page of a two-page letter bears a printed letterhead, the second should match it in quality and color but should be blank. Smaller stationery is sometimes used for extremely short letters, but paper of standard size is usually preferred even for short letters because of the great convenience for filing.

Placement of the letter on the page.   Letters should be placed on the page so that the margins are well balanced. The result should be approximately as follows: (1) there should be more white space at the bottom than at the top unless the length of the letter and the size of the letterhead makes this impossible; (2) the side margins should be approximately equal and should ordinarily be no larger than the bottom margin—often being smaller; (3) regardless of the length of the letter, the body should fall partly above and partly below the center of the page. A short letter will look better if slightly more than half the body is above the center of the page.

When a letter is written on a blank sheet, the minimum margins all around are 1 inch—preferably 1½ inches at the top and bottom. The maximum margin at the top is about 2½ inches, and at the sides is 2 inches. The space between the heading and the inside address may vary from a double space in the longest letters to six spaces in the shortest. In the shortest letters, the bottom margin is obviously controlled by the length of the letter.

When the stationery bears a printed letterhead, the point to consider when setting the top margin is the first line of the inside address. *In the shortest letters* this may be as much as 3½ inches (21 lines) below the top of the page. This would normally leave room for six or seven lines of the body of the letter above the center of the page. The side margins would, of course, be the maximum for such a short letter. *For the longest letters,* the inside address may begin only a double space below the date line, which would itself be only a double space below the bottom of the printed letterhead. For shorter letters, the date line would be placed either (1) two spaces below the printed letterhead or (2) halfway between the letterhead and the inside address.

Spacing within the letter. The best procedure, normally, is to single-space within the paragraphs of all business letters and to double-space between paragraphs. The space between the heading and the inside address varies. There should be a double space, however, between any other two parts of the letter, namely (1) between the inside address and the salutation, (2) between the salutation and the body, (3) between the body and the complimentary close, and (4) between the complimentary close and the signature if the first line of the signature is typed. More will be said later about spacing the signature.

Indentation and Punctuation

In modern practice the following system of indentation is perhaps the most widely used: The dateline (and all the lines of the heading when there is no printed letterhead), the complimentary close, and the typed part of the signature, all begin at the center or just to the right of the center of the page. If a letterhead is designed to provide a place for the dateline, it is used. Every other line begins at the left margin except that when an element that ordinarily calls for one line needs two, the second is indented two spaces. Also, although it is ordinarily unnecessary, indentation of five spaces at the beginning of a paragraph is optional—in fact, it is necessary only on those unusual occasions when a letter is double-spaced. Terminology varies, but this form is usually called the block or semiblock system.

Some organizations, however, use what is usually referred to as the full block system. That is, every line begins at the left margin. This form is simpler but the result is an unbalanced appearance that some people dislike.

Less used is the "simplified system." All lines in this system start at the left margin but the salutation is omitted, being replaced by a subject line, and the complimentary close is omitted. Unless and until this form is far more widely used, however, it cannot be recommended because its appearance might distract readers' attention from its message.

The modern system of punctuating the heading and inside address is the open system. That is, no punctuation is used at the end of any line unless, of course, it is necessary to use a period after an abbreviation.

Figures 16 and 17 show business letters in correct form, well placed on the page.

Education Commission of the States
300 LINCOLN TOWER • 1860 LINCOLN STREET
DENVER, COLORADO 80203 • (303) 893-5200

April 11, 1974

Mr. Kenneth Plate
Counselor
Belpre Middle School
200 Rockland Avenue
Belpre, Ohio 45714

Dear Mr. Plate:

Thank you very much for your recent letter to Dr. Hazlett and for your interest in National Assessment. Dr. Hazlett has asked me to reply to your several questions.

As you probably know, NAEP assesses the knowledge, skills and attitudes of young people in two learning areas each year and reports its findings on an item-by-item basis and on several variables, one of which is geographic location. Thus, we do have data on 9, 13, 17 and 26 to 35 year olds for the central region of the nation (which includes Ohio); but these data cannot be regarded as "norms." Rather, they are statistics showing performance on a particular skill, piece of information, behavior, or attitude and can best be interpreted by the reader as to whether they are good or poor.

I am enclosing a NAEP publication, which I think will be of real help in understanding NAEP's method of procedure, entitled, "What Is National Assessment?" and a copy of our "Publications List," which shows and describes all of our reports, booklets, monographs, and similar publications. Possibly after reading these, you will think of additional materials which would be of help. In that case, feel free to write me again with your request.

Best wishes for the success of your reevaluation, and let me know if we can be of further help.

Sincerely,

Ken Seaman

Kenneth W. Seaman, Director
Utilization/Applications

KWS/lm

Encls: "What Is National Assessment?" Publications List

cc: Dr. James Hazlett

NATIONAL ASSESSMENT OF EDUCATIONAL PROGRESS

Figure 16.   Letter of Approximately Average Length, Well Placed on Letterhead. It shows use of a full block form.

741 Kearney Street
Colter Bay, Wyoming 83001
March 8, 1982

Mr. John Walters, Chairman
Springfield Environmental Council
2414 Walnut Boulevard
Springfield, Colorado 81073

Dear Mr. Walters:

Thank you for your letter of March 2 inviting me to participate as a panel member at the April 25 meeting sponsored by the Springfield Environmental Council.

Much as I would like to accept this invitation I am unable to do so. As you may know, I retired at the end of January, so I am now free to do some travel that my work schedule formerly prevented. Consequently, I am planning to leave Colter Bay for a six week trip early in April.

Otherwise, I should enjoy attending your meeting, for I admire your work and am sure that the meeting will be interesting.

                    Sincerely yours,

                    Arthur M. Perkins

                    Arthur M. Perkins

Figure 17.   Extremely Short Letter, Correct in Form and Well Placed on a Blank Sheet.

Form for Individual Parts
of the Letter

The heading. If the stationery does not bear a printed letterhead, the heading contains from two to four lines, usually three. The first line, plus the second if necessary, tells the local address from which the letter was written; the next line tells the town, state, and zip code; and the last line tells the date. The name of the writer does not appear in a typewritten heading. The heading is placed so that its lines all begin at the center or slightly to the right of the center of the page. The top line of the heading is typed 1 inch to $2\frac{1}{2}$ inches from the top of the page, the exact distance depending on the length of the letter.

If the letter is written on stationery with a printed letterhead, the typed heading consists of the dateline only. This may begin at the center or slightly to the right of the center, halfway between the bottom of the letterhead and the inside address, or it may be centered and placed two spaces below the letterhead. A printed letterhead is often designed to provide a natural space for the dateline.

EXAMPLES OF TYPEWRITTEN HEADINGS

1316 Twenty-seventh Street   The Associated Engineers
Alameda, California 94501   University of Idaho
May 15, 1982   Moscow, Idaho 83843
                                February 20, 1982

Abbreviation, capitalization, and use of numbers are illustrated in the preceding headings; but since these are matters that apply to other parts of the letter also, they will be discussed as the final point under *Form*.

The inside address. The examples of inside addresses given later illustrate usage in the most common cases. As indicated, considerable flexibility is necessary to prevent some of the lines reaching an unwieldy length or to prevent the number of lines from becoming so great as to be awkward. Usually, however, three or four lines of reasonable length are sufficient.

When a letter is addressed to an individual, some title or the equivalent must always be used. To omit it is discourteous. The titles *Mr.* and *Mrs.* are the most common, but they may be replaced by *Doctor, Captain, Professor, Rabbi,* or any other term one might use before the person's name if one were actually speaking to that person. The terms *The Honorable* and *The Reverend* are sometimes used in place of titles. *Honorable* is appropriate when one addresses an important governmental official. *Reverend* is appropriate when one addresses Protestant or Catholic clergy. (Full coverage of all the possible forms appropriate for all clergy may be found in books of etiquette and collegiate dictionaries.) It is also correct to indicate the reader's degree, for example, Ph.D. or M.D., after the name rather than placing a title before the name. There should be no duplication, however; *Dr. Ralph Morgan, M.D.* would be incorrect.

For letters addressed to women, *Miss* and *Mrs.* were traditionally the only titles available unless, of course, it was appropriate to use some other form such as *Dr.* or *Professor*. In recent years, however, *Ms.* has come into wide use, in newspapers and magazines as well as in letters, so there is no reason not to use it.

There is some tendency at present, when a letter does not concern a woman's personal business, to omit the title entirely and address her, for example, merely as *Mary L. Brown*. Doing so, however, would still leave the problem of choosing a title for use in the salutation, unless you simply repeat the entire name, as in *Dear Mary L. Brown*.

If you choose to use a title, it is best, of course, to use the title that the addressee prefers, if you know it. In the past, authorities on letter writing gave a long list of options based on the woman's marital condition, but today such concerns are felt to be irrelevant to business and an invasion of privacy. Our best advice, therefore, is to use *Miss* or *Ms.* if you do not know the woman's preference, and if you do know, use whatever form she indicates.

The salutation. When a letter is addressed to an organization, the salutation should be *Gentlemen* or *Ladies*, whichever is appropriate. *Dear Sirs* is out of date except in a letter addressed to more than one individual. Though some companies omit all punctuation after the salutation, long standing convention calls for a colon.

For letters to individuals, note the following salutations, arranged in the order of decreasing formality:

Sir: Madam:	too formal for most letters
My dear Sir: My dear Madam:	rather ceremonious
Dear Sir: Dear Madam:	very common
My dear Mr. Jones: My dear Mrs. Jones: My dear Ms. Jones:	often appropriate
Dear Mr. Jones: Dear Mrs. Jones: Dear Ms. Jones:	most widely used of all terms

It is often appropriate to replace *Mr.* or *Mrs.* by some other title, such as *Doctor, Professor, Captain, Governor,* or *Mayor.* It is more complimentary to refrain from using any abbreviation for a title, except *Mr., Mrs., Ms.* and *Dr.* The terms *Honorable* and *Reverend* are not titles, and though they sometimes replace titles in the inside address, they should not be used in the salutation. Many other terms, such as *Superintendent,* are sometimes erroneously used as titles. When there is doubt whether some particular term should be used as a title, it is advisable to refrain from using it.

As for the choice between the recipient's name or *Dear Sir* or *Dear Madam,* it used to be customary to use the name only when the writer knew or at least had received a letter from the reader. At present, this convention is usually disregarded, for business letters are becoming less formal and more personal. To use *Dear Sir* or *Dear Madam* may make a letter sound cold and remote. The modern trend is to use the reader's name if you know it unless high rank or the occasion of the letter would cause informality to be in bad taste. (Obviously *Dear Sir* or *Dear Madam* or even *Dear Sir or Madam* must be the salutation when the letter is addressed to a position, as for example, Director of Personnel, if you cannot learn the person's name.) There is a tendency, however, to avoid *Dear Madam* and use *Dear Mrs. (Miss* or *Ms.) Jones,* or *My dear Mrs. Jones.* In the last analysis, your own sense of appropriateness must determine your choice of salutations. There is no reason not to use the first name of a reader if you would use it in face-to-face conversation and if there is no special reason informality would be unsuitable to the occasion.

The salutation must harmonize with the inside address. If a letter is addressed to an individual, the salutation must be to an individual and the same title must be used. If the letter is addressed to an organization, the salutation must be to an organization. This holds true even when an "attention line" comes between the inside address and the salutation.

To show how inside addresses and salutations can be handled on many occasions, the following examples are provided:

Mr. David M. Jones, President
Western Testing Laboratory
Olympic Building
South Bend, Iowa 57087

Dear Mr. Jones:

Western Testing Laboratory
Olympic Building
South Bend, Iowa 57087

Gentlemen:

Ms. Mary S. Stevens
Secretary of the Chamber of Commerce
Box 101
Caldwell, North Dakota 58203

Dear Ms. Stevens:

Messrs. William Smith and H. R. Brown
1148 Harrison Boulevard
Twin Cities, Arkansas 92116

Dear Sirs: (or Gentlemen:)

Dr. Frederick C. Moore	Frederick C. Moore, M.D.
1415 West Grant Street	1415 West Grant Street
Tulsa, Oklahoma 74199	Tulsa, Oklahoma 74199
Dear Dr. Moore:	Dear Dr. Moore:

Dr. Norma C. Church	Professor Norma C. Church
Professor of Mathematics	Department of Mathematics
University of Kansas	University of Kansas
Lawrence, Kansas 66025	Lawrence, Kansas 66025
Dear Dr. Church:	Dear Professor Church:

Director of Personnel	American Association of University Women
Hammond and Haskins Milling Company	Cascade Building
Fourth and Main Streets	761 Main Street
Sandpoint, Utah 84336	Sandpoint, Utah 84336
Dear Sir or Madam:	Ladies:

The Honorable Mary R. Randolf	The Honorable John R. George
United States Senate	The House of Representatives
Washington, D.C. 20515	Washington, D.C. 20515
Dear Madam:	Dear Sir:
(or My dear Senator Randolf:)	(or Dear Mr. George:)

The Honorable John R. Kelly	The Honorable Jay G. Walker
Governor of (State)	Mayor of (City)
(Capital City and State)	(City and State)
Dear Sir:	Dear Sir:
(or My dear Governor Kelly:)	(or My dear Mayor Walker:)

The complimentary close.   The complimentary close is placed a double space below the final line of the body of the letter. It usually begins slightly to the right of the center of the page in line with the heading.

The forms used most frequently are *Very truly yours, Sincerely,* and *Sincerely yours. Yours truly* is more formal and somewhat less friendly. *Cordially yours* is a gracious form if one is doing the reader a favor or in a position to do favors or merely wishes to be friendly, but it seems inappropriate in a letter soliciting a favor. *Respectfully yours* is used mainly when the reader's position calls for special respect or when you are formally submitting material to a superior in your own organization. It is too formal and too subservient for ordinary use. Participial endings—that is, *hoping, trusting,* and similar terms ending in *ing*—are out of date. The best attitude is to use *Sincerely, Sincerely yours,* or *Very truly yours* unless there is some reason for deviation.

The signature. The place and contents of the signature are shown in the various examples following. Whenever a letter purports to be personally written, the name of the person who wrote it must be signed in longhand, even if the primary signature shows the letter to come from an organization. The name of the writer must always be typed, however, as well as signed in longhand.

A title should never precede the writer's name in either the typed or the longhand signature except in parentheses, as shown below. When the information is relevant, however, a writer may place *M.D.* or *Ph.D.* or other such an abbreviation after the typed name or may add such a typed line as *Professor of Chemistry* or *Captain, United States Navy.*

When a woman feels that the person to whom she is writing needs guidance about how to address her in an answer, she should indicate her preference in the typed signature line by including the title *Mrs.* or *Ms.* in parentheses before her name. If no title is included, the assumed title is *Miss.* A married woman or widow may type her name, *Mary L. Brown,* followed by *Mrs. John J. Brown* on the line underneath.

By using parentheses around the title, the woman signing the letter provides the information needed; but since the parenthetical material is not regarded as part of the actual signature, she is conforming to the convention that one should not place a title before the signature.

The first example below indicates that the writer is acting merely as an individual; the second and third both show that the writer acts in behalf of an organization; the fourth shows the letter as being primarily from an organization, even though the name of the person who actually wrote the letter is also given. Note that in Numbers 2 and 3 the organization is not named, the assumption being that it was named in the letterhead. Otherwise, the name of the organization would appear in the signature. In all of the examples, the complimentary close is shown above the signature in order to show the relative positions of these two parts of the letter. At least a triple space would be left for the writer's longhand signature. If the first line under the complimentary close is typed, as in Number 4, a double space is used between the two.

(1)	(2)
Very truly yours,	Sincerely yours,
Joan M. Jones	*John M. Jones*
(Mrs.) Joan M. Jones	John M. Jones, Secretary

(3)

Sincerely yours,

Joan M. Jones
Joan M. Jones
Production Manager

(4)

Very truly yours,

THE REED PAPER COMPANY

Robert L. Brown

Robert L. Brown
Sales Manager

The attention line.   In a letter addressed to an organization, an attention line, placed as shown below, is sometimes used in addition to the inside address. By using it one can indicate the name, position, or both, of the individual from whom the letter should receive attention. Note that the salutation matches the address in spite of the attention line.

(1)

The H. C. Capwell Company
Broadway and Twentieth Streets
Oakland, California 94610

Attention of Mr. H. L. Klarnet

Gentlemen:

(2)

The H. C. Capwell Company
Broadway and Twentieth Streets
Oakland, California 94610

Gentlemen: Attention of the Credit Manager

The subject line.   A good opening will identify the subject of a letter without a special line being necessary. Sometimes, however, a subject line is desirable, as, for example, when you have been requested to refer to a file number. It is most frequently placed to the right of the salutation and on the same line or a double space below. When it is not on the same line as the salutation, it either may be centered or may align with the paragraph openings. Note the following illustration:

The Prudential Insurance Company
The Paulsen Building
Spokane, Washington 99201

Gentlemen: Subject: Policy No. 6,499,313

As you requested in your letter of July 23, I am sending you more
complete information on ...

Stenographic references, enclosures, postscripts.   When a letter has
been transcribed by a stenographer, that person's initials and sometimes those
of the person who dictated the letter are placed at the left margin, either on
line with the bottom of the signature or slightly lower. If enclosures are to be
sent with the letter, that fact is indicated either one or two spaces below the
initials, and if copies of the letter are sent to others, that information is shown
by including the letters *cc* (for carbon copy) and the names of the recipients.
The following illustration shows these details in one of the many acceptable
forms.
We shall hope to receive your consent in the near future.

                              Very truly yours,

                              THE CENTRAL ELECTRICAL COMPANY

                              W. I. Bowers

                              W. I. Bowers

WIB/lc
Enclosures: 2
cc: Will James

If a postscript is added, it is placed a double space below everything else
in the letter. It is typed in the same form that has been used for the paragraphs
in the body of the letter. It may or may not be preceded by *P.S.* Obviously,
it should be short.

As used in modern business letters, a postscript is not a means of adding
some statement that has been omitted by accident, for letters should be written
well enough that such accidents do not happen. Rather, it is a device used to
throw emphasis on some matter of special importance, or else to add something
that it seems desirable to mention but that is not really related to the main
message of the letter. Since such a device loses its effectiveness if overworked,
postscripts should be used sparingly.

Additional pages.   When a letter must consist of more than one page,
the stationery used for all pages should be exactly alike except that as mentioned

previously, only page 1 bears a letterhead. The side margins of extra pages should be the same as those on the first page. The top margin should be equal to the left margin. Often, in order that the second and later pages can be identified, the name of the person or organization addressed, the page number, and the date are typed at the top of the second and later pages. (Sometimes, however, common sense indicates that this is unnecessary.) The first line of the body of the letter comes three or four lines lower. Several forms may be used for this reference information, one of the commonest being as follows:

Dr. Irvin B. Steadman -2- October 3, 1982

Abbreviation, Capitalization, and Numbers

Abbreviation. In the body of a letter, the usage in abbreviation is the same as in any other writing. In all other parts of the letter, the main point to remember is to refrain from overabbreviating. The first rule is: In case of doubt, don't. Do not abbreviate the name of a person or organization unless the person or organization does so. Do not abbreviate months. Many organizations do not abbreviate the names of states unless they use the two-letter system adopted by the postal service, nor do they abbreviate such words as *Street, Avenue, East,* and *West* in addresses, such as *2315 West Walnut Street,* except in cases where the compass direction follows the street, as in *1914 Pine Street S.E.* Some latitude is permissible if abbreviation is necessary to prevent a line from being so long as to appear awkward. Yet even then you should use a consistent style in any given letter. That is, it would be wrong to abbreviate a term such as *Street* in the heading and write out the same term in the inside address.

As mentioned in the discussion of the salutation, it is more courteous not to abbreviate any title except *Mr., Mrs., Ms.,* and *Dr.* In the inside address, the terms *Honorable* and *Reverend* are sometimes seen abbreviated, but it is more courteous to write them out. The compliment that you imply by using one of these terms is canceled when you attempt to save time by reducing the term to an abbreviation.

Capitalization. In the body of the letter, capitalization is the same as in any other writing. In other parts of the letter, the examples that have been given are a sufficient guide.

Use of figures or words for numbers. In the body of a letter the choice between figures or words for numbers is in general the same as in any other writing, but dates and addresses call for special comment.

The normal form for a date in the United States is *June 17, 1982.* In the body, the year may be omitted if it is obvious, but such forms as *the seventeenth* or *the 17th* should not be used unless the month has already been named.

Years and days of the month are always shown as figures, but the month should not be indicated by a figure as in the form *8/11/82*. The form *11 August, 1982,* is used in the armed forces and in many countries, but not in the United States.

In addresses, the best usage calls for the house number in figures and the number of a street written out. Many organizations, however, use figures if the number is larger than ten, as in *12th Avenue* or *114th Street*. Some organizations omit the *th* or its equivalent when figures are used. It is also permissible to use figures for the number of a street if a word such as *East* or *North* appears between it and the house number, as in *12 East 42nd Street*.

THE SUBSTANCE OF LETTERS

Learning how to use correct form is no more than a starting point in the study of letter writing. The real opportunity to profit by such a study comes when attention is turned to the substance of letters.

To be sure, some letters deal with routine situations and call for no special qualities beyond the completeness, clearness, and accuracy that are essential to any type of writing. Yet even a routine letter is a personal contact; and in letters that are not matters of mere routine, the personal element becomes extremely important. Thus it is well worthwhile to learn something about the attitudes and techniques that give letters the best chance of obtaining satisfactory results.

At the outset, you should understand the difference between a letter that merely is not bad and one that is really good. A letter may not be bad if it is correct in form, is free from serious errors in English or content, and does the obvious job without giving offense. A letter that is really good, however, is one *that uses to the maximum, for the benefit of the writer or his organization, the contact that the letter represents.*

Standards of Appearance, Correctness, and Accuracy

You should make your letters neat for much the same reason that you would try to present a neat appearance in a personal contact. The appearance of a letter influences its reception, for the reader forms some sort of impression before even starting to read. A messy-looking letter can create an impression of carelessness that lessens the reader's confidence in the accuracy of its contents. Rightly or wrongly, the reader may subconsciously assume that your standards in other matters are comparable to your standards in your letters. If you have an opportunity to look at letters sent by successful businesses, you will find that they invariably are neat and attractive.

Correctness involves both letter form and English. Correct form should become so habitual that it is taken for granted. And nonstandard English in a letter, like nonstandard English anywhere, is so damaging to prestige that it is not tolerated by modern industry. The standards of correctness, however, should be those of liberal up-to-date authorities. These authorities base their judgment on the way language is actually used by literate people; thus many forms that were branded incorrect by textbooks years ago are now entirely acceptable except on extremely formal occasions.

Accuracy should include both the accurate reading of the letter you are answering, and accuracy in the information your own letter contains. A surprising amount of confusion, waste of time, and annoyance is caused by writers' failure to notice what is really said in the letters they attempt to answer. Frequently, an additional exchange of letters is the result. And as for making sure that their own letters are accurate—that is a matter that demands more care than most writers realize. Such matters as dates, amounts, the numbers of orders or statements, and file numbers demand methodical checking. Errors such as a reference to *Monday, May 18*, when May 18 falls on a Tuesday provide another example of the need for close attention to detail. For that matter, errors in ordinary words are almost as easy to make as errors in figures and can create confusion and sometimes ill feeling. If you write, "We are *not* making an exhaustive study of the cause of the trouble" when you intend to write "We are *now* making an exhaustive study," you convict yourself of a fault you have not committed. And again, if you write "*Neither* the manager *nor* the field agent would have been seriously concerned about this matter" when you intend to write "*Either* the manager *or* the field agent," you seriously misrepresent the facts and may lose a great deal of good will.

Language in Business Letters

The first point to realize about the language to be used in business letters is that there is no special language for business letters. However, many writers whose conversation is perfectly natural fall into an unnatural jargon as soon as they start a business letter. They use long, flowery expressions such as *I beg leave to inform you*. They use wordy phrases such as *at an early date* for *soon* and *at the present time* for *now*. They make excessive use of the passive voice, writing, for example, *Your letter has been received by us* and *It would be greatly appreciated by us* instead of *We have received your letter* and *We should appreciate it*. They use pompesities such as *It would be in accordance with our desires if* . . . for the simpler *We should like to have you*. . . .

A good way to avoid unnatural language is to ask yourself, "Would I feel ill at ease if I talked this way in conversation with the person to whom I am writing?" It would be going too far to say that a letter should exactly duplicate the language of ordinary conversation, but it is certainly true that language that would sound pompous or unnatural in a conversation is bad style in a letter also.

Additional examples of undesirable business-letter jargon are:

beg to advise, state, inform, acknowledge, etc.

your letter of recent date (Refer to the date specifically; or if the date is not known and the situation does not justify looking it up, merely use *your recent letter.*)

same (as in "Fill out the questionnaire and return *same* to us.")

party (with the meaning of *person.* This use of *party* is appropriate only in legal documents.)

favor us with

kindly for *please*

enclosed please find a copy of (Use *enclosed is a copy* or *a copy is enclosed.*)

We wish to *take this opportunity* (Unless the phrase has some meaning, avoid it. If you were mentioning something incidentally when writing upon another subject, the phrase might be appropriate, but most of the time it is meaningless.)

Sentences in Business Letters

Sentences in business letters tend to be shorter than sentences in most writing. This results from the fact that many letters must be dictated, or written rapidly and mailed with little if any revision. Under these circumstances, long sentences are likely to become jumbled and to slow up the process of writing. You should not form the habit, however, of making all sentences so short that you write in "primer style."

Paragraphs in Business Letters

Because most business letters are not themselves very long, their paragraphs tend to be short. The first and last paragraphs are often single sentences and are seldom more than two sentences long. The other paragraphs characteristically run from two to four or five sentences. Unless a letter is longer than average, it is unlikely to contain any paragraph of more than 100 words. Single-sentence paragraphs are permissible if they do not become too numerous, but you should not fall into the habit of beginning a new paragraph with every sentence. Unless most of the paragraphs are longer than one sentence, there is no value in paragraphing at all except the psychological value of making the letter look easier to read. As a matter of fact, most long or medium-length letters that contain too many one-sentence paragraphs are found on examination to be poorly organized.

Though paragraphs in business letters are often too short to call for the use and full-scale development of topic sentences, paragraph unity is nevertheless important. The opening of a paragraph creates an instant impression about the topic that the paragraph concerns, and it is bad writing to mislead the reader by presenting information on other topics. In letters that are longer than average and are therefore likely to contain longer paragraphs, attention to paragraph unity is especially desirable.

Beginnings of Letters

The first paragraph of a letter, being mainly a contact paragraph, should usually be short—often consisting of a single sentence and rarely of more than two sentences. Nevertheless, it may need to (1) indicate the subject and purpose of the letter, (2) refer to previous correspondence if such correspondence needs to be borne in mind, and (3) establish a satisfactory tone. Obviously it is not always necessary to refer to previous correspondence, but the first and third of these functions always call for attention.

To say that the opening of a letter must always indicate the writer's subject and purpose seems to be dwelling on the obvious, yet this need is often overlooked. For example, a person inviting the reader to address some organization might tell about the organization for two or three paragraphs before coming to the point that concerns the reader—that the letter is an invitation to deliver an address. Therefore it pays to remember: Make both the subject and the purpose clear at once so that the relationship between the reader and the contents can be easily grasped.

When you answer a letter you have received, you should immediately identify it by mentioning its subject and if necessary its date. Also, you should say enough to demonstrate that you have read it accurately. Such key facts as dates and amounts of money may well be restated, so that if any error has been made in the other person's letter it will be discovered at once. Though such errors are not frequent, they occur often enough to justify the precaution. The manner in which facts are restated should not be crude and obvious. It is poor style to say, "I have received your letter of June 15. In this letter you say that you would like to have me make reservations for you for July 17, 18, and 19." Rather, let the facts seem to be mentioned incidentally, as for example, "Thanks for letting me know, in your letter of June 15, about your need of reservations for July 17, 18, and 19." Ostensibly, the sentence was written to say "Thank you." Actually, its purpose was to repeat the dates.

Mention of your own previous letters on a subject is advisable because the reader may have forgotten them and because your present letter may not reach the same person who read the earlier letters.

The establishment of tone is especially important. This subject will be covered extensively later in the chapter. Here, suffice it to say that the original impression influences the reader's reaction to all that follows. If the tone of the beginning is offensive, it will be doubly hard for the material that follows to get a proper reception. Hence the opening should usually be friendly in tone, and should never, regardless of the circumstances, be harsh or sarcastic.

At the very beginning you should try to get in step with the reader. Even if it will be necessary, later, to include material that is unpleasant, the opening should be an attempt to suggest that you are not an opponent who is trying to defeat the reader in an argument, but a person who has approached the subject with an open mind. Even if a letter deals with a situation so serious that a display of friendliness would clearly be insincere, it is always possible to be completely courteous in the opening and to avoid any display of anger.

Some might feel that it is a nuisance to have to make the opening of a letter friendly. Yet when you adopt a friendly or at least a courteous tone at the beginning, you are merely opening your letter as a tactful person usually opens personal conversation.

Endings of Letters

The best method of ending a letter depends on whether you are attempting to secure some specific action. If not, the ending should be a final brief paragraph showing friendliness and good will. Here again, there is a parallel between a letter and a personal conversation. One usually tries to end even a business conversation in a friendly manner.

A typical letter not attempting to secure action might be one giving some information that had been requested. You might well conclude such a letter by expressing hope that the information will be useful and that the reader's project will be successful. On the other hand, if you had been unable to provide the information asked for, it would still be possible to express a hope that the reader would find it elsewhere and that you could be of some service in the future.

If the purpose of a letter is to secure action, the action desired should be suggested by the ending. The suggestion should be as specific as the circumstances permit. Sometimes the request for action may be in the form of a command: "Sign up today." Sometimes, tact would make it advisable that the command be veiled: "To make sure that you receive the accommodations you desire, make your reservations today." Sometimes it must be toned down still more: "We should appreciate it if you could let us know by April 1 the dates that you prefer." The degree of urgency must always be a matter of judgment. If you need immediate action when requesting a favor, it is especially important to be tactful. In such a case your letter is less likely to arouse a negative reaction if it tells why immediate action is important—for example, "If you accept this invitation, we are anxious to give your address plenty of advance publicity. Therefore we hope you will let us know your decision at your earliest convenience."

An ending aimed at securing action need not entirely neglect the element of good will. A skillful writer can usually phrase the request for action so that good will is apparent.

Whether a letter does or does not aim at action, one fundamental point about endings should never be overlooked: The ending of a letter should never dwell upon anything that you would not want the reader to continue thinking about. A reader tends to remember whatever was mentioned last. Suppose a letter ends by saying, "In closing, let me express once more my regret that, as originally written, my report did not contain all the information you needed." The reader's mind is thus focused on the fact that the original report was unsatisfactory. On the other hand, the letter might end, "In the version I am now sending, I am sure you will find the full information that you will need as you proceed with the project." The reader's mind is thus directed to the merits of the new version rather than to the faults of the old. Although there are many occasions when a letter must express regret, the end should, if possible, be optimistic.

Need of Planning Before Writing

Unless a letter is to be extremely short, you should plan it before writing it. Your plan may be nothing more elaborate than rough notes, but it should at least indicate the points to be covered and the order in which you will cover them. As you make your notes, it will pay you to make sure that you have at hand all the facts that must be included, so that you will not have to stop writing or dictating while you look up some bit of information.

The time used in making notes will be short and will be more than compensated for later because you will be able to forge ahead with assurance when you write. Also, planning will usually improve the quality of your letter. That is, it will insure your putting related material in the same part of the letter, will result in better arrangement than you could achieve while struggling with questions of phraseology, and will permit you to concentrate on how to express each idea rather than being constantly preoccupied with the question of what you are going to say next.

To be sure, as you gain experience and deal with situations that have become familiar, you will eventually be able to base many a letter on a plan you have used on similar occasions in the past. Even so, it will usually be advisable to have in mind or on paper a checklist of points to be covered; and it will also be advisable, before you write even a routine letter, to pause long enough to make sure that a routine letter will really cover the situation. Nothing is more annoying to your reader than to receive a letter that ignores some of the relevant facts.

Arrangement of Material in Letters

In writing a letter, as in writing anything else, you will need to be governed by two considerations as you plan the arrangement of the material—the subject matter and the reader. Whatever may be the message of a letter, there are two basic methods of arranging the material. These might be called "the order of saying *Yes*" and "the order of saying *No.*" In any specific situation, one or the other should definitely be preferable.

When you are saying "yes" to a request, you should say it at once—sometimes even in the opening sentence. This is appropriate because a reader who learns that a request has been granted is in a more receptive state of mind to hear anything else you wish to say. Perhaps you must explain that granting such requests cannot be made a regular policy, or that careful consideration was necessary before consent could be granted. Perhaps you need to explain the background. All such material should be held back, however, until the affirmative answer has been given.

Some writers neglect this principle. They tell all the reasons that might justify a refusal before informing the reader that the request is granted. Thus they make the reader antagonistic, or, at best, secure only superficial attention to this material while the reader skims over the page looking for the definite answer. Moreover, by delaying the consent too long, they seem to give it grudgingly, and thus lose the good will that the consent might otherwise have gained. In view of all this, when you intend to say "yes," you should say it at once and say it graciously, then offer any explanations that are necessary, and finally, bring the reader's attention back to the fact that the request has been granted.

When saying "no," however, you should not blurt out your refusal at the beginning, but should hold it back until you have given the reasons for the decision. Thus the decision seems to be the outgrowth of the facts. If the "no" comes before the reasons, the reasons seem to have been scraped together to justify an answer that had been decided upon at the outset. The proper method is to try to make the reader see that a negative answer is indicated by giving evidence and leading to that answer gradually. When you make your decision known, you should do so as tactfully as possible. Although no letter should be long-winded, the very fact that you have taken pains to write a full explanation suggests that you have not rejected the request peremptorily.

Naturally, it is not possible to generate so much good will with a refusal as with a consent, but the arrangement of material recommended above at least makes the best of the situation. A refusal that is tactful will create much less resentment than one that is tactless.

Not every letter, to be sure, forces you to grant or refuse a request, but the basic principle can be applied in almost every situation. If your letter bears good news, let the reader know it at once; if the news is bad, break it gently; if it is partly good and partly bad, first tell the good news, then tell the bad news, and next direct the reader's attention toward whatever aspects of the situation are most cheerful.

As you apply these principles, it is a good idea to decide very early how you want a letter to end. You need not go to the extreme of writing the last sentence of a letter first; but it is only common sense to decide what idea you want that last sentence to express. If you do so, you can move toward that ending more smoothly. Also, you can avoid the misfortune of using, in the middle of a letter, the idea that would be the best possible ending—therefore having either to use an ending that is not as good as it might be or to repeat ideas that you have already expressed.

Negative, Neutral, and Positive Approach

When we say that an approach to a letter is negative, we mean that by the power of suggestion the writer creates in the reader's mind an undesirable impression about the writer's attitude or about the way in which a situation should be interpreted. When we say that the approach is neutral, we mean that there is nothing in the writer's manner that creates either a desirable or an undesirable impression. When we refer to the approach as positive, we indicate that the power of suggestion operates in a favorable manner.

The difference between the negative, the neutral, and the positive approach can be seen in the following sentences, all of which convey the same basic fact.

1. I was surprised to discover in your letter of June 26 that you claim I have made an excessive charge for investigating your lighting needs.
2. I have received your letter of June 26 concerning my charge for investigating your lighting needs.
3. I have been glad to do as you requested in your letter of June 26, and look into the question of whether I have made the correct charge for investigating your lighting needs.

The first sentence sounds indignant; it suggests that the writer's purpose is to prove the reader's letter was not reasonable; it shows that the writer regards the reader as an opponent in an argument. The second is completely noncommittal as to the writer's attitude. The third, without implying that a concession is to be granted, makes it appear that the writer welcomes the opportunity to correct a mistake if one has been made or to explain how the charge was arrived at if it is correct. A positive approach is always best when it is possible, and in any event the negative approach can always be avoided.

One cause of the negative suggestion is the use of words with a negative flavor. For example, "We have received your letter in which *you claim* that the last shipment was improperly packed," implies that the writer questions whether the statements of the reader are accurate. (Even if such were the case, it would be undesirable to make that fact apparent at the outset.) Other characteristic negative terms are "you neglect," "you fail," "your complaint," and "your error."

The negative tone may also be a matter of basic outlook—a tendency to see the dark side of a situation rather than the bright side. One writer may write, "We regret to say that unfortunately we cannot comply with your request until . . ." Another and better writer will say, "We shall be glad to comply with your request as soon as . . ." The difference is that one sounds like a refusal and the other sounds like a consent. One stresses the fact that immediate compliance is not forthcoming; the other stresses the fact that eventual compliance is possible. Again, one writer may write, "My experience is limited, consisting of nine months as assistant to . . ." Another writes, "My qualifications have been increased by nine months' experience as assistant to . . ." The first suggests that the reader consider nine months as very slight experience. The second encourages the reader to think of how much a person may know at the end of nine months that was not known before.

Finally, certain kinds of material may in their very nature be negative. Consider the sentence, "Now that it's fall and you're getting warm clothes for the children, hadn't you better make sure that your automobile chains are in good condition?" This is negative because it makes the reader think of other expenses, which may be so hard to meet that economy would recommend making the old chains serve. Another example of negative material is this ending to a letter: "I hope that our decision will not interfere with the pleasant relationship that has always existed between our companies." The writer showed by this ending that the decision is one that *might* interfere with the pleasant relationship. A better version might have been, "I believe that the reasons I have offered to explain our decision show it to be appropriate, and I shall look forward to a continuance of our pleasant relationship."

Tone

A discussion of tone is really an extension of what has been said about the negative, the neutral, and the positive approach, for it concerns the impression created in the mind of the reader about the writer's personal attitude. Just as in a personal conversation, this impression can either increase or decrease the likelihood of satisfactory results.

The tone of letters can be infinitely varied. A letter can sound friendly, cheerful, courteous, serious, formal, informal, cordial, hostile, ingratiating, blunt, sarcastic, indignant, flippant, whining, evasive—in fact, it can suggest any attitude that can be suggested by a person's manner in a face-to-face conversation. It is possible, of course, for a letter to have a purely impersonal tone. The best that can be said for such a tone, even on a routine occasion, is that it does no damage. But on any occasion that is not entirely routine, there is often a danger that it will create a damaging impression of coolness and indifference.

Obviously some of the tones named above—sarcasm, bluntness, hostility—are undesirable. A warning against them is nevertheless called for because many a person who would agree that they are undesirable will, under the stress of emotion, write letters that possess them.

As for letters that sound apologetic or ingratiating, their tone is often the result of carrying such qualities as reasonableness and friendliness too far. No quality is so good that it cannot be overdone.

What tone, then, should you try for? So far as it is possible to generalize we might say: Your letters should almost always sound friendly; and on the few occasions when friendliness would obviously be insincere, they should as an absolute minimum be entirely courteous. They should usually be cheerful. And the tone of modern letters is strongly in the direction of informality except when informality would suggest lack of suitable respect.

None of this means that a letter need be unbusinesslike or that it cannot be intensely serious when seriousness is called for, any more than a personal conversation would need to lack seriousness because one's manner is friendly.

How is such a tone achieved? Skillful letter writers, especially in letters that go outside their own companies, form a mental picture of the image they want people to hold of their companies, and they bear this image in mind as they write. Eventually it becomes practically automatic for them to adopt a suitable manner when they write letters. This is comparable to the way in which a person in military service automatically adopts a manner suitable to the occasion when talking in the performance of duties. It is comparable to a salesperson's habit of talking to a prospect in a certain manner, regardless of his or her own state of mind at the moment. Unless you are fortunate enough to possess instincts that will always guide you to a tone that is appropriate, you should keep reminding yourself to adopt a desirable tone until it finally becomes second nature for you to do so. If you fail to do this, you are likely to form habits that produce a contrary result.

The "You Attitude"

The "You Attitude" means merely that the writer attempts in every way possible to show thoughtfulness about the reader's interest. A good example is one letter written by a manufacturer refusing to accept an order from an individual consumer. This manufacturer might have offered as a reason the nuisance it would be to handle such small orders. Instead, the letter stressed the advantage to the purchaser of being able to see the various sizes and colors and of obtaining products without delay when served by a retail merchant.

This "You Attitude" can be achieved even in letters that would seem to offer little opportunity for it. For example, a letter written to collect back payments on a refrigerator purchased on the installment plan would have as its actual purpose the securing of the payment; yet the writer might indicate that the letter was being written because the company did not want the reader to lose the refrigerator, as would happen if payments were delayed; or the writer might stress how desirable it would be to the reader to protect a credit rating.

Good Will in Business Letters

Whatever its message may be, a letter should present that message in a manner that generates or retains as much good will as possible. The way to create or retain good will, so far as you can do so by means of writing techniques, is mainly to express good will. It is hard for a reader to remain unfriendly in the face of consistent friendliness on your part. It bothers people more to say "no" to someone who has been pleasant than to someone who has been indifferent or antagonistic. Showing good will pays off by making the reader want to see things as you see them.

Also, making good will permeate your letters is a sensible precaution for it can prevent inadvertently offensive remarks from doing damage. If a statement in a conversation is taken in an unpleasant way that you did not intend, you can notice the effect at once and correct your slip. Moreover, once said the remark passes by at once. But when a letter gives offense, you are not present to observe its effect and have little opportunity to set the matter right. Also, what you have said is there to be read repeatedly. Consequently, it is necessary, when you are writing, to take extra precautions to avoid giving offense unintentionally. There is no better way to do this than to make good will so evident that the reader will realize an unfortunate slip could not have been meant as it sounded.

The time when it is hardest to express good will is when you must answer an unpleasant letter from an angry writer. It is entirely natural to feel an impulse to answer such a letter sharply, but the impulse should be resisted. In refusing to lose your temper you are not necessarily giving evidence of weakness. And one point not to be overlooked is this: Having blown off steam, the writer may no longer feel as belligerent as before and may even feel a little ashamed at having lost self-control. If you answer in kind, you relieve the writer of worry about going too far; but if you are courteous as well as firm in your manner—if you refrain from entering a contest in recrimination—you have a good chance of making the writer regret the remarks and wish to show a less quarrelsome and unpleasant nature. This may go far toward making the person easier to deal with.

If you are tempted to regard anything said above as soft or emotional, remember that industry spends millions of dollars annually in the belief that good will is profitable. Expressing good will in your letters will increase the chance that they will be effective both in their immediate results and in their long-range contribution to a favorable image of you or your company. And just as it is worthwhile to include material for the sake of good will, it is often sensible to omit remarks that, though justified by the facts, are not really essential and would sacrifice good will. It is important to remember that you cannot assume every reader to be a fair-minded, unprejudiced judge. Therefore your manner and material should be determined not by what you would be *justified* in saying but by what will produce the result you desire. The satisfaction of "telling off" a reader who has been unreasonable is a luxury you cannot afford. Constant attention to good will is by no means sentimentalism or a shady practice. We may hope, of course, that it will usually be a sincere expression of your actual feelings; but it is definitely called for by intelligent self-interest.

EXERCISES

Exercise 1

Change the following addresses, salutations, and complimentary closes so that they conform to the best modern usage. Do not change anything that is already correct. You may assume that where abbreviations appear in company names, the forms used are correct official forms.

(1)

Mr. John O. Clark, president
Cascade Paper Company
1289 Burk dr.
Box 687
Cascade, Washington 98498

My Dear Mr. Clark:

Sincerely,

(2)

E. A. Thompson, Inc.
1224 Grove Avenue
Millwood, Arkansas

Attention of Personnel Director

Dear Mr. Thompson:

Sincerely yours,

(3)

Dr. Doyle M. Hobbs
Walters Dental-Medical Building
3958 Sherman St
Springfield, Ohio 44903

Dear Mr. Doyle,

Very truly yours,

(4)

Mr. Henry R. Norton
4223 Thirteenth Street
Middleton, Vermont 05982

Gentlemen:

sincerely,

(5)

The Honorable Walter P. Pigeon
The House of Representatives
Wash. D.C. 20515

Dear Honorable Pigeon

Very truly yours

(6)

Collins Pre-Mix Company
Tenth and Franklin Streets
Caulfield, Mississippi 38925

Attention of the President

Dear Mr. Jackson:

Sincerely yours,

(7)

Richard I. Dormer
Upland Chemicals, Inc.
Box 71
Cranston, Texas 78201

Gentlemen:

                                             Sincerely,

(8)

Prof. Chuck C. Baker
Dept. of Chemical Engineering
Kiley Technical Institute
Kiley, Penn. 15695

Dear Dr. Baker,

                                           Sincerely yours,

(9)

Ms. Jayne C. Patterson, President
Ames and Coope, Incorporated
2363 Bristol Way
Longview, Mass. 01033

Dear Sirs:

                                           Very Truly Yours,

(10)

Millicent C. Milbrook
Principal of Westbank Senior H.S.
Westbank, Nebraska 68569

My dear Principal Milbrook,

                                           Yours,

Exercise 2

Rearrange the material in each of the following items so that it is an inside address in acceptable form, with an attention line if one is called for. Add a suitable salutation after each address.

1. The purchasing agent of Campbell and Kelly, Inc. The company is located at 37 Winchester Street, Houston, Texas 77018.
2. Daniel G. Burk, director of the Central Steel Corporation. The main offices are located at 12 Rockefeller Plaza, New York City. The zip code is 10022.
3. Edgar H. P. Manning, an associate professor of physics at the Atlanta Technical Institute, Atlanta, Georgia 30364. He has a Ph.D. and works in the Physical Sciences Department.
4. The Northern Division of Universal Plastics, located at 522 Thirty-second Street, Tacoma, Washington 98468.
5. Two men, Dr. Carl N. Curtis and Oliver Edwards, M.D., who work at the Kettleman Medical Clinic in Denver. The mailing address is Box 3022, and the zip code is 80375.
6. The Business and Professional Women's Club, Toledo, Ohio. You wish to address someone in the speaker's bureau. The club meets at the City Center, Box 37. The zip is 43609.
7. The Pacific Marine Supply Company, 14 Wharf Street, San Francisco, California 94109. You wish to address the sales manager.
8. Judith P. Smith, who is director of research at the Zenith Manufacturing Company, Minneapolis, Minnesota. The street address is 18256 Central Avenue. The zip code is 53414.
9. The purchasing agent for the Gary Pump Company, Box 737, Gary, Indiana 46444.
10. Central Missouri Agricultural Bureau, 324 Davis Way, Kansas City, Missouri 64195. You wish to address the director.

Exercise 3

Each of the following sentences occurred, as indicated, at or near the beginning or the ending of a letter. Most of them are faulty in some respect, either because they contain unsuitable ideas for the positions that they occupy or because their tone, ideas, and language would be unsuitable regardless of their position. Write an improved version of each sentence that needs improvement. If it is necessary to interpret a situation more definitely or to supply additional facts, feel free to do so.

BEGINNINGS

1. I wish to draw attention to the fact that you have been consistently overcharging me every time I order spare parts.
2. The computer system that you have provided for us has given continual trouble since it was put in operation and we insist that you provide us

with a simpler program that the operators can handle or replace the entire system.

3. I realize that a person in your position has heavy demands, timewise, but I hope the following facts will persuade you to join one more committee.

4. We have been in business for 20 years and have never had a complaint like yours about our service before.

5. Though questionnaires are sometimes regarded as a nuisance, I hope that the one enclosed will not be consigned to your wastebasket.

6. The report you wrote about icing conditions during high-altitude flights is unacceptable because you neglected to address one of the points we wanted covered in the investigation.

7. Your request for a refund cannot be approved until you provide us with more information.

ENDINGS

8. In closing, let me once more express regret that our mechanic failed to make the proper adjustments to your solar system, and assure you that he will try harder in the future.

9. We now believe we have ironed out the difficulties in our delivery system so that you will no longer experience the delays you have called to our attention.

10. While we recognize that our plant sometimes emits noxious odors, we are careful to keep within the limits set by the Department of Environmental Quality and thus have no plans to reduce the emissions because the reduction would not be cost-effective.

11. Since our investigator discovered that you have been using our batteries in commercial vehicles, which is expressly prohibited by our guarantee, we cannot make the adjustment that you requested.

12. If you do not replace the entire control panel without delay as we have requested, we shall be forced to refer your shoddy work to our lawyers for action.

13. We hope that the reset valves you have installed in our air conditioning system function without further problems so that we can feel justified in referring further work to your company when we begin our expansion program this June.

Exercise 4

Study the following letters to discover the respects in which the principles explained in this chapter have not been complied with. Try also to think of ideas that might, if they had been added, increase the letters' effectiveness. Criticize or rewrite letters designated by the instructor.

(1)

Dear Mrs. Hortlandt:

Thank you for your letter inviting me to come for an interview on May 30. Unfortunately, I cannot get away on that date because it in the middle of exam week.

I could come for an interview after June 3 and would appreciate your letting me know whether you would be able to talk with me then instead of May 30.

I would be pleased if you would also give me a little more information that would affect my decision if you offer me a position. I would like to know whether you would consider paying my moving expenses. Also, could you let me know whether I would be asked to move very often if I joined your corporation.

You expressed an interest in when I could start work if I am employed. I could start after June 15, which would give me enough time to pack and give my wife time to give notice at her job.

In addition to the information I requested above, I would be interested in learning about the possibilities of my wife's finding work in your community, or what the possibility is of her finding work with your company. She has a degree in recreational education and is currently employed by the Park Department here, teaching various outdoor skills classes. Would such a position exist in Riverview? Would it be possible that a similar position might exist in the school system?

I shall hope to hear from you soon so that I can know what date to plan on coming after June 3. However, if it would be inconvenient for you to postpone the interview so long, I could come the week of May 20 to 25, if that would be better for you.

Sincerely yours,

(2)

Gentlemen:

We have received your letter telling us that you accept our offer to move the heavy equipment from your old science building to the new building that is nearing completion.

We shall be glad to do this work for you, but cannot possibly complete it by September 15, which is the date you specify. As we indicated in earlier correspondence, moving this equipment will be a four-week job. The earliest date when we could begin would be September 4, and there is a strong likelihood that the date might be from one to two weeks later. Consequently, the earliest date when we could guarantee to have everything moved would be October 15.

If you would review the case, you will not blame us for this delay. We submitted our offer on June 15, and on July 1 we wrote to ask that you hasten your decision. After waiting ten days we took on another job and went to work on it on July 20. We cannot do both jobs at once, and cannot hope to complete the one we are now working on before September 1.

We realize the importance to you of getting moved before the opening of the school year, but since it took you six weeks to reach your decision, we can hardly be blamed for not being able to start work only two weeks after you did so. We would have liked to give your job top priority, but we could not risk having the idle time that would have existed if we had not accepted the other job when it came along.

There are not many outfits in this area who are really competent to move heavy scientific equipment, and we doubt that you will find any of them able to serve you on such short notice. We would like to do the job for you and know that you will let us know immediately whether our offer to do it beginning some time in the first half of September and completing it by October 15 is acceptable.

If the acceptance is long delayed, we might again be forced to take on other work and be tied up during the dates that are now open in September and early October. We therefore hope to hear from you soon.

Very truly yours,

(3)

Dear Sir:

I am writing to complain about the manner in which the county road crews have for a long time consistently neglected the condition of Derby Road between Highway 93 and the bridge over Cottonwood Creek.

This stretch of road is so full of chuckholes that it looks like it had been bombed by an airplane, but the road crews ignore its existence. Not long ago they went over the entire part of Derby Road that lies east of the highway. A little earlier they had patched the black top on the portion that lies west of the bridge. I am unable to understand why the half-mile stretch between the two should go without attention when it is rougher than either.

As a taxpayer who helps to foot the bill for the upkeep of all the roads in the county, I see no reason to tolerate the neglect of a stretch which I and others who live upon it must use every time we leave home. I trust that you will put an end to this discrimination.

Yours truly,

(4)

Dear Sir:

I have received your letter in which you complain about what you claim is neglect by county road crews of a stretch of Derby Road west of Highway 93.

There is one fact, however, that you ignore. The stretch of bad road lies within a tract that was annexed to River City over a year ago.

I agree that the condition of this stretch of road is bad and that the county road crews have done nothing to make it better, but we have our hands full caring for our own roads, and cannot be expected to do the city's work.

Since the city is now responsible for the stretch of road you complain about, I recommend that you address your complaints to the city government.

16 Special Types of Letters

Not every letter can be classified according to type, and the inclusion of this chapter is not intended to imply that such is the case. Rather, its purpose is merely to show how the general principles apply in a number of characteristic situations. No effort is made to include all the types that are generally recognized. Credit letters, for example, are omitted because they are unlikely to be written by anyone in a position that calls for technical writing. Sales letters are omitted for the same reason, though some of the sales letter techniques are explained in the discussion of letters urging action. The types included justify the attention they receive not only because they illustrate the application of general principles but also because you may some day occupy a position where any of them may have to be written.

INVITATIONS

Though invitations of a social nature are beyond the scope of this book, there are many times when invitations are essentially business letters. For example, anyone who is active in a professional organization may need to write an invitation to attend or participate in a meeting.

In writing an invitation, make the purpose of your letter clear very early. Don't leave your reader wondering, "Why am *I* being offered this information?" Also, in or immediately after the opening, include something to catch your reader's interest. One good way to do this is to refer to something your reader has done or something you have read or heard about him or her.

Try to persuade your reader that the expenditure of time involved in the acceptance of your invitation will be justified. The importance of the occasion, the number expected to attend, facts about the organization, names of others who will be present—any such information can be useful in this connection.

Make a subtle appeal to self-interest if you can plausibly do so. Many a person accepts an invitation, partly at least, in expectation of gaining good will or publicity for the organization the person is associated with. Take care, however, not to be crude and obvious in appealing to self-interest. Your tone should make it clear that though you hope your reader will somehow benefit from accepting your invitation, you consider that his or her acceptance will render you a service.

Make sure that you include all the facts your reader will need in making a decision—especially the date, the time, and the place of the meeting. Make it clear what you expect or hope your reader will do—deliver a major address, participate in a panel discussion, attend a social function. In an invitation to deliver a major address, indicate about how long it should be and either suggest a subject or say that a subject of the reader's choice will be satisfactory. One possibility is to suggest a possible subject but make it clear that use of this subject is optional. If the occasion calls for any comment on financial matters, say what is necessary.

When all these points, or as many of them as seem appropriate, have been covered, end an invitation by suggesting action, that is, by referring to your hope to receive a letter of acceptance. The main difficulty in writing the ending is the question of how to induce a rapid response without seeming to rush the reader unreasonably. (In this connection, be considerate. Don't delay so long in sending your invitation that you must ask your reader to answer in haste.) Perhaps the mildest way to request early action is to use the phrase, "at your earliest convenience," which is not jargon, because it is the simplest way of expressing the idea. If it is necessary to ask for action by a specific date, mention the reason—for example, the need of printing a program or beginning publicity.

Make your ending pleasant, in fact make the letter pleasant, by showing a desire to be hospitable and to make the occasion enjoyable for the person you are inviting. An offer to make hotel reservations or to meet your reader at the airport is one way of showing hospitality and consideration. In any event, your ending can easily be written so as to be complimentary as well as being a request for action.

Though all the preceding suggestions are worth considering, you should not try to follow so many of them in one letter that the letter grows unusually long. Usually, a letter from two-thirds to three-quarters of a page is long enough. If the reader accepts and needs further information, you can supply it later. In the original invitation it is better to select the ideas that best suit the occasion rather than to try to use every device you can think of.

The following letter illustrates the application of these principles.

Dear Mr. Kiley:

Like other snowmobile owners in Centerton I have been greatly interested in the work that the Mountainview Snowmobile Club has done under your leadership to improve the opportunity of snowmobilers to enjoy their hobby.

Several snowmobilers in Centerton have formed a committee to organize a similar club in our area. As a first step we have called a meeting at the

Community Center, to be held at 8:00 p.m. on December 3. We would like to extend to you a cordial invitation to attend this meeting and speak to us about the way you organized your highly successful club and about its accomplishments since it was organized.

We believe that if we direct our efforts wisely, we can develop a successful organization. The group I mentioned, which has asked me to serve as its chairman, has been sounding out snowmobile owners and their response has been enthusiastic. The local snowmobile dealers are helping us to arouse interest, and I am confident that at least fifty people will attend the meeting— probably more if we are able to announce that you will be there.

Since the existence of a strong organization here would help your own organization to accomplish its purposes on a statewide level, I hope that you will feel it worthwhile to accept this invitation.

If you can do so, I also hope that you can spend enough time here that some of us on the committee I am speaking for will have a chance to extend our personal hospitality. We shall be very much pleased if we receive word of your acceptance in the near future, so that we can spread the word among potential members.

Sincerely,

LETTERS GIVING INSTRUCTIONS

Letters giving instructions, as we will discuss them here, are those primarily intended to tell the reader to do certain things and to explain, if necessary, how to do them.

The first requirement, if those who receive instructions are to follow them properly, is absolute clarity. As a Prussian general once remarked, "Anything that can be misunderstood will be misunderstood." True, it may take extra time to make instructions unmistakably clear; but the extra time devoted to clarity is usually less than the amount of time it takes to repair the damage caused by having things done wrongly.

Clearness is partly, of course, a matter of writing. Everything about clearness said elsewhere in the book, especially in the discussion of style, applies to clearness in writing instructions. Completeness is also an element in clearness. In giving instructions you will find the time well spent if you think the subject through, visualize every contingency, and cope with it if you think it might arise and cause uncertainty. And in dealing not with contingencies but with regular essentials, you will have to use good judgment in deciding which details would be useful and which can be omitted.

Usually instructions will be clearer if at least a minimum amount of information is provided about the situation that makes them necessary. A reader who is aware of reasons and purposes will usually grasp instructions faster, remember them better, and follow them more intelligently than would be likely if expected to follow them blindly. But the actual orders should stand out sharply and distinctly rather than merging with the explanatory matter.

A letter of instructions can often be improved by the use of mechanical devices such as numbering, headings, and extra indentation. These devices were discussed in Chapter 6, in the section dealing with instructions. To be sure, a letter is seldom long enough to permit the use of very many mechanical devices, and also they may take more time to work out than it is practical to spend on a letter to a single individual. But they should not be overlooked in a letter that goes to a number of people.

Finally, tone can be important in a letter of instructions as in other letters. Instructions are not always given exclusively to subordinates. For example, instructions on such matters as safety procedures, accounting methods, or use of cars from a company car pool might be issued to people in other lines of authority. And even in giving instructions to subordinates, a pleasant and courteous tone increases the chance of securing willing cooperation.

Following is a letter giving instructions:

Dear Ms. Kennedy:

We are glad to know that you have decided to add an index to your book about the early history of your home town of Pavelle, which we are publishing for you.

For your assistance we are enclosing a specimen page of a good index. If you use it as an example, you can avoid any uncertainty about what is customary in regard to such matters as capitalization, punctuation, indentation, and wording.

We think it will also be helpful if we explain how experienced index makers go about their work to avoid confusion and unnecessary effort. Basically, it calls for the use of 3–5-inch cards, each of which bears a single entry. In using them, proceed as follows:

1. When you receive the page proofs, look over each page; and when you find an item that a reader may wish to look up, record it near the upper, left-hand corner of a card, with the page number following it. If the item is extensive enough that the reader might be helped by reference to separate parts of it or to related items, make a subordinate card for each such item. Each subordinate card should show the main item where it would appear on the main card and the subordinate item below it but more deeply indented.

2. When you make out a card for a main entry, pick out the key word—the word that best identifies the actual subject. Start the entry with this key word, put a comma after it, and then add what is necessary. For example:

Wild Rivers, Areas Classified as

3. If information on some point might be looked up under either of two words, make a card for each. For example you might have a main-entry card for Camp Grounds followed by cards naming streams or other locations where camp grounds are available; and you might also have a main card for each of several streams, followed in each case by subordinate cards—one referring to camp grounds on the stream and others to other information about it.

4. When you have made out cards for everything you believe a reader is likely to look up, arrange them so that the main cards are in alphabetical order, followed in each case by the subordinate cards. Then alphabetize the subordinate cards in each set on the basis of the first important word in each—

which means, for example, ignoring a preposition even if it is the first word in a subentry.

5. Cross out the main entry on each subordinate entry card—lightly enough that it is still legible. (The reason for showing the main entry on a subordinate card is to identify it if it gets out of place.)

6. When you have made and arranged your cards in this manner, you need only copy them onto your ordinary manuscript paper and your index will be complete. We advise, however, that you look over the copy you have made so as to eliminate needless duplication by means of cross-references.

If you use this method of working with cards you will find that it facilitates alphabetizing and prevents confusion. And if you write the entry on each card exactly as it should appear in the finished index, you can copy the cards without being concerned about anything except copying correctly.

We hope that these instructions plus the enclosed page will make it easier for you to produce your index so that your book can be published as soon as possible.

Sincerely yours,

INQUIRIES AND ANSWERS TO INQUIRIES

Letters of Inquiry

The task of writing letters of inquiry may fall to the lot of a person in almost any position. When an inquiry is sent to a person or organization that has a responsibility for supplying information—for example, a chamber of commerce or a governmental bureau—it poses no special problems. The same is true when the receiver may have a chance to profit by supplying information because it provides a chance of selling goods or services. But when it goes to a recipient who is under no obligation to answer and who does not profit by doing so except in good will, it must be written with more care. Most organizations, to be sure, answer legitimate inquiries as they answer other correspondence. But whether they supply the desired information depends in part on whether the writer succeeds in making them really want to be helpful and in making the answer as easy as possible to write. Our discussion will concern mainly inquiries to be sent when the circumstances call for tact and care.

The resemblance of one such inquiry to another is great enough that the following instructions may be applied in most occasions.

1. Open your letter by making it clear that your purpose is to obtain information, and identify the subject that the information concerns. Don't leave your reader wondering about the reason for concern about the situation you are dealing with.

2. Tell your reader why the information is needed. Though your explanation should not be long-winded, it is neither courteous nor reasonable to ask your reader to use time for your benefit unless you show that your purpose is neither trivial nor contrary to the reader's own interests.

3. Unless it is obvious, explain why you chose the reader as a source of information. As you do this, you will often be able to say something complimentary that will increase your reader's desire to be obliging.

4. Ask the questions that you hope your reader will answer. These questions, of course, are the actual reason for your letter's existence and will be discussed below in more detail.

5. Offer to return the favor by supplying similar information to your reader, or perhaps by supplying information on the results of your investigation if you are writing to a number of other people. Of course, no such offer should be made if it would be an empty gesture, but there are many times when a mutual exchange of information is natural among individuals or organizations engaged in similar activity.

6. End with a tactful suggestion of action combined with the assurance that you will appreciate whatever help you receive. In expressing your appreciation it is best to avoid saying "Thanks in advance," either directly or by implication, unless your personal relationship with the reader justifies your taking assistance for granted. There is a real though subtle difference between saying "I shall appreciate whatever information you send me" and saying, "Thank you in advance for answering these questions" or "Thank you for this information."

The action requested should be made as easy as possible to perform. For example, you might encourage your reader to jot down answers at the bottom or in the margin of your letter and enclose a stamped, self-addressed envelope. If you do this and if your questions can be answered without looking up the information, your reader may answer you at once rather than waiting until some later time when writing letters.

The actual questions, as mentioned above, are the heart of the letter. Unless your needs are so simple that you are asking only one or two questions, both courtesy and self-interest dictate that your questions be worked out carefully. They should be well-organized, with those dealing with a single aspect of the subject placed consecutively. Each question should be as brief as clarity permits. The answers called for should be brief facts or estimates rather than general discussions, for a discussion is likely to be vague and filled with hedging and qualifications.

If there are several questions, it is advisable to number them. When there is room, it may be desirable to make them stand out as a single block of material by changing the side margins. The numbers make it easy for the reader to check up on precisely what facts are needed in order to answer. Also, they make it difficult to overlook questions. And finally, they encourage answers in the exact order of the questions, which is a great help if you are tabulating responses from many persons.

Though an inquiry should request information mainly by asking specific questions, it is usually advisable, also, to invite the reader to add comments or other information that might be considered relevant. Especially, the reader should be invited to supply any additional information that affects the interpretation or the significance of the answers.

Closely related to an inquiry is a questionnaire, which of course would be accompanied by a covering letter. In fact, a questionnaire is merely the result of removing the questions from the letter, placing them on a separate page, and providing room for answers. Unfortunately, though a questionnaire must be used when questions are numerous, many people who answer straight inquiries disregard questionnaires. If a questionnaire seems necessary, the letter that accompanies it should merely be a brief, normal letter of inquiry without the questions. The page bearing the questions should contain enough information, however, to be answerable if separated from the letter.

A letter of inquiry follows:

Dear Professor Saunders:

I have heard about your system of teaching students in your technical writing classes how to write reports by having them help faculty members in the professional schools produce reports on their research projects. Since I teach technical writing, I am very much interested. I would be really grateful if you would help me by supplying answers, based on your experience, to the following questions:

1. Have you received satisfactory cooperation from the faculty members of the professional schools?

2. Has the benefit to the students depended to any great extent on the faculty member's performance rather than on their own efforts?

3. How large a percentage of the student's total work in the course is faculty-related?

4. Does the system provide an opportunity for the student to demonstrate his judgment as to what the report should include and what it should omit?

5. Does the system test the student's sense of responsibility for the accuracy of data?

6. Does the system force students to do their own thinking—for example, to interpret the significance of data?

7. Since two to four students work on the same project, and join in the production of the same report, is it possible to pass judgment on their individual contributions?

8. In the same vein as number 7, have you found a way to tell whether, in each student team, some single student who is a natural leader or unusually competent does the lion's share of the work though the others get equal credit?

As I have indicated above, I can see that the system might solve some of the problems of teaching the course, but I am uncertain about some of its aspects. If you will give me the benefit of your experience by answering my questions, I will really appreciate your assistance. I shall of course be pleased if you add any further comments that you think would be of interest.

Sincerely yours,

Answers to Inquiries

The following suggestions apply when the answer to an inquiry is sent to a reader who is not a potential customer for goods or services. The answer to a potential customer is essentially a sales letter and is therefore beyond the scope of this discussion.

When your answer to an inquiry contains most or all of the information requested, the opening should make it clear that the reader's request is being complied with, and of course it should cover all the points that must be covered in any opening. Then should come the information requested, as far as it can be provided. If part of the information cannot be provided, a statement to that effect comes next, accompanied by a suitable expression of regret and whatever explanations are called for. The ending is usually an expression of good will. For example, it can be a statement of your hope that the reader's project will be successful or an offer to provide additional information if such an offer is suitable.

When your answer to an inquiry does not provide the information requested, its contents should be arranged in the pattern suggested for saying "No" under *Arrangement of Material in Letters*, Chapter 15. Rather than blurting out the fact that the request is being refused, you should first attempt to show that you would like to comply with it. Then should come a brief statement of the reasons that you cannot do so, leading to a tactful but not evasive statement that the information requested cannot be supplied. In stating reasons you should not be unduly apologetic but should avoid bluntness.

After the actual refusal, you should include at least one or two sentences so that the final impression will be your good will rather than your refusal. You can create such an impression by saying that you hope the reader can find the information elsewhere, that the project will be successful, or that you can be of service in the future. Many times, when you cannot answer exact questions, you may be able to send printed material or suggest other sources of information. A well-written refusal can at least prevent the reader from feeling rebuffed.

The following letters are examples of how the answer to an inquiry may provide information or may say that the information requested cannot be provided.

Gentlemen:

We have received your letter asking about the Mohawk garbage compressor and are glad to provide the following information in answer to your questions about how it has worked out in the Excelsior Apartments.

1. The Excelsior Apartments consists of about two hundred units housing about 450 to 500 residents.

2. The garbage to be disposed of averages about 1000 pounds per day. We estimate its original bulk to be great enough to fill forty garbage cans.

3. The compressor handles all kinds of garbage—cans, bottles, food waste, and paper.

4. The compressor reduces the garbage to about one-sixth of its original bulk.

5. The operation is automatic. When a certain amount of garbage reaches the container, the compressor is turned on by an electric eye. Very little attention is necessary except for removing the compressed blocks.

6. The system has been trouble-free since we installed it 18 months ago.

7. We estimate that the saving in labor resulting from our use of this system will pay for it within two and one-half to three years.

Since you ask us to add any information that we think might be helpful, I will mention that our building was designed with the use of a compressor in mind. The tenants on each floor can dump their garbage into chutes that carry it directly into the compressor, located in the basement. Thus it is handled only once, and then in the form of compressed blocks. The system would be less efficient if the tenants had to put their garbage into cans and we had to move these cans to the basement, dump the garbage, and then return the cans to their former locations.

To be sure, the compressor has a door through which garbage can be inserted, but only a trivial fraction of our waste material needs to be handled in this way. I do not know whether the apartment you are planning will provide chutes such as those that we use, and I believe that the advisability of your using a compressor like ours may well depend on the answer to that question.

I hope that this information is helpful and will be glad to answer any further questions that may occur to you.

Sincerely yours,

Dear Miss Jones:

We have received your letter asking about our cloud-seeding operation during the early part of the year on the Warren River drainage—its extensiveness and our appraisal of the results.

As you may know, this operation was carried out under a contract with Central Light and Power Company. Consequently we do not feel free to release information about it to anyone except our client.

In the hope that it will be helpful, however, I am enclosing our brochure that contains a substantial amount of information about our methods and the results that we have obtained on other occasions.

Central Light and Power is of course free to release any information we developed for them. If you send them the same set of questions that you sent to us, it is possible that they may provide you with answers.

In any event, since we are naturally interested in seeing the public made aware of the potential benefits of weather modification, I hope that you can find enough material so that you can write the article you are working on.

Sincerely yours,

COMPLAINTS AND ADJUSTMENTS

Complaints

A letter of complaint is one of the types most likely to be badly written. It is likely to be written by a person who is irritated—perhaps justifiably—by both the facts to be complained about and the necessity of writing the letter; and with the reader at a distance, the temptation to write a harsh, angry letter may be hard to resist. Many a person who would be moderate if face-to-face with someone throws restraint aside and is unpleasant in a letter.

The purpose of a letter of complaint should not be to relieve your emotions. Rather, it should aim at affecting the attitude and perhaps the actions of the reader. Thus, even if harshness would be justified, you defeat your own purpose by being harsh, you anger the reader, and you make fair judgment on the merits of your case improbable. All in all, you are unlikely to gain any desirable result by a display of irritation that you could not gain without it and yet retain good will.

A letter of complaint should, at the least, be completely courteous, and it is often more effective if the tone is friendly. You should address the reader as a fair and reasonable person. Thus your letter will seem intended to make sure the reader becomes clearly aware of certain facts and realizes their significance—facts that you would presumably wish to know if the case were reversed.

This does not mean that you need to be weak and apologetic, nor that you need to minimize the seriousness of a serious occasion. But it is possible to be direct and intensely serious without being quarrelsome or sarcastic. Even when a situation is so serious as to call for a clear warning that drastic action may become necessary, the warning can be phrased so that it does not sound like a threat. In keeping with the "You Attitude," it can be expressed in such a way as to show your desire to avoid an action that presumably would be to the reader's disadvantage.

Another attitude to guard against is self-righteousness. Though it may be desirable to present yourself as fair and reasonable, you are not likely to obtain the reaction you desire if you seem convinced that you are perfect and your reader is highly reprehensible.

One may hope that in avoiding personal unpleasantness in a complaint you are not falsifying your feelings; but whatever your feelings may be, your letter should not be written to express them. It is written to produce a desired effect on the reader, and a letter along the lines recommended is most likely to secure the effect desired.

When a complaint must contain a request for some specific action—for example, grant an adjustment—you should assume that the reader accepts the moral and legal responsibilities, state the facts on which your request is based, and ask for whatever action you think the situation calls for. The tone of such a letter is improved by an expression of regret that the request is necessary. The conclusion should of course be the specific request for action.

Following is a letter of complaint:

Dear Mr. Zimmerman:

For the last three years your garage has been doing the service work on our trucks and company cars. During most of this period we have been pleased with the results.

For the last two or three months, however, there have been a good many occasions when the service received has not been as good as before. The latest incident occurred last week, when the cap was left off after brake fluid had been added to one of our trucks.

Earlier incidents included such slip-ups as failure to replace the cap on a gasoline tank, leaving tires underinflated, and a good many other and similar cases of carelessness. Also, there have been occasions when work has been done that we had not authorized—for example, replacing spark plugs during a tune-up when the ones in the car had not been in service long enough to need replacement, and installing a full tail pipe behind a muffler when an extension was all that was needed.

We have brought these matters to the attention of the head of your service department when they have occurred, but have not noticed any improvement.

I would really regret the necessity of taking our work elsewhere, but I shall be forced to do so if we have further cause for concern. I am bringing this matter to your personal attention in the hope that you can bring the service we receive up to its former high quality, so that we can continue to bring our work to you in the future.

Very truly yours,

Answers to Complaints
and to Requests for Adjustments

Three kinds of letters are covered in the following discussion: answers to complaints that have not asked for any specific adjustment, answers that refuse adjustments, and answers that grant adjustments. All three have a good deal in common; in fact, everything that is said about the first applies to both the others.

Answers to complaints not involving adjustments. The key to writing a good answer to a complaint is the same as the key to success in writing anything else: Form a clear picture of the result you hope to accomplish, and write in a manner designed to accomplish that result. You will find it is easy to forget this, however, in answering a complaint. The complaints you will answer will not always be justified, and whether justified or not they will sometimes be unreasonable or unpleasant. Consequently, you will be tempted to say, and may even be justified in saying, things that would be better left unsaid. You may constantly have to remind yourself that you are not writing to express your emotions or to win an argument. You are writing in order to affect your reader in a certain manner, usually to make the person feel less injured or antagonistic. The tone and content should be settled upon with this definite purpose in mind.

This general advice is outlined in the following suggestions:

1. Even if you are answering an angry, unreasonable letter, your tone should be pleasant, or at least completely courteous. This does not mean that your letter

must be subservient, but only that it should be a calm, reasonable attempt to reach an understanding.

2. You should avoid the role of an opponent who is trying to defeat the reader in an argument. To accomplish this result it is well to mention very near the opening some point or principle on which you and your reader will agree.

3. Detailed restatement of the complaint refreshes the reader's memory of the original annoyance and revives anger. Therefore you should limit your restatement of the complaint to the minimum that is necessary to show you have read it with care and understand it correctly.

4. If the complaint contains misstatements, do not repeat them for the express purpose of proving that they are false or erroneous. Of course you must state the facts as you know them to be, but it is better to refrain from calling attention to the fact that what you are saying contradicts what the reader said. In doing this you give the reader a chance to back down with a minimum of embarrassment.

5. Expressions of regret should be proportionate to the seriousness of the occasion. They should be neither so perfunctory as to suggest indifference nor so excessive as to magnify a small matter into a large one.

6. Explanations should not be too long and involved. To have them so creates a risk of seeming unduly apologetic or unduly argumentative. If the tone creates a satisfactory impression, a simple explanation of what went wrong will usually be sufficient.

7. The ending should not revert to the cause of the complaint. It should be based on the assumption that the matter at hand has been cared for and that all will be well in the future.

Some of the preceding suggestions picture your reader as a person whose complaint has been unpleasant or immoderate. This emphasis is justified by the fact that unpleasant complaints are the hardest to answer. Your problem will of course be simpler when you answer a complaint that is free from a quarrelsome tone. The opening should make it clear that you are glad your reader let you know about the complaint. It should show that you have been glad to look into the case open-mindedly and are anxious to set matters right if you have been at fault. (There is no reason, however, that on every occasion blame must be fastened upon someone.) Include whatever explanations seem relevant and express an appropriate degree of regret. If the complaint called for action, say what you have done or propose to do. End as recommended in the suggestions above.

The main consideration here is to convince your reader that you are really concerned about whatever it is that caused the complaint. It is possible, of course, to magnify a small matter into a large one; but even a moderate complaint should not be answered in such a perfunctory manner that you appear indifferent.

The following letter is an answer to a complaint.

Dear Mrs. Parr:

Thank you for your letter of May 10 letting me know of your concern about the manner in which our trucks are driven in your neighborhood. We are sorry that they must use the residential area where you live to reach the construction work on the new levee.

We certainly do not want to have our trucks driven in any but a safe and legal manner. Consequently I have brought your letter to the attention of all of our drivers and made it entirely clear to them that they are to drive properly in the future.

I think it likely that most of them have been doing so already, and would be glad to receive information that would help me to single out the exception or exceptions. Each of our trucks can be identified either by its license number or by the number painted on its door. If the offenses continue, and if you can supply me with one of these numbers on the truck that is at fault, and tell me when and where the offense occurred, I will take the matter up with the individual driver concerned.

I believe, however, that what I have already said to all drivers will remedy the situation and that you will have no further cause for concern.

Sincerely yours,

Letters refusing to make adjustments. Included in Chapter 15, on business correspondence in general, was a discussion of the order of saying "no." What was said in that discussion applies, as might be expected, to a letter refusing to make an adjustment.

In brief, the order of saying "no" calls for holding back the actual statement of a refusal until the facts and reasons for refusing have been presented. This arrangement conveys the impression that the request was received by a person who was willing to consider it with an open mind and grant it if the facts showed that an adjustment was due. Having the actual statement of refusal precede the statement of reasons conveys the impression that the writer decided to refuse and then looked for reasons to justify the decision. This, plus the fact that being refused does not put the reader into a receptive state of mind, makes it undesirable for the refusal proper to be placed near the opening.

The opening of a refusal to make an adjustment should of course acknowledge the letter you are answering and should include enough facts to show you understood it correctly. It should convey the impression that your first reaction was, "Let's look into this matter and see whether we should grant an adjustment." Even if you know before you start writing, as of course you will, that you are not going to comply with the request, you can open your letter by saying that you are glad your reader felt free to bring the case to your attention, or that you are sorry about the trouble. Often it will be useful if you state the principle on which you base your decisions on matters of the kind in question. For example, you might mention your willingness to make repairs without charge if the product you had sold developed trouble within the guarantee period and had been used for the purpose intended.

Next should come the facts that you consider relevant and perhaps your reasoning as you considered these facts. In refusing to grant a request that is based on incomplete or inaccurate statements, present the facts that justify your decision but do not go out of your way to prove that the reader is guilty of misstatements. Nothing is gained by making your reader feel embarrassed. If you show up the weakness of a case too devastatingly, the reader may dislike dealing with you in the future, or may become angry at you in order to escape self-condemnation. Likewise, in trying to demonstrate that you are not to blame for something that has gone wrong, do not make a point of putting the blame on the reader. Most of the time, the less said about blame, the better.

Having stated the reasons for your decision, you may logically go on to state the decision itself. Be tactful as you do so. Avoid the word *refuse* if possible. It would be better, for example, to avoid such phraseology as, "We must therefore refuse to grant your request for a reduction of our charge from $275 to $200" and say instead, "We believe that in view of these facts, our charge of $275 is correct." The effort to be tactful should not be carried so far, however, that it makes your answer unclear or evasive.

When your decision has been stated it is usually desirable to add a few lines turning attention to the future. You can say that you look forward to a continuance of pleasant and mutually rewarding relations, that you would like to be of service in the future, or whatever else the specific circumstances permit you to say. Though any idea or attitude can be carried too far, the general principle of ending a letter on a cheerful, forward-looking note is sound.

It is possible, of course, that a letter constructed along these lines may not have the desired effect; but after all, letters cannot achieve miracles. Sometimes no amount of skill in saying "no" can prevent the reader from feeling resentment. But when refusal is necessary, the method explained should hold resentment to a minimum.

The letter that follows is an example of a well-written refusal to make an adjustment.

Gentlemen:
I have received your letter telling us that the sidewalk we laid for you, leading from your parking lot to the rear door of your building, is not holding up as well as you had expected.

We make it our policy to stand behind our work, so I sent one of our people to inspect the walk that you refer to. He agrees that it is in poor condition. He also learned, however, that loaded trucks often use it to get closer to the building to unload.

As you requested at the time, we designed and laid an installation suitable for any amount of pedestrain traffic. We could, of course, have provided a driveway, but its cost would have been substantially higher. Neither the preparation of the subgrade nor the thickness and reinforcement of the concrete for a sidewalk make it strong enough to be used by trucks.

We feel that the deterioration of your sidewalk has occurred not because our work was defective but because it was used for a purpose for which it was not designed. Under the circumstances, I believe you will understand why we cannot agree to repair it at our own expense. We would be glad, however, to do a repair job—or to build you a driveway if trucks must use it, at the lowest cost that will permit us to do a satisfactory job.

If you would like to have us submit an estimate, let me know and we will do so immediately.

Sincerely yours,

Letters granting adjustments. Sometimes you may need to grant an adjustment on an occasion that is important enough to call for more than a routine consent. In this event, the letter differs from a refusal in one important respect: The action being taken should be indicated extremely early in the letter. Thus the letter will contain: (1) an acknowledgment of the complaint (though the term *complaint* would not be used); (2) a statement that the request for an adjustment is being granted (these two parts may be combined); (3) whatever expressions of regret are called for unless regret has already been expressed in the opening; (4) whatever explanation seems necessary; (5) an attempt to turn the reader's thoughts away from the occasion for complaint and focus them toward the future.

This order is psychologically sound. The reader will listen to explanations more understandingly after learning that the request has been granted.

When as a matter of policy you must grant an adjustment to which you do not feel the claimant is entitled, you should grant it graciously. Some writers will offer all the reasons that might justify a refusal before finally saying that an adjustment has been granted. By adopting this grudging, argumentative method, they sacrifice the good will that they are making the adjustment in order to retain. A letter will not have the desired effect if it shows that although the request is being granted, the reader is considered unreasonable to have made it.

The letter that follows is an answer to a complaint in which an adjustment was granted:

Dear Mrs. Smith:

I was sorry to learn from your letter of March 10 that the tile drain we installed for you has not been functioning properly, and agree that it is our responsibility to take care of the trouble.

When I received your letter I sent one of our foremen out to see what was the matter. By means of a probe he discovered that the tile was 15 inches under the surface of the ground. Apparently, it froze during the cold spell last month; and when the weather changed and the heavy rain began, it did not thaw out and drain away the water as fast as it should have done.

It is highly unusual for the ground to freeze so deep in this area, and I think that a recurrence of the trouble is very unlikely. However, we stand behind our guarantee, and if you feel it necessary, as soon as the weather permits, I will send out a crew to dig up the tile and bury it deeper. Though this means that the work we did last summer will have to be done over again, which adds to our expense, there will be no charge for this service.

I am really sorry that the drain tile did not solve your problem when the rains came; but since I have never known of the ground in this area freezing to a depth of two feet, I believe that there will be no trouble in the future.

Very truly yours,

LETTERS URGING ACTION

Many letters written for the purpose of urging action do not fit into any of the types discussed elsewhere. There are letters intended, for example, to urge the reader to contribute to some fund, to join a program for smoke or erosion control, or to join some organization. The writing of such letters may be demanded in any business or profession.

Letters of this type resemble sales letters closely enough that they can often be based on the sales letter pattern. This pattern involves four steps: (1) attracting attention and securing interest; (2) arousing desire; (3) convincing the reader that the urged action will satisfy that desire; (4) securing action. The key words are *interest, desire, conviction,* and *action.* Though the lines that mark off one step from another need not be sharply drawn, the four functions must always be included in any letter that attempts, under one cover, to persuade the reader either to make a purchase or to perform an action of some other kind.

To obtain *interest* it is best to get the letter under way with a minimum of the preliminaries that usually occur at the opening. The reader should immediately encounter something of self-concern rather than material that concerns the writer. If the opening can suggest that the reader's own welfare or the welfare of the reader's organization is involved so much the better.

One of the poorest openings encountered recently was in a letter by a social fraternity attempting to raise funds. It read: "It seems as though every time we write to the alumni we are asking for something." This showed that the writer was uncertain of the appropriateness of the request. Moreover, it put the reader on guard to refuse. The writer should have aroused enthusiasm about the fraternity and its project before letting the reader realize that a request for money was coming. In other words, the writer should have adopted the "You Attitude."

In attempting to arouse *desire*, remember one fundamental point. You do not attempt to *create* desire. Rather, you try to think of some desire that the reader already feels, fan it into intensity, and then associate the action aimed at with its satisfaction. Technical writers can profit by seeing how advertisers use this principle. For example, a person may not desire an air-cooling system; but everyone desires comfort, so the desire for comfort is exploited in a letter selling an air-cooler. Similarly, a person may not desire membership in an organization, but may be eager for professional advancement, or a company may be induced to join a smoke-control program not so much to provide smoke control as to obtain good will. The starting point in arousing desire is to ask yourself, "What are some of the desires that the reader already feels?" and then, "How can I connect the idea I am trying to sell with one of those desires?"

This is not to urge you to attempt anything specious or far-fetched. The connection you attempt to show should really exist. In fact, if the connection between the action you try to obtain and the desires the reader feels is not genuine, the letter has little chance of success.

In arousing desire you do not emphasize every idea you can think of. You should make one single "appeal" dominate. Other possible appeals may be mentioned but should be kept subordinate. If a second letter is sent, a different appeal may be emphasized; but in any single letter, one incentive is emphasized and the others are kept in the background—visible, perhaps, but not conspicuous.

To secure *conviction*, simply use whatever evidence is best suited to the occasion. No generally appropriate methods can be suggested, as have been suggested for attracting interest and arousing desire. Whatever evidence you offer, you should carefully avoid an argumentative tone. Try to make it appear that you and your reader are both following certain facts to their logical conclusion.

In the effort to secure *action*, three points must be borne in mind: (1) The action suggested should be complete and specific. Rather than "Let us know your decision," you should write, for example, "Check the enclosed card to indicate your decision." (2) The action should be made as easy as possible. For example, it might be helpful to supply an envelope or card that is already stamped and addressed, or perhaps you can supply blanks to fill in rather than forcing the reader to write a complete letter. (3) There should be an effort to make the action immediate, even if delay would do no harm. Once readers start postponing action, they are likely to do so indefinitely, for the impulse to act is stronger at the moment of reading the letter than it will be later, when its contents have faded from memory. Therefore it is well to offer some plausible reason to act at once, preferably citing some benefit to the readers, if they do.

Certain final cautions are needed in regard to a letter aimed at action. First, the opening paragraph should be brief. This will make it look easy to read and thus encourage the receiver to start reading. If your action letter is a form letter, the hardest problem, quite often, is to induce the reader to read it at all. Second, the length of the letter needs careful consideration. If it is too long, a reader may not complete it—in fact may not even start it; but if it is too short, the reader finishes it before it has time to make any impression. Finally, you should not discuss any expense that you are asking the reader to incur until you have aroused enthusiasm about the project concerned. Talking about expenses too early arouses the reader's defenses and dampens enthusiasm.

An example of a letter urging action is in the section that covers form letters.

FORM LETTERS

Everyone is familiar with form letters, yet they pose some problems in writing that you may not realize when reading them.

A form letter is reproduced in quantity, so it is less impressive than a personally written letter, for the reader immediately realizes that the message is of no more personal concern than it is to many other people. It is frequently necessary to accept this disadvantage, however, when the same message must be sent to many people and when the time and expense of individual letters would be prohibitive. Moreover, the disadvantage can be somewhat reduced by obtaining a good reproduction job on a good grade of paper and by sending the letter as first-class mail in a sealed envelope. Sometimes, also, individual addresses and salutations can be inserted by typewriter so that the effect is less impersonal. If available, a word processor might be used. This allows the same letter to be typed automatically with only individual details, such as names and addresses, being changed, thus making the form letter almost indistinguishable from a personal letter. As any of these measures is considered, its expense must of course be taken into consideration.

Unless individual inside addresses and salutations are inserted, they are the parts of a form letter that pose the first problem. Sometimes it is possible to improvise two or three lines to replace the inside address, as for example:

To the Officers and Advisers of
Student Chapters of
The American Society of Civil Engineering

Ladies and Gentlemen:

More frequently, however, the inside address is omitted, and except for the heading, the letter begins with the salutation. This can be *Dear Sir* or *Gentlemen,* or *Dear Madam, Ladies,* or another appropriate salutation if the letter reaches both men and women. Sometimes instead, you may be able to use more specific salutations such as *Dear Purchasing Agent, Dear Subscriber, Dear Member, Dear Motor User,* or *Dear Friend.* These variations should be carefully chosen, however, for they are in danger of seeming too coy. Still another device is to replace the salutation with such a line as *To All Users of Drafting Instruments* or *To All Users of Power Saws.*

The only other question of form arises in connection with the signature. Sometimes, in form letters, an effort is made to make the letter seem more personal by reproducing the writer's handwritten signature. Often, however, the primary signature (that is, the first line below the complimentary close) is typed, usually capitalized throughout.

There is no reason that the body of your form letter really needs to differ from the body of individual letters, either in form or content. However, since a form letter is not likely to be dictated and sent without revision, you can often spend more time planning its layout. For example, you can emphasize a paragraph by changing the margins, or you can use the device of having the first paragraph end with three double-spaced periods in the middle of a sentence, and thus tease the reader into the second paragraph. Sometimes these eye-catching devices help to offset the disadvantage a form letter suffers in comparison with an individual letter, but they should not be so extensive that they make a letter look cluttered.

The soundest way, however, to compensate for these disadvantages is to take more care with a form letter than it is possible to take with most individual letters. Having more opportunity to revise, and being free to write a form letter when you can do it justice, you can often produce a much better letter than you can dictate as part of the day's routine.

One more caution is needed: The message of a form letter must be expressed so that it is appropriate for *every* person who will receive it. For example, if some of the receivers are in universities and others in business, or if some have been members of an organization and others are not yet members, the phraseology must be appropriate for either group. And if the letter is to be sent at different times of the year as need arises, it must avoid seasonal allusions.

Following is a form letter urging action.

Dear Sir or Madam:

As you know, the southern entrance to our city of Lakeland is unattractive because the highway runs parallel to the levee that was constructed twenty years ago as protection against high water that occurs during the spring runoff of the Canterbury River. Since you are a member of one of the service clubs that have always been interested in civic improvement, we are writing to solicit your help in getting rid of this unsightly feature.

The Army Corps of Engineers has scheduled a meeting in room 116 of the courthouse, to begin at 8:00 p.m. on Thursday, April 16. Their purpose is to determine whether our community would be seriously in favor of having the height of the levee reduced, having its present slope made more gradual, and having it landscaped where it runs beside the highway.

The Corps now has the authority to do this work as a pilot program like the one that has already led to a similar undertaking at Longchamps on the Weaver River.

Reducing the height of the levee would not create a danger of flooding. The two dams that have been built upstream since the levee was constructed have ample storage capacity to hold back excessive runoffs.

Whether this project is placed high on the Corps' list of priorities depends a great deal on whether the citizens of Lakeland show a keen desire to have it undertaken. If we let this opportunity slip by, there is little likelihood that we will have another like it for many years.

Feeling sure that you would like to have a conspicuous and unsightly stretch of rocks and dirt replaced by a parklike stretch of green lawn, shrubs, and trees, we urge you to attend the meeting and speak out in favor of this new project.

Sincerely yours,
Jane R. Miller, President
Lakeland Civic Improvement Society

ASSIGNMENTS

In writing the letters called for in the following assignments, make up headings, addresses, and signatures. When such an assumption is reasonable, assume that the letterhead occupies the top two inches of the page and use only a dateline as a heading. Be sure to type your letter and place it properly on the page.

In developing the details of a situation, you may need to assume that the date is different from the date on which you actually write the letter. You may also need to make up additional details. However, you should *not* make up facts that would change the essential nature of the problem. In particular, do not yield to the temptation to imitate the specimens that appeared earlier in the chapter. Rather, try to develop your own skills.

1. Write a letter to the United States senator in your district inviting him or her to speak at a meeting of an organization to which you belong. The organization could be a club, your school, or a civic group. Be sure to include the name of the group and explain its purpose. Indicate that you would like the senator to speak on the state of the economy or about recent legislation as either affects your organization, but assure the senator that you would be happy to listen to a talk on any subject. Although the talk will be for your organization, you plan to advertise the meeting and will open it to the public.
2. Write a form letter to be sent to numerous people inviting them to attend an event of the kind described in Assignment 1. (If this is to be an invitation to a banquet, do not offer complimentary tickets.) Though this is a form letter, it is to have an individual name and address on each copy, so provide what is necessary. If study assignments have not already covered form letters, you should read that section before writing the assignment.
3. Assume that you have received the letter called for in Assignment 1 but are unable to accept the invitation. Write an answer. If it seems appropriate, offer to have someone else in your organization come in your place. Remember that you wish to keep the good will of the organization involved so assure them of your continued interest in their goals.
4. Write a letter of instructions to be sent to a substantial number of people. The group might be homeowners in your community or students in

your school. The instructions are the steps to be taken to reduce the possibility of becoming a victim of crime. In the letter, give specific steps that your readers should take, but be careful not to frighten them with the specter of rampant violence since a too lurid approach might offend the recipients and cause them to throw the letter away unread. Your goal is to assure the readers of the general safety of your area and at the same time to educate them about what individuals can accomplish to increase their safety.

5. Write a letter of instructions to be sent to an individual who will proctor a writing test for an applicant to a graduate school. The proctor's task is to put the candidate at ease, provide the writing materials, time the test, and collect the papers and mail them to the school. In addition, the proctor is to interview the applicant concerning the reasons for applying to the school and the goals the applicant hopes to achieve by going there. The proctor is to write a brief report about the interview and submit it along with the test by a certain date.

6. Write an inquiry asking an organization that has been using some product, machine, or equipment about whether it is receiving satisfaction. You wish to know because your own organization is considering purchase of the same item. Though you invite general comment, include several specific questions.

7. Write a letter answering an inquiry and providing most or all of the information requested. This may be an answer to the inquiry called for in Assignment 6 or to some other inquiry that you have the information to answer.

8. Write a letter answering an inquiry—in this case an answer that contains little if any of the information requested. The specific situation may be similar to the one suggested in Assigment 6.

9. Write a letter asking for an adjustment based on the situation that follows or on a similar case of your own choice. The organization you belong to has ordered a product, machine, or equipment at a price that included certain options. After the item arrived and has been in use, it was discovered that several of the options have been omitted. For instance, you have ordered several automobiles for your company's fleet; the vehicles were to have special ratio differentials, electric overdrive and "gas saver" engines for fuel economy. With use, however, it was determined that these items were not the fuel-efficient options. Not only has your company paid for options it did not receive, the cost of running the automobiles will be substantially higher than planned. There are several ways adjustment could be made: a refund of the difference, since the vehicles have been purchased on time payments; trading the cars in for the vehicles ordered, with payment for their use; a refund plus a cash settlement for the difference in operating costs. Pick one or a combination of the suggested adjustments for your letter. Remember that you wish to make an amicable settlement even though you may have grounds for a legal suit.

10. Write a letter based on the following case or on one that is similar to indicate that action will be taken unless an undesirable situation is

remedied. Your company is located in an industrial park with winding roads, expanses of lawn, and trees and shrubs. Recently a number of people have begun using the main road through the park as a shortcut between two arterial roads and have been driving faster than is safe. Your company feels that if the city does not enforce the speed limits, the roads must be closed to through traffic by means of a gate with a guard. You would rather not close the road because a number of people, attracted by the beauty of the landscaping, enjoy driving through the park on weekends and in the evenings. Your company appreciates the good will involved while the community enjoys the benefits of a park. However, you fear the consequences of a serious accident and feel that unless action is taken by the city, you will be forced to close the road. You have already spoken to the police department, who say that their jurisdiction is questionable because the road is on private property and they are understaffed and cannot spare a car to patrol the area except intermittently. Address your letter to the city council. Don't make your proposed action sound like a threat, because you wish to retain the council's good will, but the problem is serious and requires immediate attention.

11. Write a complaint—one that does not ask for an adjustment—based on the following case. Your company has been remodeling its main office building. Work crews arrived as scheduled, set up scaffolding, blocked parts of hallways and stairs, appropriated one elevator, and generally created problems for the smooth working of your company. Then the remodeling work seemed to stop for several weeks, with only a few men showing up part of the time to do what seemed like minor jobs. The work has now been completed, but you feel that your company was inconvenienced for an unnecessary length of time. You are aware that contractors sometimes bid on a job when in fact they are not available, leave the job they are currently working on long enough to set up on the new job, then return to the original job to finish it before actually starting work on the new one. The reason for this practice is to ensure work for the contractor's crews. You suspect you have been a victim of such a practice, but have no proof. Write a letter letting the contractor know of your dissatisfaction but use reasonable restraint—don't accuse anyone of shady practices, but you wish to prevent such inconvenience in the future.

12. Write an answer to the letter called for in Assignment 11. Do what you can for the sake of good will. Your company subcontracts much of the actual work and all companies engaged in this kind of work frequently have similar problems. Although time and materials can be estimated in advance, often difficulties arise during the actual work that cannot be foreseen. Sometimes previous jobs take longer than planned, sometimes workers cannot be found to fill out crews. Indicate that you will call the complaint to the attention of your subcontractors and will do your best to see that similar inconveniences do not recur. Point out that your primary concern is in the quality of the work completed, and

since no complaint was made about it, you assume that it was fully satisfactory.

13. You are connected with a company that manufactures electric motors and sells them wholesale. A distributor writes to complain that he purchased an order from you only to discover a short time later that a competitor made a similar purchase at a lower price. The distributor wants a refund. You check your records and discover that the distributor made his order in August at a point when the price of copper was at the highest it had been in several years so that the price of wire you buy for windings in the motors was up. His competitor had ordered several motors earlier while the price was lower, which allowed you to order wire at a reduced price. Since you resupply based on orders, your price must be based on the current market price of your raw material. While in individual cases, such a practice may appear unfair, in the long run it allows everyone to purchase your motors at the lowest possible price and still allows you to make enough profit to remain in business. Write a letter declining to make a price adjustment.

14. You are a general contractor operating on a small scale. A store owner writes to tell you that water still leaks into his basement though one of your men supposedly plugged the leaks. Some of the merchandise kept in the basement has been damaged and will have to be sold at a loss. He also says that your statement charges him for three and one half hours of labor, but the man was on the job an hour less than that amount. The writer feels that you should do the job over again without charge. He is also considering going to court to recover the loss of his merchandise.

    Facts not in his letter are as follows: You had told the man that you could not guarantee an inside job; but outside repairs were prevented at the time by the weather and, besides, would cost substantially more. You were given the go-ahead under such circumstances. Also, the workman who did the job on an emergency basis had to leave another job to do it, had to go to a store for materials, and had to drive back the five miles, afterwards, to the job that he had left. This accounts for the apparent overcharge for time. Write a letter explaining these facts, but saying that in an effort to be fair, you will reduce the labor charge to the amount your customer says it should be. You will not do the job over without charging him. Try to sell the customer on the idea of waiting until the weather is good and then having the leak patched from the outside, which will cost more but can be guaranteed.

15. Write a form letter 250 to 300 words long in which you try to convince the readers to perform some action such as promoting a project desired by your school, supporting or opposing a zoning change, changing the season for hunting—anything of current interest in your school, your community, or your profession. As an alternative you might consider soliciting pledges to support some program or activity.

    The date may need to be different from the date on which you actually write. Those who receive the letter should have something in common, such as being farmers, parents, sportsmen, people interested in public health or environment. Supply an inside address that would be suitable for all who receive the letter. (Unless the sort of group to which the letter is sent is obvious from the letter itself, add a note giving this information to the instructor.)

17 Letters Concerning Employment and Resumés

For a person about to graduate from college, the problem of securing employment varies from year to year depending on the current job market. Even in poor years, however, large corporations and governmental units send representatives to colleges and universities for the purpose of recruitment to such an extent that a letter of application is not always the first contact between the employer and the prospective employee. On the contrary, when graduating seniors today write letters in connection with employment, they often have been interviewed already, have received an application blank, and may have filled it out and submitted it. As a result of an interview, a company representative will have formed and recorded impressions about the applicant's appearance and personality to supplement the facts about qualifications obtained from the applicant and from other sources.

Under such circumstances, the traditional instructions about how to write letters of application are not entirely applicable. Rather than starting from scratch and writing a full-scale appplication, applicants will base letters on the way they size up the situation as a result of the interview and on any leads they may have picked up regarding what they should emphasize among the many things they may feel justified in saying about themselves.

Still, the full-scale letter of application is not a matter to be neglected. There are still numerous occasions when it may be written by students seeking their first employment after graduation, and it is as important as ever to the person who already has a position but wants to apply for something better in a different company. Therefore this chapter will first cover the traditional letter of application, because even those who do not need to write the full letter may need to use parts of the material it contains. The chapter will also discuss how to construct resumés that really present your qualifications, rather than simply present statistical data. Next, attention will be centered on letters written after an applicant has been interviewed. And finally, some concrete suggestions that apply to letters written under either circumstance will be presented.

THE CONVENTIONAL
LETTER OF APPLICATION

The Opening

When you write a conventional letter of application for a specific position, start with a short paragraph indicating the position you are applying for and telling, perhaps, where you learned that the position was open. If necessary, identify yourself—for example, as a graduating senior or as the occupant of whatever job you hold. Make it clear that your letter is definitely an application, if such is the case, rather than just an expression of interest that includes some facts about yourself to justify a request for information. In general, let your opening paragraph merely establish contact and provide the information that will enable your reader to judge your qualifications without being annoyed by unanswered questions.

Try to put something that will arouse your reader's interest either in the first paragraph or in the opening of the second paragraph if you can think of anything that will serve such a purpose. This might be, for example, a reference to some extraordinary qualification or to an unusual combination of qualifications. If someone well known to the reader told you about the opening or is willing to recommend you, the information should be placed here. Not everyone is fortunate enough to be able to include such a bid for interest, but if you can somehow, at the outset, set yourself apart from the run-of-the-mill applicants, what follows is more likely to receive attention.

The Central Section

The central section should follow smoothly after the opening. Since modern practice involves the inclusion of a resumé with almost every application letter, the old sections in the letter on education, experience, personal information, and references no longer apply. In fact, many writers are at a loss about saying anything after the opening paragraph, assuming that their resumés include all pertinent information. Such an attitude is a mistake, for a number of reasons. First, all resumés tend to look alike to personnel managers and others looking through them for applicants. Second, what may seem obvious in the resumé to the writer may not be obvious to the reader. And all the information in your resumé appears equally important, unless you draw attention to it.

It follows, then, that the central section of your application letter should draw attention to specific points in the resumé, explaining or amplifying items that might be unclear or overlooked. The judgment you make about which items to select for amplification depends on your ability to analyze your readers, for an application letter is probably the most closely focused writing you will do. Naturally, you should discuss those items that will make you a more desirable candidate, and if there are items that might disqualify you if left unexplained, you should discuss those.

For example, suppose your overall grade average in school is low, but you received high grades in the specific area you are now applying for. Rather than assuming that the reader will look over your transcript to discover this fact, you might point out, not that your grades were poor, but that you did exceptionally well in courses related to your major interest. Or perhaps you spent time on particular projects that turned out well. Even though your overall work suffered, you could discuss the nature of the projects. In short, you must pick those points in your background that would be most important to your reader so that they reflect well on you, then point out how they do so. Do not assume your reader will be able to figure it out without your help. Like others in business, those people hiring for jobs are busy and appreciate information that makes their tasks easier.

If you have experience in the specific area you are applying for, you naturally should discuss some aspect of the experience that would not normally be part of the information on the resumé. For instance, if you advanced more rapidly than was normally the case in the organization you worked for, you might point that out and explain some of the reasons. If you were involved in a successful project, even if you were not in charge, you might discuss the project, and perhaps point out what you learned from the experience.

Above all, in the application letter, you must be straightforward and base your information on material that could be verified if the reader cared to follow up by contacting your references. But you cannot be falsely humble. You must assess your best qualities from the reader's perspective and draw attention to them. People in the business of reviewing job candidates expect you to put your best self forward. If you base your assessment solidly on verifiable information, you will avoid sounding like you are bragging and yet sound like someone worth hiring.

The Ending

You will improve your letter if you can say something to give it a lift at the end. A single-sentence summary of your qualifications might serve the purpose. Another possibility would be a comment on the work you have done to qualify yourself for such a job and the effort you will consequently put forth to make good. Whatever you may say to strengthen your ending must be the outgrowth, of course, of material that your letter has included, and it is therefore impossible to offer a formula for your guidance.

The major function of the ending is to suggest action. This action may be the offer of a job, the invitation to come for an interview, or a letter providing information or telling you what to do next. You will need to appraise the situation, decide what action is the next likely development, and phrase your suggestion accordingly.

Try to phrase your effort to secure action in such a manner that it avoids negative suggestion. When you write, "*If* my qualifications are of interest, I shall be glad to come for an interview," you imply uncertainty whether the reader will be interested. Though you should not appear to take too much for granted, you might avoid this implication by writing something such as, "I shall appreciate your consideration of my qualifications and shall be glad to come for an interview on any date that you suggest." Such an offer gets rid of the *if* without seeming overconfident.

Unless your potential employer is hard-pressed for help, it is best to refrain from trying to hasten action. If it is essential, however, that you receive an early response and if you are tactful, you might be able to speed matters along without giving offense. No resentment would be aroused, for example, by such a sentence as, "Since I shall be leaving town for one month on June 10, I should appreciate it if I might have an interview before that date." Most employers are normally considerate and are not likely to resent a reasonable request tactfully made. But pressure must be applied with care or your effort to obtain early action may reduce the chances that action will be favorable.

RESUMÉS

The resumé is a complete picture of the facts. It should bear your name, address (future as well as present if a change is imminent), and telephone number, should be dated, and may include the position you are applying for. It should contain routine personal information, and it should give full information on education, experience, and references.

A clear understanding of the relationship between the resumé and the letter is important. The function of the resumé is to free the letter from certain routine information that is uninteresting but necessary for completeness. In the resumé you provide the full record. In the accompanying letter you focus the reader's attention on the particular facts that are most important and bring out their full significance.

A resumé cannot perform all the functions of a letter; rather, it performs the functions of an application blank. However completely the resumé may present the facts, it remains for the letter to create a personal interest, for it is only in the letter that you seem to be talking to the reader personally.

There is no standard form for the resumé, but it should be a tabulation rather than a discussion in sentences. To decide on a system of headings, jot down as rough notes all the facts to be covered, and then work out a plan of headings to cover them. Arrange the material on the page so it has a pleasant appearance, and make it as neat and accurate as the letter itself.

Types of Resumés

The traditional resumé, now often called the chronological resumé, is likely to have such headings as *Vocational Goals, Personal Information, Education, Experience,* and *References.* The education and experience sections are stressed, with pertinent data listed in reverse chronological order, that is, the most recent information first, then the second most recent, and so on. Such resumés are preferred in conservative industries where jobs are well-defined and the skills required are fairly standard. Chronological resumés are particularly effective if they show you have advanced steadily in your field.

In recent years, however, other types of resumés have found wide use, especially in fields with less defined job areas and skills, such as the service and personality-enhancement industries. Applicants in these fields feel that the traditional resumé hides their real skills behind a screen of dates and job titles. Thus the "functional" resumé has developed, which highlights skills and de-emphasizes specific job assignments. Such a resumé might briefly summarize education and jobs under a heading such as *Summary of Background* and go on to stress skills the applicant has acquired. These resumés might include such headings as *Leadership Skills, Research Skills, Machine Operating Skills,* and the like. Functional resumés are particularly useful for those individuals who have changed careers, and thus have a long list of jobs that appear irrelevant to current employers, and for individuals with a wide variety of jobs with different employers. A functional resumé might also be useful for students with little or no job experience of a traditional type.

In addition to the two named "types," resumés come in various "mixed" types, which use some elements of one type and some of the other. The goal, of course, is the creation of a document that makes the applicant look like a good candidate for the job and somehow stand out from all the other candidates. Our advice is to use whatever elements suit you best, but remember, you will stand out best not by trying gimmicks but by analyzing your potential employer and presenting your credentials in a way that shows that you best fill the employer's needs.

Contents of Resumés

Whichever type you decide to use, the following information is expected by people examining resumés.

Vocational Goals. Most resumés include the kind of work being applied for. This area can be as short as "Chemical Engineer" or as long as a detailed description of the kind of job atmosphere you are trying to find. Often this section can include a statement about long-term goals, such as "Chemical Engineering position with opportunity to use administrative skills; eventual management position with international firm allowing chance for travel and work in other cultures." In a period of plentiful jobs in your specialty, such a description may help you find a more congenial position than you might otherwise find. Functional resumés tend to include longer descriptions in this section.

Education.   As a minimum, name the higher educational institutions that you have attended and the degrees you have earned or are about to earn. If you have an outstanding academic record, mention it. Do not hesitate to include extracurricular activities, for they help to round out the impression about the kind of person you are.

The amount of detail about education should be governed by the amount of other information you expect to present, the strength of your scholarship, and the extent to which your academic training is directly related to the work you would do on the job. Facts about education that one applicant should mention—for example, courses taken—might well be omitted by an applicant whose other qualifications are stronger.

Experience.   Nothing is better evidence of your fitness for a job than successful experience in a job of similar nature. If you have had such experience, include it in detail. Do not feel, however, that experience in a job of any kind is unimportant. In almost any kind of work, you have had a chance to demonstrate good personal qualities, and it is well known that personal qualities are often a decisive factor in an employee's success. In general, the picture of your experience should be complete. In the chronological resumé, if anything is omitted, there will be noticeable gaps of time in your record, and these will make a reader wonder whether you have something to conceal.

Personal Information.   As a minimum, tell your date of birth. Although discrimination because of age is illegal, employers rarely hire a 20-year-old to supervise older colleagues. Other data, such as sex, weight, height, marital status, and freedom from physical handicaps were once routinely included but now tend to be omitted unless important to the job. Reference to leisure interests will sometimes help the reader to think of you as a person, especially if your hobby skills can be related to job skills. Additional facts about your background help to make the picture complete and are especially likely to be valuable if your background is in any way related to the field you are seeking to enter. Of course the value of personal information, like the value of any information, is relative. Whether certain facts are worth including may depend on how much else you have to say, and it might be helpful for one applicant to include information that another could omit.

References.   Practice varies today on the inclusion of references. Many resumés simply state, "References available upon request." Others list the references. In either case, you should have available three to five individuals who can speak knowledgeably about your qualifications. With this in mind, you may use one friend of long standing, and one teacher, but the majority of your references should be employers since the person doing the checking will want to know about your working skills more than any other point. If you are just completing your work in college, include at least one reference from the institution that is awarding your degree. The number of references should of course depend on how many are necessary to confirm what you have said in your letter. Courtesy and self-interest make it advisable to secure the consent of anyone whom you name as a reference. And by all means make sure that you have the name (spelling included), address, telephone number, and position of each reference absolutely correct. Unless it is obvious, make sure that the reader knows why a person whom you name as a reference is qualified to judge your fitness for the job you are seeking.

LETTERS FOLLOWING INTERVIEWS

As mentioned earlier, for a student who is about to earn a degree in a large or middle-sized institution, the routine of obtaining a position may begin with an interview on the campus and filling out of an application blank. If such a student writes a letter at all, it will be only to supplement the considerable information already in the hands of the potential employer, or to cope with some later development. The person who writes under these circumstances may be able to apply some of the preceding instructions but the letter will not stand independently as the major instrument for obtaining employment.

After an interview, you can help yourself by writing a good follow-up letter. Of course you should write to say "Thank you for the interview." even if you have decided that the job is not one you want. In a time of mobility, you never know when, or where, you may meet that interviewer again. If you are interested in the company after the interview, however, you should expand the thank you by the addition of other information that would be helpful. What kind of material could you include? Some of the possibilities are as follows:

> You can answer or supplement your earlier answers to questions that were not entirely settled.

> You can describe the growth of your interest in some matter that you had not thought about intensively before the interview—for example, a kind of work that is different from what you had had in mind.

> You can dwell upon facts about the company that increase its appeal to you.

> You can say that you have been giving additional thought to what qualifications you can offer in some particular respect, and supplement what you said earlier on that point.

> You can say that you have been thinking a great deal about some remark made by the interviewer, and review some aspect of the case in the light of this added consideration.

Admittedly it takes imagination to translate these general ideas into definite, concrete terms. There is no guarantee that every applicant can produce a letter that will help. But it is probably true that most job seekers, after an interview, think of things they wish they had said, or wonder whether they made themselves entirely clear on certain points, or think of questions they wish they had asked. The exercise of ingenuity will often make it possible to use some such afterthought as a point of departure for a letter.

And what can such a letter do for you beyond supplying concrete facts not supplied before? The answer to this question depends largely on your adroitness as a writer. Assuming that you write it skillfully, such a letter can help you to gain credit for initiative, originality, imagination, and aggressiveness. It can convey the impression of a lively interest in the company, beyond the fact that you want a job.

Also, it can demonstrate your literacy and your skill in writing—a skill that improves your potential value in any large organization. Just as there are some people who write badly but make a good personal impression, there are others who can present themselves better when they write than when they talk. If you are one of these, your letter may help to counterbalance your conversational shortcomings. And even if you are fortunate enough to make a good impression in a conversation, you will strengthen that impression by showing that you write well in addition to talking well.

Of course if you are weak in writing, there is no reason to feel that you can help your prospects by writing when it is not essential to do so. The best short-range policy under these circumstances is to keep every letter as simple as possible and at least to avoid the pitfalls that are pointed out in the remainder of this chapter. The only suitable long-range policy is to improve your skill in writing rather than reconciling yourself to a weakness in one of the most basic skills in science and industry. Poor writing abilities are a handicap that will pursue you during your entire working life.

GENERAL SUGGESTIONS

Neatness

Neatness is a must. Before the people who receive your letters begin to read, they form an impression about you on the basis of the appearance that affects the reception of the contents. If sloppiness and carelessness are qualities of your letters, they associate you, at least subconsciously, with those qualities, and you are handicapped from the outset.

Further, a failure to make your letter neat is not very flattering to those to whom you send it. You force your readers to conclude either that you regard them and the position as unimportant or else that you are satisfied to turn in an inferior performance on an important occasion. You can hardly expect them to believe you will do your utmost to serve the interests of their company if you are careless even when working directly in your own interest.

In view of these facts you should no more send a slovenly looking letter than you should appear for an interview with uncombed hair and dirty clothes.

The "You Attitude"

Don't let your desire for a position make you overlook the fact that a potential employer is less interested in your eagerness for the job than in the question of whether you can become a valuable employee. Your tastes, ambitions, and desires are irrelevant unless they will cause you to do better work. Sometimes, to be sure, they will improve your work. The fact that a job is in line with your long-range ambitions might make you try harder to do it well. Your liking for a certain region might reduce the likelihood that you would resign because of a desire to live elsewhere. Still, you are wasting space if you talk about such matters without relating them to the way you will function as an employee.

In this connection, before you apply for one specific position you will do well to consider carefully the demands of the job as well as your own qualifications, so that when you write your application you can make it clear to the reader that you have done so. The effort to create such an impression must be made carefully; you do not wish to appear to be telling the reader what he or she wants. But if you do the job well, you can suggest that before you decided to apply, you carefully considered the question, "Is this the right kind of job for a person with my qualifications?" All of us know things about ourselves that no one else knows. Your potential employers realize that try as they will, they cannot discover all of your limitations. If you can convince them that you have coolly analyzed your strong points and your weak points and are applying for the job that your abilities qualify you to perform, they will take your application far more seriously.

Objectivity

Keep your letter objective. You will gain nothing from making unsubstantiated claims that you are self-reliant, dependable, tactful, and the like. Possibly you are, but such claims are just as likely to appear in the letters of applicants who have no justification for making them. Anyone who has had much experience in hiring employees knows this, and nothing will lose interest faster than claims that are not supported by objective evidence.

Tone

A satisfactory tone is usually the result of striking the right balance between extremes. You should aim to appear self-confident but not conceited; obliging but not ingratiating; earnest but not pompous.

The tone of your letter will result in part from your choice of language. Many a person whose manner in a conversation is entirely normal becomes stiff and pompous when writing a letter of application. Phrases such as *earnest desire, intense determination,* and *it is my wish to* are unlikely, however, to increase your chance of obtaining a position. On the contrary, they suggest that you are so eager to be impressive that you have been unable to express yourself in a natural manner. Everything said about the language of business letters in Chapter 15 applies with special emphasis to letters of application; and if it is not fresh in your mind, you would do well to review the section dealing with that point.

Excessive Use of "I"

Since you are telling about yourself when you write a letter of application, you will probably find it difficult to avoid using "I" at the beginning of every sentence, thus making your letter monotonous. It is better, however, to use *I* liberally than to try so hard to avoid it that your sentences become involved and unnatural, or that you sacrifice parallel structure where it might be appropriate. The best way to solve this dilemma is to write the first draft of your letter without worrying about how often *I* appears and then, as you revise it, change a sentence here and there until the use of *I* is no longer conspicuous.

Discussion of Salary Expected

Much of the time you will know whether a position pays an acceptable salary before you apply for it; and even when this is not the case, you are usually in a better strategic position if you let the employer be the first to bring up the subject. If you feel it necessary, however, to mention the salary you would expect, the place to do so is near the end of the letter. Your first job is to convince your reader that your services would be of value, for a potential employer who has not yet been convinced that your services would be of value is not interested in what it will cost to obtain them.

Making Sure All Questions
Are Answered

Before you consider your letter complete, make sure that it will not leave any unanswered questions lurking in the reader's mind. For example, make it easy to see that there are no substantial periods of time left unaccounted for in your history. Also, unless it is apparent that the position you apply for is an advancement, explain why you prefer it to the one that you already hold. If you are not working, let the reader know the reason unless it is obvious, as it will be if you are just finishing school.

Failure to make the picture complete by the inclusion of such information may delay a decision about whether to hire you until there has been another and needless exchange of letters, or may even result in the job's going to someone else who seems to be qualified and who has provided all the desired information.

Seeing Things as the Reader Sees Them

One final suggestion is in order. When you are keenly interested in a job and sincerely convinced that you could do it well, you will be strongly tempted to make statements about yourself that sound less impressive to someone else than to you. You will probably feel that these statements are justified, and you may be right in thinking so. The reader, however, does not see things through your eyes. It is therefore a good idea to imagine yourself receiving the letter—from someone whom you do not know any more about than the reader knows about you. Ask yourself, "If I were the reader, would these statements impress me as I hope they will impress the person I am writing?" If you are capable of using your imagination in this manner, you may eliminate some lines that sounded good to you when you wrote them, but that later seem less appropriate. By doing so you will produce a letter that is not diluted by ineffective material and will therefore be more likely to carry conviction.

SPECIMEN LETTERS AND RESUMÉS

Following are two letters of application, each accompanied by a resumé. These letters are in line with the principles described in the preceding discussion; but they are specimens, not models. If a letter of application is assigned, do not borrow phraseology from the specimens unless it fits your own circumstances.

1149 Park Hill Road, Apt. C
Corvallis, Oregon 97330
April 10, 1982

Personnel Director
Environmental Protection Agency
National Field Investigations Center
Denver Federal Center
Denver, Colorado 80225

Dear Sir:

I would like to apply for a position as Field Investigator with your agency. I shall be receiving my Bachelor's degree in Forestry from Oregon State University on June 10, and shall be available at any time after that date.

As indicated in the attached resumé, I have worked two summers as a Fire Guard. Between fires, I was assigned to a variety of duties, but the one most applicable to the kinds of tasks I would undertake with your agency was the Camp Ground Patrol duty. For this duty, I was required to check the two camp grounds in the Big Creek district to make sure campers were obeying regulations and were not causing needless damage. This duty involved contact with the public, and frequently required a great deal of diplomacy in convincing campers to obey the rules. I wore no uniform and had no real power except to report to the ranger, so I had to use persuasion. Throughout the period I had this duty, I had to resort to calling the ranger only once. It seems to me that the skills I used would be similar, though of course not on the same scale, as the skills I would need as a member of your agency in discussing possible violations with individuals being investigated.

With this in mind, besides the regular courses leading to a Bachelor of Science in Forestry, I have taken courses in speech, persuasion, group dynamics, and psychology. They have taught me a great deal about people, and as a result they have made my contacts with the public more enjoyable for me, as well as, I think, more successful.

I have always enjoyed the outdoors, and have for a long time planned to find work in an area where I could help so that others could enjoy it as I have. I feel that a position with your organization would help me achieve this goal, and I feel I could contribute to your agency's goals as well. I would be available for an interview at your convenience any time but the first week in June, when I take my final exams. Thank you very much for your consideration.

Sincerely,

John Q. Carter

John Q. Carter

Figure 18.   Well-Written Application Letter. The writer has not done the exact kind of work applied for but makes good use of other material.

RESUMÉ

John Q. Carter

School address: Permanent home address:
   1149 Park Hill Road, Apt. C   Box 67, Route 4
   Corvallis, Oregon 97330   John Day, Oregon 97845
Phone: 503 754-4266   503 693-7659

Date of Birth: July 10, 1961   Place of Birth: Bend, Oregon

Vocational Goals:
   Work that combines contact with people and strong environmental interests, where I can use my background in science and
   forestry with my study in psychology.

Education and Work Experience:
   B.S. in Forestry, Oregon State University.   B average.
                Expected June 10, 1983
   Graduated John Day High School.   B average.
   Fire Guard, Big Creek, Montana; Summers 80, 81, 82.
   Ranch work, part time and summers, 1974–79.

Skills:
   Communications. Written papers and reports throughout college
      including major reports in technical report writing, forest
      engineering, and advanced psychology classes. Personal
      communications were necessary in my Fire Guard assign-
      ments and in my supervisory positions.
   Supervisory. Head of ten-man planting crew for Forest Service,
      and overseer of a haying crew on my grandfather's ranch.
   Forestry. Bachelor of Science degree plus three summer's experi-
      ence in all areas; including forest management, recreation
      management, and forest ecology.
   Independent work skills. Most of the jobs in the Forest Service
      involved working alone or with one or two others. On the
      ranch I often worked alone. In all these, I had to show
      self-motivation and stick-to-itivity.
   General outdoor skills. I have done everything there is to do on
      a ranch including set fences, drive heavy equipment, re-
      pair vehicles and farm machinery, nurse sick animals.
      Fire Guard work requires all outdoor skills often under ar-
      duous conditions.
   Personnel skills. Have always gotten along with fellow workers,
      both superiors and those working under me. Was active
      member of Outdoor program in college, was volunteer Big
      Brother, campaigned for student government, was on in-
      tramural basketball team.

References:
   Available from Oregon State University Placement Office, Corvallis, Oregon, 97331.

Figure 19.   Specimen Functional Resumé (to be sent with letter shown in Figure 18).

739 Circle Drive
Pullman, Washington 99163
April 10, 1982

Mr. Eugene M. Johnson
The Asphalt Institute
College Park, Maryland 20740

Dear Mr. Johnson:

During a recent visit to Pullman, Mr. Harvey Darrow, who was associated with The Asphalt Institute for many years before his retirement, told me that you usually have openings at this time of the year for two or three chemical engineers. In June I will receive an M.S. degree in Chemical Engineering from Washington State University, and I would like to apply for one of these openings.

I believe I could do satisfactory work for you because both my education and my work experience have been in line with your operations as Mr. Darrow explained them.

My graduate study has been in a field that involved work with asphalt. I participated, for example, in studies at the Riedesel Testing Facility—studies including both the design and testing of various asphalt mixes to determine their suitability for highway surfacing and other uses.

Before I returned to the university to earn an advanced degree, I was employed for one year by the Superior Paving Company of Richland, Washington and supervised the laying and inspection of highway surfacing. Then for a year I was a laboratory assistant in the asphalt laboratory of the Corland Oil Company in Martinez, California.

I have talked at length with Mr. Darrow about your work and have read several of your reports describing one or another of your projects. The more I learned, the more interested I became. I would welcome an opportunity to join your staff because your research work is the kind I have been preparing to do and because the atmosphere in which it is carried out, as described by Mr. Darrow, would make my work enjoyable.

To make more information about me immediately available, I am enclosing a resumé, but I would be glad to fill out an application blank if you will send one.

I would greatly appreciate any word you might send me about my prospects.

Sincerely yours,

Edwin A. Neal

Edwin A. Neal

Figure 20. Well-Written Application.

QUALIFICATIONS OF EDWIN A. NEAL
Applicant for a position as a chemical engineer

<u>Address</u> 739 Circle Drive
Pullman, Washington 99163
509 632-8971   <u>Date</u> April 10, 1982

PERSONAL INFORMATION

Date of birth: June 24, 1955   Health: Excellent
Weight: 165   Marital status: married,
Height: 5 feet 10 inches   two children
                                       Leisure interests: golf, chess

EDUCATION

Cheney High School, Cheney, Washington. Graduated in 1973
Eastern Washington State College—1973 to 1975
Washington State University—1975 to 1977
                    1980 until present
    Degrees: B.S. in Chemical Engineering—June, 1977
              M.S. in Chemical Engineering—will receive in June,
              1982
    Academic record: undergraduate 2.9, graduate 3.1
    Activities: member of American Society of Chemical Engineers
                 (student chapter)

EXPERIENCE

<u>Experience related to professional objective</u>

    July, 1979 to September 1980—Laboratory assistant in asphalt
    laboratory of Corland Oil Company, Martinez, California
    June, 1978 to July, 1979—Superior Paving Company, Richland,
    Washington. Duties: to inspect and supervise laying of asphalt
    surface on roads and airport runways.
    September, 1977 to June, 1978—Assistant (part-time) Mate-
    rials-Testing Laboratory, Washington State University

<u>Experience—General</u>

    Summers of 1975, 1976, and 1977—Truck driver, A & S Produce
    Company, Kent, Washington

REFERENCES

Duane V. Kellogg   Martin A. Driscoll
Professor of Chemical Engineering   Associate Professor of Chemi-
Washington State University   cal Engineering
Pullman, Washington 99163   Washington State University
509 729-3651   Pullman, Washington 99163
                                       509 729-5444

Donald R. Newsome, Manager   Frederick S. Whitcomb
Superior Paving Company   Director, Asphalt Laboratory
1860 Front Street   Corland Oil Company
Richland, Washington 99352   Box 895
509 324-6278   Martinez, California 94553
                                       805 622-3333

Figure 21.   Chronological Resumé (to be sent with letter shown in Figure 20).

ASSIGNMENTS

Assignment 1

Construct a chronological resumé, using your own job experience and education. Do not make up or pretend to have had experience that you do not have, but present the information in the best light possible. If you are in school and will not graduate in the near future, plan the resumé as if you were going to use it in applying for a summer job as a trainee, or for a part-time position while continuing your education.

Assignment 2

Construct a functional resumé. Although you may use the same or similar headings as the example in Figure 19, make sure the headings you decide to use reflect your true abilities. As in Assignment 1, plan the resumé to fit your current situation as closely as possible. When you have finished the resumé, compare it with the one you constructed in Assignment 1 to determine which represents you better, and which you are more comfortable using.

Assignment 3

Write a letter of application setting forth your own qualifications for a position that you might plausibly apply for. Let it be assumed that you definitely know that an opening exists and that the reader knows little if anything about you. In stating your qualifications limit yourself to facts except that if the instructor permits, you may assume that you are about to complete or have just completed the course you are taking. Obviously, this assumption would include your having gained any experience you are likely to gain before graduation. Do not make unrealistic offers, such as an offer to travel farther for an interview than you would really be willing to travel under the circumstances. Plan your letter to include a resumé.

Assignment 4

Assume that you had been scheduled for an interview with some potential employer's representatives who were on your campus or in the town where you are located. For some reason, such as an attack of influenza, you were forced to cancel the interview. You have learned that in the near future a representative of this employer will be holding interviews in some other city perhaps fifty or one hundred miles away. Write to see if you may be interviewed there.

You may assume, if you like, that you have already submitted an application blank or, if you prefer, that you are sending a resumé with your letter. If other assumptions would make the assignment more plausible in your particular case they are also permissible. The basic situation, however, should remain unchanged. In this letter you are attempting to demonstrate your interest in the company concerned and seize the opportunity to say enough about yourself to arouse interest.

Assignment 5

You have been negotiating with a potential employer and have reason to believe that the offer of a position is likely. Another employer with whom you have also been in contact makes you a definite offer. You feel that a position with the first employer mentioned would be preferable, but the one you are offered is good enough that you do not wish to lose your chance for it.

Write a letter to the employer who made the offer. You should probably tell him that you have been negotiating for another position but the major purpose of your letter, at least ostensibly, should be to ask some questions you would like to have answered before you make a definite decision.

Assignment 6

Under the same conditions described in the preceding assignment, write a letter to the potential employer who has not offered you a job, telling about the offer that you have received. Without letting your letter sound like a threat, indicate that you will soon have to accept or decline the offer you have received. Try to make your reader understand that though a position with the reader's company would be your first choice, the other position is good enough that you are reluctant to turn it down unless you can feel confident of something better. Try in a tactful manner to obtain a definite decision as soon as possible, or at least to obtain definite information about when a decision will be made.

Assignment 7

Assume that you have been interviewed by the representative of a potential employer four or five days ago. Write a follow-up letter of the kind described in the chapter.

IV

HANDBOOK OF FUNDAMENTALS

The Handbook of Fundamentals provides answers to many of the common questions about writing that might arise during the production of any paper or report. In addition to making the book a more useful reference work, the Handbook allows the text proper to treat material that concerns technical writing specifically, unencumbered by many questions that concern all writing—questions that may be of major importance if overlooked, but that nevertheless would make the treatment of technical writing too complex or lengthy. Another reason for including the material here rather than in the text proper is the wide range of abilities of the readers of the book. Many users seldom need to be reminded of a point of grammar or usage and thus are ready to deal with the advanced problems of technical writing, report writing, or proposal writing. Others need to spend time with more basic matters while pursuing the advanced. For the former, a casual glance now and again at a dictionary or handbook may be enough; for the latter, a reference work needs to be constantly at hand. For both groups, questions are more likely to be pursued if answers can be readily found. Of course, no handbook can substitute for adequate training in English; however, to gain the most value from this handbook, we suggest that readers study the material enough to become familiar with the presentation so that when a question arises, they will know where to look for the answer. The Handbook includes material on standard usage, punctuation, spelling, and broader matters such as unity and coherence in sentences and paragraphs.

PARAGRAPHS

General Comment

The following discussion is concerned mainly with paragraphs that comprise the bulk of most writing, technical or otherwise. It also contains brief comments on paragraphs of isolated statement. To be sure, there are other kinds of paragraphs: introductory, transitional, and concluding paragraphs, and paragraphs of dialogue; but the first and third have been covered in the text, the second does not give rise to much trouble and the fourth is rarely seen in technical writing.

The function of a paragraph is to group together sentences that concern the same topic and combine to form a thought unit. Some of the points to bear in mind in producing paragraphs that perform this function are as follows:

1. Effective paragraphs do not result from merely beginning a new paragraph whenever the one you are writing has grown long enough, nor even from merely beginning a new paragraph whenever you take up a new topic. Your paragraphs will be effective only if you know, each time you begin one, what point you intend it to make or at least what ground it will cover. That is, as you begin a paragraph you should have a definite idea about where you will be at its end.

2. A paragraph is likely to be more effective if its basic idea is expressed in a topic sentence. A topic sentence is usually most useful near the beginning of the paragraph, but it may be placed at the end; occasionally, for the sake of emphasis, it may be placed at the beginning and restated at the end in different words. Wherever it appears, it should dominate the paragraph, and the rest of the paragraph should serve mainly to develop it.

3. "Developing" a topic sentence means providing the additional facts and ideas that are needed to make it clear and acceptable, and possibly to indicate its importance. If a sentence does not need such development, then either the idea it conveys is not big enough to form the basis of a paragraph or else—much less frequently—it should stand alone as a single-sentence paragraph. If an idea cannot be developed in a paragraph of reasonable length, then it should be broken up into ideas of paragraph size.

4. If the topic sentence is merely implied and the reader must grasp the point of the paragraph from the details themselves, the first sentence should never fail to create an accurate impression about the topic and the basic idea of the paragraph. Readers who skim rapidly should not be misled about the contents of a paragraph when looking at its first sentence.

5. Sometimes it is necessary, for the sake of fairness and completeness, to include material that runs counter to the main point of the paragraph. This material should not be placed at the end, for the end should strengthen, not weaken, the impression the paragraph is supposed to make. If some facts in a paragraph amount to concessions, the best pattern is likely to be: (1) a topic sentence making the main point clear, (2) the facts that must be conceded—introduced in a manner that shows them to be concessions, and (3) the material that establishes the main point of the paragraph despite the concessions.

6. If it is necessary to use transitional material to bridge the gap between two paragraphs, such material should open the second of the two rather than close the first. Such placement avoids weakening the emphasis of the first, and helps to introduce the topic of the second. Also, since most writers do place transitional material at the beginning of the new paragraph, that is where the reader will expect it to be.

Length of Paragraphs

One need only look at contemporary writing to see that paragraphs vary widely in length. The special kinds that were mentioned at the beginning of this discussion are often extremely short. The same can be said of the paragraphs in popularized treatments of technical subjects, and of paragraphs in publications with columns so narrow that long paragraphs would appear difficult to read.

Nevertheless, normal paragraphs serve their purpose best when they contain from three or four to seven or eight sentences—which would be from 75 to perhaps as much as 200 words. Experience has shown that paragraphs of such length usually present the material to the reader in thought units that can be grasped most effectively at one time. Occasionally it may be desirable to have a paragraph consist of only one or two sentences, but when a substantial piece of writing consists mainly of one-sentence paragraphs, the reader receives no help in determining which sentences are related to the same topic.

Paragraphs of Isolated Statement

Paragraphs of isolated statement are statements that do not need to be expanded upon but do not merge into the development of a single larger idea. For example, in presenting a series of conclusions each consisting of a single sentence, it is often desirable to treat each item as a separate paragraph. The same treatment is often suitable for some instructional material and for various other kinds of material, including much of the material in this Handbook.

Writing paragraphs of isolated statement obviously does not call for any special technique. They should be recognized as a distinct and acceptable type, but their existence does not justify excessive use of one-sentence paragraphs when the subject matter calls for related rather than independent statements.

STANDARD USAGE

The following discussion has only one purpose—to help you to avoid the problems in usage that occur most frequently. Consequently, its contents are limited to the information that will best serve that purpose. A definition of grammatical terms appears later in the Handbook.

There are times, of course, when literate people disagree about whether some form may be used. Many a usage that was once considered incorrect is now acceptable to liberal authorities. The point of view in this book is liberal, but in many cases both sides are presented when a form is disputed. This has been done on the grounds that if there are still many literate people who consider something to be an error, those who use this book are entitled to be warned. In this connection, you should note that many liberal authorities concede—in fact take pains to point out—the acceptability of forms that they avoid in their own work.

Case of Nouns and Pronouns

Nouns and pronouns may be nominative, possessive, or objective in case. Nouns are identical in nominative and objective cases, but change form when they become possessive. Pronouns vary in form as shown by the following list:

NOMINATIVE	POSSESSIVE	OBJECTIVE
I	my, mine	me
you	your, yours	you
he	his	him
she	her, hers	her
it	its	it
we	our, ours	us
they	their, theirs	them
who	whose	whom

All pronouns except those listed above and others derived from them, for example *whoever*, which becomes *whomever* in the objective case, are identical in nominative and objective cases and form the possessive, if at all, just as nouns form it.

The possessive forms *your*, *its*, and *whose* should not be confused with *you're* (you are), *it's* (it is), and *who's* (who is).

The three cases are used in accordance with the following rules:

1. The nominative case is used for a subject of a finite verb.

   VIOLATION: We shall hire *whomever* applies.
   CORRECT: . . . *whoever* applies. (The word in question is the subject of *applies* rather than the object of *hire*. The object of *hire* is the entire clause *whoever applies*.)
   VIOLATION: *Us* draftsmen need better light.
   CORRECT: *We* draftsmen . . . (*We* is correct, for it is in apposition with *draftsmen* and thus is a subject of *need*.)
   VIOLATION: He is older than *me*.
   CORRECT: He is older than *I*. (*I* is correct for it is the subject of the implied verb *am*.)

2. The nominative case is used for a subjective complement.

   It is *I*.
   It was *we* workers who objected.

   VIOLATION: It was *him*. It was *me* whom you saw.
   CORRECT: It was *he*. It was *I* whom you saw.

Note. In spite of the rule, such forms as "It was *me*" and "This is *him*" are widely used in conversation even by literate people, because the forms that are formally correct sometimes sound stilted. It is still best, however, to follow the rule in any piece of writing where formal style is called for; and even in conversation many discriminating people try to phrase their remarks so that they do not need to choose between stiffness and violation of the rule.

3. The possessive case is used to indicate possession.

VIOLATIONS: A violation of this rule almost always results from the fact that the possessive and some other form are pronounced exactly alike. (Examples: *it's* and *its*, *who's* and *whose*, *ladies* and *lady's*. *It's* is correct only for *it is*, and *who's* for *who is*. *Ladies* is plural, not possessive.) Thus the errors do not indicate ignorance of the need for the possessive case, but rather ignorance of (or carelessness about) how the possessive case is formed. Information on forms that are correct for the possessive case appears in the list of pronouns at the opening of this discussion and in the discussion of the apostrophe in the section on punctuation.

4. The possessive case is used for a noun or pronoun that precedes and modifies a gerund.

VIOLATION: They objected to *him* altering the records.
CORRECT: They objected to *his* altering the records.

Note. When a pronoun has no possessive form, for example *this* or *that*, Rule 4 must of course be disregarded. It is also disregarded if the presence of other words between the noun or pronoun and the gerund modified would make the possessive sound unnatural.

CORRECT: There is no record of *this* being done before.
CORRECT: The chance of *anyone* in the vicinity wanting to buy the property is slight.

5. If a noun or pronoun stands for something inanimate, the possessive case should be avoided, if possible.

UNDESIRABLE: The *furnace's* grates; the *street's* surface.
PREFERRED: The grates *of the furnace;* the surface *of the street.*

Note. There are many exceptions to this rule, especially when the noun in question involves time, as "an *hour's* work," or "a *day's* pay." Exceptions are also permissible to avoid awkward constructions.

6. The objective case is used for the object of a verb, verbal, or preposition.

Direct object of verb: The letter encouraged *her*.
Indirect object of verb: The letter gave *him* useful information.
Object of verbal: Whom do you intend to hire? (*Whom* is the object of the infinitive phrase *to hire*.)
Object of verbal: Answering *them* was difficult. (*Them* is the object of the gerund *answering*.)
Object of preposition: Three of *us* were chosen. (*Us* is the object of the preposition *of*.)

COMMON VIOLATIONS: They offered my partner and *I* a good contract. (*Me* would be correct, for the term is used as the indirect object of *offered*. This type of error results from excessive concern about avoiding *me* as a subject, as in "John and *me* are going." Uncertainty as to what form to use after *and* can be settled quickly by eliminating *and* along with the term that precedes it. "They offered *I* a contract" would obviously be wrong.)

All of *we* draftsmen were questioned. (*Us* would be correct, for *us* along with *draftsmen*, is the object of the preposition *of*. The subject of the sentence is neither *we* nor *draftsmen*, but *all*.)

He is a man *who*, in spite of his youth, we can trust. (The parenthetical phrase *in spite of his youth* obscures the fact that the objective *whom* rather than the nominative *who* is needed, for the term is the object of *trust*.)

Who do you give the answer to? (*Whom* would be correct, for the term is used as the object of *to*.)

Note. In conversation and in informal writing there is a strong tendency to permit *who* at the beginning of a sentence despite the rules. In formal writing, however, it is still desirable to follow the rule.

7.   The objective case is used for either the subject or object of an infinitive.

CORRECT: They expect *him* to do well. (*Him* is the subject of the infinitive *to do*. The object of *expect* is not *him*, but the entire infinitive phrase, *him to do well*.)
CORRECT: We knew *him* to be competent.
CORRECT: We knew it to be *her*.

Agreement of Verb and Its Subject

A verb must agree with its subject in number, person, and gender.
Practically all of the violations of this rule have to do with disagreement in number. Person and gender will therefore not be discussed here. Rules involving agreement in number are as follows:

1.   When the subject of a verb is compound and its parts are joined by *and*, it ordinarily takes a plural verb.

The time and the place *are* uncertain.

2.   When the subject of a verb is compound in form but singular in meaning, it takes a singular verb.

Blue and gold *is* a pleasing color combination.

3.   When the subject of a verb is compound and its parts are joined by *or*, it takes a singular verb if its parts are singular and a plural verb if its parts are plural. If its parts differ in number, it agrees with the part that is closer.

CORRECT: Either the oak or the elm *is* to be removed.
CORRECT: Either the chairs or the tables *are* to be refinished Monday.
CORRECT: Either the cow or the pigs *are* to be slaughtered.

4. The word *number*, numerical quantities, and fractions take either singular or plural verbs according to their meanings.

The number of complaints *has* been increasing.
A number of changes *have* been made.
Three days *is* a long wait. (*Three days* refers to a single period of time, even though *days* is plural in form.)
Three days *have* passed since your letter was mailed. (Each day is conceived as a unit rather than as part of a single period of time.)

5. A collective noun takes a singular verb when the group it refers to is regarded as a unit, but takes a plural verb when the statement concerns the members of the group as individuals.

The crowd *was* breaking up.
The crowd *were* going to their homes.
The committee *has* adjourned.
The committee *have* taken their seats.

Note. When you feel that following the rule would result in an awkward sentence, revise the sentence. For example: "The *people* in the crowd (rather than *the crowd*) were going to *their* homes."

6. Some nouns that are plural in form take singular verbs when singular in meaning— especially nouns ending in *ics*.

CORRECT: The *news is* encouraging. *Mathematics is* difficult. *Tactics wins* battles.
BUT: Our *tactics are* proving successful.

7. Intervening words should not be allowed to interfere with the agreement of a verb with its subject. Many of the errors in agreement are caused by intervening words.

CORRECT: Good management plus favorable economic conditions *was* responsible for the improvement. (The phrase introduced by *plus* is not part of the subject.)

8. A verb agrees with its subject even though a subjective complement that follows is different in number.

The border *was* shrubs of several species.

Note. When you feel that following the rule would result in an awkward-sounding sentence, revise the sentence. For example: "The border contained shrubs of several species."

Agreement of Pronouns
with Antecedents

The rules. The basic rule is that a pronoun and its antecedent should agree in person, number, and gender. When a pronoun has a compound antecedent the following additional rules apply:

1. If the antecedent consists of two or more nouns connected by *and*, the pronoun should be plural.

   The table and the chair showed *their* age.

2. If the antecedent consists of two nouns connected by *or*, the pronoun should be singular if both nouns are singular, and plural if both nouns are plural. If the nouns differ in number, the pronoun should agree with the nearer.

   You may pay the bill if the manager or the auditor will give *his* permission.
   You may use either bolts or screws if *they* are large enough.
   Either the bolts or the clamp must have been loose in *its* place.
   Either the clamp or the bolts must have been loose in *their* places.

3. If the antecedent is two or more nouns connected by *nor*, the pronoun should be singular if both nouns are singular, but *plural if either noun is plural*.

   Neither the manager nor the auditor gave *his* consent.
   Neither the bolts nor the screws had fallen from *their* places.
   Neither the employer nor the employees were willing to change *their* attitudes.
   Neither the employees nor the employer were willing to change *their* attitudes.

The problems. When a pronoun does not agree with its antecedent, the disagreement, almost always, is in number. The problem does not usually result from ignorance of the rule, but from uncertainty about whether the antecedent itself should be treated as singular or plural. Numerous though the problems are, a surprisingly large proportion of them result from one of three conditions. Any person who can master these three conditions will thenceforth have little trouble with the reference of pronouns: (1) when the antecedent is a *collective noun;* (2) when the antecedent is *a noun of common gender;* (3) when the antecedent is not a noun but *an indefinite pronoun.*

1. *Reference of a pronoun to a collective noun:* A collective noun is sometimes treated as singular and sometimes as plural. It is treated as singular when the statement where it appears applies to the group as a group, and plural when the statement applies to the members of the group as individuals.

> CORRECT: The board of directors makes *its* report.
> CORRECT: The board of directors took *their* seats.

The forms that are undesirable occur when a writer first treats a collective noun as singular, then as plural—usually making it the subject of a singular verb and the antecedent of a plural pronoun. The best way to avoid such inconsistencies is to change the construction of the sentence.

> DOUBTFUL: The orchestra *waits* with *their* instruments ready.
> IMPROVED: The orchestra waits with instruments ready.
> DOUBTFUL: The team *was* photographed wearing *their* new uniforms.
> IMPROVED: The team was photographed wearing new uniforms.
> IMPROVED: The members of the team were photographed wearing *their* new uniforms.

2. *Reference of a pronoun to a noun with common gender:* Writers sometimes use a plural pronoun (*they* or *their*) to refer to a singular antecedent because they don't want to use the singular *her* or *him* in referring to a noun such as *spectator, student,* or *employee.* The rule that has been traditionally followed by literate writers and speakers calls for the masculine unless the gender of the person referred to by the noun is known to be feminine.

> WRONG: A person should be able to rise above *their* environment.
> CORRECT BUT STILTED: A person should be able to rise above *his or her* environment.
> CORRECT: A person should be able to rise above *his* environment.
> IMPROVED: A person should be able to rise above environment.
> IMPROVED: People should be able to rise above *their* environment.

3. *Reference of a pronoun to an indefinite pronoun:* Sometimes the antecedent of a pronoun is not a noun, but an indefinite pronoun. Most of the indefinite pronouns are listed below. Each word on this list is treated as singular, even though some of them are plural in meaning. Hence any pronoun referring to a word on this list should be singular in number.

INDEFINITE PRONOUNS

one	everyone	somebody
anyone	anybody	nobody
someone	everybody	each
no one		

All the words on the list are common in gender. Hence, in an effort to achieve agreement in gender, writers use the plural *their*, which can be common gender, in order to avoid *he* or *she*. This tendency is intensified by the fact that some of the words, for example *everyone*, are actually plural in meaning. Nevertheless, all the words listed are treated as singular (no one would think of writing "Everyone *are* here"), and except as noted below are not to be referred to by the plural *their*. Unless the feminine singular is known to be appropriate in the individual case, the masculine singular (*he, his,* or *him*) should be used. *His or her* is of course correct but is so stilted as to be undesirable. If you wish to avoid using *he, his,* or *him*, you should recast the sentence to avoid the singular rather than attempt to violate the convention.

> DOUBTFUL: No one should neglect *their* responsibilities.
> CORRECT: No one should neglect *his* responsibilities.
> DOUBTFUL: Everyone should pay *their* bills promptly.
> CORRECT: Everyone should pay *his* bills promptly.
> WRONG: Each of those hired will be required to supply *their* own tools.
> CORRECT: Each of those hired will be required to supply *his* own tools.
> IMPROVED: All those hired will be required to supply *their* own tools.

Exceptions.   The rules just explained and those under No. 2 above apply to speaking or writing in a formal style, and many people who are careful with their language follow them even when speaking or writing informally. There is a strong contemporary tendency, however, to relax the rules. Some of the more liberal authorities would accept the use of *their* to refer to a noun of common gender or an indefinite pronoun—especially a pronoun with a plural meaning.

The tendency has been intensified by the emergence of the feminist movement, but even before the movement emerged, at least one President (John Kennedy) and more than one senator was heard to use *their* when the antecedent was such a word as *everyone* or *anyone*. More recently, systematic listening to television has shown that even among literate speakers, the rule is disregarded as often as it is obeyed. Among these speakers have been former President Carter, Walter Cronkite, Dick Cavett, and Laurence Olivier. In print, *their* has been used by financial columnist Elliot Janeway, a former presidential press secretary, a distinguished psychiatrist, the director of public affairs of a large corporation, the Associated Press in nationally circulated dispatches, and even by "Dear Abbey."

Agreement of a Demonstrative Adjective with its Object

A demonstrative adjective (*this, these, that* and *those*) should agree with the word it modifies.

> WRONG: *These* (or *those*) kind of brakes may give trouble.
> CORRECT: *This* (or *that*) kind . . .

Note. In the error illustrated, the demonstrative adjective should agree with *kind*, not *brakes*.

Adjectives and Adverbs— General Comment

1. An adjective may modify a noun or pronoun directly (in which use it is called an attributive adjective) or as a predicate adjective—so called because it appears in the predicate and applies to a noun or pronoun by assistance of a verb.

> The *old* building should be abandoned. (Attributive modifier. *Old* modifies *building*.)
> The building is *old*. (Predicate adjective. *Old* relates to *building* by assistance of the verb *is*.)

2. An adverb is used to modify a verb, an adjective, or another adverb.

> The snow had melted *rapidly*, so what might have been a *very* difficult trip was made *quite easily*. (*Rapidly* modifies the verb *melted; very* modifies the adjective *difficult; quite* modifies the adverb *easily*—which itself modifies the verb *was made*.)

3. Use of an adjective instead of an adverb to modify a verb is usually a serious error.

> WRONG: The car runs *good*. (The adjective *good* should not modify the verb *runs*.)
> CORRECT: The car runs *well*. (*Well* is correctly used as an adverb. In some meanings, *well* is also an adjective; but *good* is never correct as an adverb except when the phrase *as good as* appears in such a sentence as "He was *as good as* elected.")
> WRONG: I can do it *easy*.
> CORRECT: I can do it *easily*. (*Easily* is the adverbial form of easy.)
> WRONG: I can do it *easier* than he can.
> CORRECT: I can do it *more easily* than he can.
> WRONG: Among all those who protested, he talked the *angriest*.
> CORRECT: Among all those who protested, he talked *most angrily*.

Note. The error of using an adjective to modify a verb is especially prevalent in the comparative and superlative degrees as shown in the last two examples of errors. (For discussion of degree, see Rule 7 below.)

4. Use of an adjective to modify another adjective or an adverb (as contrasted with an adverb modifying a verb, discussed in Rule 3) is not formally correct, but sometimes may pass as dialect rather than being considered a serious error.

> DIALECT: The car was running *considerable* better. (*Considerable* is an adjective and should not modify the adjective *better*.)
> DIALECT: The next day, I felt *some* better. (*Some* is an adjective, not an adverb. It should be replaced by the adverb *somewhat*.)

DIALECT: It was *real* nice of you.
CORRECT: . . . *very* nice of you.
DIALECT: It was made of *pretty* good material.
CORRECT: . . . *rather*, or *quite*, or *fairly* good material.
DIALECT: He was late *most* every morning.
CORRECT: . . . *(almost)* every morning.

Note. It would be better, except on informal occasions, to replace *real*, *pretty*, and *most* with words that are recognized as adverbs. There is no way, except by familiarity with the specific word, to determine what adjectives are considered merely colloquial rather than wrong when used as adverbs.

5. Adverbs are rarely misused for adjectives as direct modifiers, but use of an adverb when a predicate adjective is needed is an extremely common problem. A predicate adjective follows the verb but does not modify it. Rather, it relates to the noun or pronoun used before the verb as a subject as shown in No. 1. This use should not be confused with the use of an adverb to modify the verb. A predicate adjective characteristically follows a "linking" verb such as *be, become, seem, appear, smell, feel, look, sound, taste*.

CORRECT: The glue is *sticky*. (Predicate adjective. *Sticky* refers to the noun *glue* rather than modifying the verb *is*.)
CORRECT: The man looked *angry*. (Predicate adjective. *Angry* describes the man rather than changing or adding to the meaning of the verb *looked*.)
CORRECT: The man looked *angrily* at the intruder. (Adverb modifying verb. *Looked*, in this sentence, has a different meaning than in the example above. It is not used as a linking verb, but indicates an action. *Angrily* tells the manner in which the action was performed.)
WRONG: The dinner smells *well*. (*Well*, except in the sense of healthy, is an adverb, not an adjective. An adjective is needed here, for *smell* does not indicate an action and *well* tells a quality of the dinner rather than modifying *smells*.)
CORRECT: The dinner smells *good*. (Predicate adjective.)
CORRECT: The dog smells *eagerly* at the bone. (Note the different meaning of *smells*, and note that *eagerly*, an adverb of manner, modifies *smells*.)
CORRECT: The clerk felt *bad* about the error. (Predicate adjective.)
CORRECT: The motor was running *badly*. (Adverb modifying verb.)

Note. Those who make this type of error do so because they have been corrected for such errors as "The motor is running *bad*," or "He reads *good*." In an effort to avoid using an adjective when an adverb is needed, they are reluctant to use an adjective after a verb, even when an adjective is correct. One use of *badly*, however, calls for special comment: There is a strong tendency at present to accept "The clerk felt *badly* about his error" as correct despite the rule. *Bad*, of course, is preferable.

6. When the question of whether to use an adjective or an adverb has been answered, it may still be necessary to determine whether the particular word one wishes to use is an adjective, is an adverb, or may be used as either. Often, the adverbial form of a word may be distinguished from the adjective form because of its *ly* ending, as in *strong, strongly; clever, cleverly; sure, surely*. The *ly* ending, however, is not always a means of identification, for there are some adjectives that end in *ly* (*friendly, manly*) and many adverbs that do not end in *ly*. In case of doubt, it is advisable to consult a dictionary to determine the part of speech of the word in question.

7. Most adjectives and adverbs may be positive, comparative, or superlative in degree. The change in degree is achieved by more than one method, as illustrated in the following examples:

ADJECTIVES

Positive	*Comparative*	*Superlative*
easy	easier	easiest
large	larger	largest
curious	more curious	most curious
good	better	best
bad	worse	worst

ADVERBS

Positive	*Comparative*	*Superlative*
slow or slowly	slower or more slowly	slowest
easily	more easily	most easily
quickly	more quickly	most quickly
well	better	best
badly	worse	worst

Note. There are some who point out that terms such as *perfect* and *unique* cannot logically be used in the comparative or superlative degrees. *More perfect* is a contradiction in terms, as are *more unique, more empty*, and so on. It would be logical, however, to use such expressions as *more nearly* perfect, unique, full, or empty. The case, as stated above, is logical, but English is not always a logical language. The Constitution of the United States refers to a *more perfect union*.

The comparative rather than the superlative degree must be used to indicate comparsion of two objects. The superlative degree is used when three or more objects are compared.

WRONG: His was the *best* drawing of the two drawings that were submitted.
CORRECT: His was the *better* drawing of the two that were submitted.
CORRECT: His was the *best* of the three drawings that were submitted.

UNITY AND COHERENCE
IN THE SENTENCE

Unity of Thought

In this discussion we are concerned only with the declarative sentence—the sentence that makes a statement. Such a sentence may lack unity either because it fails to include all the material needed or because it contains extraneous material. The first of these two weaknesses may be prevented by the avoidance of primer style. The second may be guarded against by being alert to the question of whether all the material in the sentence really merges to make a single statement. Consider, for example, the sentence, "Mack trucks, which are now being produced in three foreign countries, would be the best kind for us to buy for hauling logs and lumber." There is no clear reason for placing the fact about foreign production in a sentence with the purpose this one seems to have. There are many times, of course, when certain facts may be presented as either one sentence or more than one, but this does not justify inserting into a sentence facts that are not related to its purpose.

Sentence unity also demands that we do not end one sentence and begin another at an illogical place. Consider these sentences:

> The Ellsworth machine has ample power and those who have used it say it is easy to operate, but it is extremely noisy. Also, it is somewhat bulky.

The first of these two sentences names two good features and one bad. The second mentions another bad feature. If we assume that two sentences are desirable, sentence unity would be improved by telling the good features in one sentence and the bad in the other. A general rule to follow is: For the sake of sentence unity, construct your sentences so that they draw together the facts that are most closely related.

Unity of Structure

When a sentence lacks unity of structure it may be faulty in either of two respects: (1) It may be a mere fragment because it lacks some element necessary for grammatical completeness. (2) It may contain the error variously called a "comma splice," "comma fault," or "comma error." (A "run-on sentence" is basically the same error as a comma splice and is covered under that heading.)

Fragment treated as a sentence. A declarative sentence, to be complete grammatically, must contain a subject and a finite verb. Thus a participial or gerund phrase, a prepositional phrase, or a dependent clause is not a complete sentence, regardless of its length or complexity. Actually, it should not be necessary to subject a sentence to grammatical analysis in order to determine whether it is complete. If it is read attentively, a fragment may be distinguished from a complete sentence by the simply fact that it does not actually make a statement.

The error may usually be corrected by either of two methods: joining the fragment onto a sentence with which it is naturally connected, or changing it so that it may stand by itself as a complete sentence. The following examples show how fragments may be joined to other sentences or rewritten so that they are structurally complete. Whether the corrections indicated would be the best corrections possible would depend on the context.

FRAGMENT: The volume grew lower and lower. *Finally becoming so slight as to be inaudible.* (Participial phrase rather than sentence.)
CORRECT: The volume grew lower and lower, finally becoming so slight as to be inaudible.
FRAGMENT: He was employed by the government for six years. *First in Washington, and later in New York.* (Prepositional phrases.)
CORRECT: He was employed by the government for six years, first in Washington and later in New York.
FRAGMENT: The company was losing money steadily. *Although sales were as high as in previous years.* (Subordinate clause rather than independent clause.)
CORRECT: The company was losing money steadily even though the sales were as high as in previous years.
FRAGMENT: Some of the expenses were increasing. *Overhead, for example, and the cost of raw materials.* (The italicized expression has no verb; it is used in apposition to *some of the expenses* in the preceding sentence.)
CORRECT: Some of the expenses were increasing. Overhead, for example, and the cost of raw materials were higher than before.

Note. Skilled authors do sometimes use incomplete sentences as a device to achieve special effects. Dickens, for example, wrote: "Dogs, indistinguishable in the mire. Horses scarcely better; splashed to their very blinkers." But this incompleteness was no accident. Until you are skillful enough never to violate the rules of grammar accidentally, however, you should not violate them for literary effect.

The comma splice. When a sentence contains two independent clauses, the punctuation between them must be a semicolon unless they are joined by a coordinating conjunction. Use of a comma instead of a semicolon on such an occasion is the error variously called comma splice, comma blunder, or comma error. It is more than a mere mistake in punctuation; it is a serious error in sentence structure—an indication that the writer does not know when a statement is complete.

The only coordinating conjunctions are *and, but, for, or, nor,* and sometimes *so* and *yet.* These should be distinguished from conjunctive adverbs such as *however, moreover, therefore, consequently, furthermore,* and many others.

A handy rule of thumb for avoiding the comma splice is found in the fact that a conjunctive adverb may be buried within the clause that it introduces. Thus, if a connective word might come elsewhere than between the clauses it connects, a semicolon rather than a comma is necessary. For example:

> I have inspected it, *but* I shall be glad to inspect it again. (*But* could come only between the clauses.)
>
> I have inspected it; *however,* I shall be glad to inspect it again. (*However* might, if one desired, be placed after *glad* rather than at the beginning of its clause; thus it is a conjunctive adverb, and a semicolon must be placed between the clauses.)

There are various ways in which a comma blunder may be corrected. The simplest but not always the best is to replace the comma by a semicolon or a period. Another possibility is to change the conjunctive adverb to a coordinating conjunction. Often, however, it is better to change one of the independent clauses to some other form.

> COMMA SPLICE: We shall be forced to find a new route, the grades on the one suggested are too severe.
>
> CORRECT: We shall be forced to find a new route; the grades on the one suggested are too severe. (Correction by change in punctuation.)
>
> CORRECT: We shall be forced to find a new route, for the grades on the one suggested are too severe. (Correction by using a coordinating conjunction.)
>
> CORRECT: The grades on the route suggested are so severe that we shall be forced to find a new route. (Correction by change in construction.)
>
> CORRECT: The severity of the grades on the route suggested makes it necessary to find a new route. (Correction by change in construction.)

Coherence

If writing is to be clear, every sentence must be coherent; that is, it must hold together. The relationship of every part to every other part and to the entire sentence must be unmistakable. To achieve this result it is necessary to consider the order of the parts, their structure, and the connectives that indicate their relationship. Lack of coherence often results from failure to apply the following principles:

1. Similar parts of a sentence should be expressed in parallel form so far as their contents permit.

    > NOT PARALLEL: Wood was used for some of the parts, but others were made of metal.
    >
    > PARALLEL: Wood was used for some of the parts, and metal for others.
    >
    > NOT PARALLEL: Students in the night school learn auto mechanics, and are also taught drafting.

PARALLEL: Students in night school learn auto mechanics and drafting.
PARALLEL: Students in night school are taught auto mechanics and drafting.

(a) There should be no unnecessary change in the subject or voice. (Change in one frequently results in change in the other.)

NOT PARALLEL: While the mechanic was tuning the motor, the tires and battery were checked by his assistant.
PARALLEL: While the mechanic was tuning the motor, his assistant checked the tires and battery.
NOT PARALLEL: Washing the equipment was one of his duties, and he was also expected to keep the supply bins filled.
PARALLEL: He was expected to wash the equipment and to keep the supply bins filled.

(b) The elements of a series should be parallel in form.

NOT PARALLEL: A technician must learn the *use, upkeep,* and *how to repair* equipment.
PARALLEL: A technician must learn the *use, upkeep,* and *repair* of equipment.

If *the* is used before any element in a series except the first (in which case it applies to all the elements), it must be used before every element.

WRONG: Information is needed about the length, the width, thickness, and weight of the sample.
CORRECT: Information is needed about the length, width, thickness, and weight of the sample.

If *a* or *an* rather than *the* is the suitable word to precede any element in a series, it should usually be used before *each element where it would be suitable*. This is especially necessary if some elements are plural.

We must bring nails, screws, *a* hammer, *a* chisel, and glue.

When the elements of a series cannot be made parallel in form without producing an awkward result, the phraseology should be changed so as to make it clear that a series is not intended.

WRONG: The chips are mixed with liquor and steam, fed into the apparatus, and the cooking process takes place as the mixture moves through the tubes.
CORRECT: After the chips are mixed with liquor and steam, they are fed into the apparatus, where the cooking process takes place as the mixture moves through the tubes.

(c) The elements that follow correlative conjunctions (*either . . . or, neither . . . nor,* and so on) should be parallel in form.

WRONG: It can either be shipped by freight or by express.

CORRECT: It can be shipped either by freight or by express.

2. Connectives should clearly and accurately indicate the thought relationship of the elements that they connect. The connectives that give most trouble are *as* and *while*.

(a) *As* is sometimes ambiguous because it might indicate either time or cause.

AMBIGUOUS: *As* the mixture was heated, its color changed.

CLEAR: When the mixture was heated, its color changed.

CLEAR: Because the mixture was heated, its color changed.

(b) *While* has two legitimate meanings, "during the time that" and "although." It is also loosely used in place of *and*. As a result, many a sentence in which it appears is ambiguous. Unless the context makes its meaning instantly clear, it should be used sparingly for *although* and not at all for *and*.

AMBIGUOUS: *While* his breathing was becoming more rapid, his fever had fallen.

CLEAR: *Although* his breathing had become more rapid, his fever was falling.

CLEAR: At the same that his breathing was becoming more rapid, his fever was falling.

AMBIGUOUS: Their truck tires are good, *while* their passenger car tires are even better.

IMPROVED: Their truck tires are good, *and* (or *but*) their passenger car tires are even better.

3. Reference of a pronoun to its antecedent should be clear and unmistakable.

(a) There should be only one word to which a pronoun may plausibly refer.

AMBIGUOUS (because of indirect discourse): The manager notified the foreman that *he* was being transferred. (Manager or foreman?)

CLEAR: The foreman received notice of his transfer in a conference with the manager. (*His* has been related to *foreman* before the manager is mentioned.)

CLEAR: The manager, as he mentioned to the foreman, was being transferred.

Note. Textbooks often suggest that unclearness caused by indirect discourse be eliminated by changing it to direct discourse. This is rarely appropriate, however, in technical writing.

AMBIGUOUS: Dairy workers should not be allowed to care for cows when *they* are ill.

CLEAR: When dairy workers are ill, they should not be allowed to care for cows.

CLEAR: Dairy workers should not be allowed to care for cows that are ill.

(b) Reference of a pronoun to a "weak" antecedent should be avoided.

WEAK REFERENCE: The engineers must give special consideration to the smoke-stacks. *They* must be sturdy enough to withstand earthquakes. (Engineers or stacks?)

IMPROVED: The smokestacks must be sturdy enough to withstand earthquakes, so the engineers must give them special consideration.

(c) A pronoun should not be used when its antecedent must be inferred by the reader because it has been expressed in some other form than a noun.

VAGUE REFERENCE: An employee who is injured should receive first aid even though *it* may not be serious. (*It* has as an antecedent the idea of *injury*; but the idea is not present as a noun and must be inferred from the verb *injured*.)

IMPROVED: An employee who is injured should receive first aid even though *the injury* may not be serious.

(d) Vague use of *it* or *they* when a specific noun could be used should be avoided. (This rule would not prevent the use of *it* as in "*It* is raining," or "*It* is apparent that . . .")

VAGUE REFERENCE: *It* says in the report that the method is efficient.
EXPLICIT: *The report* says that the method is efficient.
VAGUE REFERENCE: In restricted areas, *they* require that visitors be identified.
EXPLICIT: *The company* requires that in restricted areas, visitors be identified.
EXPLICIT: In restricted areas, visitors must be identified.

(e) Use of *which* or *this* to refer to the general idea of a preceding clause or phrase rather than to a specific word is condemned by some textbooks but indulged in by many reputable writers. When this use can be avoided easily, it should be avoided.

VAGUE REFERENCE: He has been studying every night, *which* should increase his value to the company.
IMPROVED: He has been studying every night, so his value to the company should be increasing.
IMPROVED: The study he has been doing every night should increase his value to the company.

(f) Use of *you* for *one* is often inappropriate. Unless you are using an informal style, do not write *you* unless you mean the reader. If *one* sounds stiff, change the sentence.

INFORMAL: The factory is interesting, but *you* are not allowed to enter without permission.
FORMALLY CORRECT: The factory is interesting, but *one* is not allowed to enter without permission.

IMPROVED: The factory is interesting, but *visitors* are not allowed to enter without permission.

(g) Use of the same pronoun to refer to two different antecedents should be avoided.

CONFUSING: Although *it* is less crowded on the second floor, *it* is not strong enough to support the heavy machinery. (The first *it* is indefinite in meaning; the second refers to *floor*.)

IMPROVED: Although the second floor is less crowded, it is not strong enough to support the heavy machinery.

4. Modifiers (words, phrases, or clauses used to change or limit the meaning of other words) should be as close to what they modify as it is possible to place them without awkwardness. No possible doubt should exist as to what is being modified.

(a) Certain adverbs are especially likely to be put in the wrong position. These adverbs are *even, hardly, nearly, scarcely,* and especially *only* and *not*.

MISPLACED MODIFIER: We *only* tried three times.

IMPROVED: We tried only three times.

MISPLACED MODIFIER: Every small business can*not* grow large. (This is literally a statement that *no* small business can grow large.)

IMPROVED: Not every small business can grow large.

MISPLACED MODIFIER: The boat *almost* seemed ready to sink.

IMPROVED: The boat seemed almost ready to sink.

MISPLACED MODIFIER: It *hardly* felt as if it were heavy enough.

IMPROVED: It felt as if it were hardly heavy enough.

(b) Phrases and clauses, as well as words, should be placed as close as possible to what they modify.

MISPLACED MODIFIER: It made a good impression on the inspectors carrying a full load.

IMPROVED: Carrying a full load, it made a good impression on the inspectors.

MISPLACED MODIFIER: After it was repaired, it was given a trial run by a testing laboratory that lasted one week.

CORRECTION BY CHANGE OF STRUCTURE: After it was repaired, it was given a one-week trial run by a testing laboratory.

MISPLACED MODIFIER: The factory reopened after a one-month shutdown on June 1.

CLEAR: After a one-month shutdown, the factory reopened on June 1.

MISPLACED MODIFIER: It was given an overhaul by a crew of expert mechanics that was long overdue.

IMPROVED: It was given an overhaul that was long overdue by a crew of expert mechanics.

(c) "Squinting" constructions should be avoided. (Squinting constructions are those in which a modifier is placed between two objects, either of which it might modify.)

SQUINTING: The smoke jumpers who had been flown in *immediately* put out the fire.

CLEAR: The smoke jumpers, who had immediately been flown in, put out the fire.

CLEAR: The smoke jumpers who had been flown in put out the fire immediately.

(d) Though split infinitives are no longer regarded as a serious error, they are frequently awkward. Therefore one should not *unnecessarily* split infinitives in an effort to place a modifier nearer to its object.

AWKWARD: They decided *to quickly complete* the repairs.

IMPROVED: They decided to complete the repairs quickly.

(e) In attempting to place modifiers near their objects, one should not needlessly interrupt the smooth flow of thought through subject, verb, and complement.

AWKWARD: The crew extinguished, after a long and difficult struggle, the fire in the slashings.

IMPROVED ARRANGEMENT: After a long and difficult struggle, the crew extinguished the fire in the slashings.

5. Comparisons should be logical and complete.

(a) Objects that are compared should be similar in nature.

ILLOGICAL: The power of the diesel is greater than the steam engine.

LOGICAL: The power of the diesel is greater than that of the steam engine.

LOGICAL: The diesel engine is more powerful than the steam engine.

(b) Illogical use of *any* should be avoided.

ILLOGICAL: The first design was simpler than *any of the designs*.

LOGICAL: The first design was simpler than any of the later designs.

(c) If a comparison is not completely expressed, the words that are omitted must be clearly and unmistakably implied.

INCOMPLETE AND AMBIGUOUS: The manager trusts her *more than the superintendent*.

COMPLETE: The manager has more trust in her *than in the superintendent*.

COMPLETE: The manager trusts her *more than the superintendent does*.

INCOMPLETE: The highway is *as rough, if not rougher than* the side road.

COMPLETE BUT AWKWARD: The highway is *as rough as, if not rougher than*, the side road.

COMPLETE: The highway is as rough as the side road, if not rougher.

6. Use of mood and tense should reflect a consistent point of view.

INCONSISTENT USE OF TENSE: The production of grade "A" milk *required* the use of methods that *meet* official standards.

CONSISTENT: The production of grade "A" milk *requires* the use of methods that *meet* official standards.

INCONSISTENT IN MOOD: If you *would* give me an extension of time, I *shall* appreciate it.

CORRECT: If you *would give* me an extension of time, I *should* appreciate it.

INCONSISTENT IN MOOD: If the generated voltage *drops* below "E," current *would flow* through and help to rotate the armature.

CONSISTENT: If the generated voltage *drops* below "E," the current *will flow* (or *flows*) through and help to rotate the armature.

CONSISTENT: If the generated voltage *were* to drop below "E," the current *would flow* through and help to rotate the armature.

7. Dangling participles and gerunds should be avoided. A participle is a verb form used as an adjective. A gerund is a verb form used as a noun. (See Definition of Grammatical Terms.) When a participial or gerund phrase comes at the beginning of a sentence, it should be followed immediately by some term indicating who performed the action indicated, so that the action will not be attributed to the wrong agent.

In respect to meaning, a dangling participle or gerund is undesirable because it is likely to be incorrectly interpreted when first read. Grammatically, it is wrong because it is a modifying element with nothing to modify—thus the term *dangling*—and has no grammatical connection with the remainder of the sentence.

DANGLING PARTICIPLE: *Having thought the case over carefully*, my opinion was unlikely to be changed.

CORRECT: Having thought the case over carefully, *I* was not likely to change my opinion.

CORRECT: *Since I had thought the case over carefully*, I was not likely to change my opinion.

DANGLING PARTICIPLE: It can be built with either one or two doors, *depending on the wishes of the buyer*.

CORRECT: It can be built with either one or two doors, *whichever the buyer wishes*.

DANGLING GERUND: *By rotating the crystal*, the light is directed so that only the desired wavelength is reflected back to the absorption cells.

CORRECT: By rotating the crystal, *one* directs the light so that only the desired wavelength is reflected back to the absorption cells.

8. Dangling infinitives should be avoided.

DANGLING INFINITIVE: *To conduct the test properly,* the motor must run at a constant speed. (It sounds as if the motor were conducting the test.)

CORRECT: To conduct the test properly, *one* must keep the motor running at a constant speed.

CORRECT: *If the test is to be conducted properly,* the motor must run at a constant speed.

9. When elliptical clauses or phrases are used, a sentence should be constructed in a manner that does not permit the reader to form a false impression. (An elliptical clause, phrase, or sentence is one from which something is omitted, being implied by what is expressed.)

MISLEADING ELLIPSIS: *While looking for a suitable location,* the truck broke down.

CLEAR: While *we were* looking for a suitable location, the truck broke down.

MISLEADING ELLIPSIS: *When working on the barn,* the problem of protecting the cattle hindered them.

CLEAR: When working on the barn, *they were hindered by* the problem of protecting the cattle.

PUNCTUATION

The major function of punctuation is to make writing clearer and easier to read. The rules, except for those that are mere conventions, have come to be what they are because experience has shown us how punctuation can best contribute to clearness. Thus there is no fundamental clash between the idea of punctuating for the sake of clearness and punctuating in accordance with rules. The best way to assist clearness through punctuation is usually to follow the rules.

To be sure, the existence of rules has not caused usage in punctuation to be completely uniform. The rules are flexible enough to permit a writer, on many occasions, to exercise judgment in deciding between equally correct alternatives. The lack of uniformity does not result from widespread disagreement about the rules themselves, nor from general disregard of the rules. Most of the rules are generally agreed upon by authorities, and the occasions when well-educated writers violate them are not frequent.

Three marks are especially important to anyone who has difficulty with punctuation: the period, the semicolon, and the comma. Even though familiarity with the other marks is desirable, you will find that if you can learn to use these three marks correctly—especially the comma—most of your problems with punctuation will be solved.

Finally, it should be emphasized that even though punctuation may be valuable, there is a limit to what it can accomplish. Punctuation is not a substitute for smooth, easy-flowing sentence structure. There are many times when a writer who is worried about how to punctuate a sentence should rewrite it rather than punctuating it, for difficulty in punctuation may well be the result of awkward writing.

The following rules are generally agreed upon for ordinary writing. They are not intended to cover footnotes, bibliographies, the technical parts of letters, or headings in a report. The punctuation of these special forms is illustrated where the forms themselves are discussed.

The Period

1.   Periods are used at the ends of all sentences except those that are interrogative or exclamatory.
2.   Periods are usually used after numbers preceding items in a list; for example, the numbers preceding the entries in this Handbook.
3.   Periods are used after most abbreviations.

   a.m., a.d., Fig., i.e., R.F.D., U.S.

Exceptions.   Many abbreviations made up of the first letters of words that comprise the names of organizations are written without periods (NATO, UNESCO, TVA, NASA). Periods are often omitted, also, after abbreviations that are peculiar to technical style.

4.   To indicate the omission of words in quoted material, three spaced periods are used unless the omission occurs at the end of a sentence, in which case a fourth period is added.

   "The survey . . . covered 174 of the 192 colleges now accredited in this field."

The Comma

The comma is especially important because it is the main device by which the grouping of words, phrases, and clauses is indicated. Consequently it is used, and unfortunately misused, more than all the other marks combined.

Its use, however, is not haphazard. Competent writers almost always use commas for one of two purposes: to set off some element of the sentence from what precedes, what follows, or both, or else to separate two elements as they might be separated by a pause or a rising inflection of the voice if one were speaking. The few uses that fall into neither of these categories are the arbitrary use of commas on certain specific occasions, to be pointed out below, and the insertion of a comma when its presence is necessary for the sake of clearness.

The rules that follow, though numbered consecutively throughout, are grouped in accordance with the uses referred to above.

COMMAS USED TO SET OFF

1. An appositive or a term of direct address is set off by commas.

The original factory, an old stone structure, is still standing.
Your answer, Mr. Smith, is entirely satisfactory.

Note. No comma is used when a noun and its appositive are so closely related as to join in expressing a single idea.

The invasion was led by my brother John.

2. An adverbial clause preceding the clause it modifies, or an adverbial phrase at the beginning of a clause, is usually set off by a comma.

After the achievement tests had been completed, the results were tabulated.
On all floors except the second and the fourth, the fire hazards have been removed.

Note. If an adverbial clause or phrase is extremely short, and if omission of the comma could not cause confusion, the comma may be omitted.

When he arrived he was admitted immediately.
During July the plant will be closed.

3. Independent elements, participial phrases, gerund phrases, and other such constructions at the beginning of a sentence are set off by commas.

No, the shipment has not yet arrived.
Worried by the complaints, we began an investigation.
The contract having been broken, no payment was due.

4. A conjunctive adverb (*however, moreover, therefore,* etc.) is usually set off by commas when it comes *within* the clause to which it applies. When it comes at the beginning of a clause, it may or may not be followed by a comma but will always be preceded by a period or semicolon. (See Semicolon, Rule 2.)

His objection, therefore, was ignored.
I had heard the rumor before; consequently, I did not believe it.

5. Any mildly parenthetical element is enclosed in commas if it seems desirable to set it apart from the rest of the sentence. A writer is called upon to use judgment in applying this rule, for too many commas will make a sentence jerky and hard to read.

The newer strains, to be sure, will survive the blight.
The answer, when received, was unsatisfactory.
The central section, for example, was undamaged.
The frame, they insisted, was too tight.

6. A term such as *namely* or *that is*, used to introduce an example or a list, is usually set apart from that example or list by a comma. (The mark that precedes such an expression depends on the sentence structure.)

The usual crops—that is, wheat, peas, and alfalfa—are in good condition.
Three species of tree were observed, namely, pine, fir, and cedar.

7. Nonrestrictive clauses are set off by commas. Restrictive clauses, however, are not set off.

The south side, which has been exposed to the sun, was badly faded.
He moved to Arizona, where the climate was not so moist.

but

The road went to pieces where the permafrost had been disturbed.
All motorists who drive recklessly should be fined heavily.
I have never been there when the legislature was in session.

Note. In the first two examples, the clauses introduced respectively by *which* and *where* merely add some additional facts. If they were omitted, the meaning of the remainder of the sentence would be unchanged. Hence they are nonrestrictive. The clauses introduced by *where, who,* and *when* in the other three examples are restrictive. Each is used to limit—to *restrict*—the meaning of the main statement, which would be radically changed if the clause in question were omitted.

Note. Sometimes a sentence does not make sense unless a clause is interpreted in a single way—restrictive or nonrestrictive. When this is true, an error in punctuation merely increases the difficulty of reading. There are times, however, when restrictive and nonrestrictive interpretations are equally reasonable. When this is the case, an error in punctuation leads a reader to misunderstand the meaning. Note how the meaning of the two sentences that follow depends on punctuation:

The people from Troy, who had come early, obtained choice seats.
The people from Troy who had come early obtained choice seats.

8. A word or phrase that is placed in an abnormal position in a sentence should be set off by a comma or commas.

    To a trained accountant, the problem would look easy.

9. A direct quotation is set off by a comma or commas.

    "The tires are threadbare," he asserted, "and will blow out at any moment."
    He asked me directly, "Will September delivery be acceptable?"

Exceptions. A quotation that blends into the regular structure of the sentence is not set off by commas. A title in quotation marks is not set off by commas unless some other rule makes commas necessary.

The poet's prophecy about "airy navies grappling in the central blue" has become an unpleasant reality.
The rhythm of "The Raven" is very striking.

COMMAS USED TO SEPARATE

10. A comma is ordinarily used between two independent clauses that are joined by a coordinating conjunction. The coordinating conjunctions are *and, but, for, or,* and *nor.* (*Yet* and *so* may also be treated as coordinating conjunctions when this rule is applied.)

    The trees had been damaged by fire, and the wildlife had been destroyed.
    The building is old, but it has been kept in good condition.

Note. When both clauses are short and simple, the comma may be omitted.

    It was damaged but it still is usable.

Note. If a comma is used *within* one or both of two independent clauses, the comma between them is sometimes replaced by a semicolon. (See Semicolon, Rule 5.)

11. When a sentence contains a series, the elements in the series are normally separated by commas.

    Cattle, sheep, and hogs are now selling for higher prices.
    New deposits have been found in Canada, in Africa, in Central America, and in Alaska.

Note. If a comma is used *within* any element in a series, it is often better to use semicolons rather than commas *between* the elements.

Note. Opinions differ over whether to use a comma before a conjunction (*and* or *or*) that precedes the last item in a series. In technical and scientific periodicals and in material published by the United States government, use of the comma is predominant. In journalistic and popular publications usage is divided. Sometimes a comma is essential for clearness because *and* or *or* is used within one of the items. For example:

> The panels were painted red, green, yellow, and black and white.

Without the comma after *yellow*, it would be impossible to know whether *black* belonged with *yellow* or with *white*. In view of this, it seems advisable to regard the comma as normal punctuation rather than trying to check each series to see whether a comma is needed for clearness.

12. Two or more adjectives preceding a noun are ordinarily separated by commas. (The comma before the last adjective is omitted, however, if that adjective is so closely associated with the noun that the two merge into a single thought unit.) Likewise, a comma is used between adverbs that modify the same object unless a conjunction such as *and* or *but* is used between them.

> A big, powerful truck is needed.
> He has a modest, unassuming manner.
> It was housed in a large wooden structure.
> The watchman was a feeble old man.
> Slowly and relentlessly, the stream wore away the dike.

13. Commas are variously used to separate items in dates, places, and numbers as illustrated in the following examples:

> *In dates:* Payment shall be made on September 15, 1956, at the main office of the company.
> *In places and addresses:* San Francisco, California, is an important shipping point. The company is located at 70 Fifth Avenue, New York, New York 10011.
> *To separate adjacent sets of figures:* In 1950, 675 men were added to the payroll.
> *Between the digits of numbers:* 10,984. 234,617. 1,856,445. (The comma is omitted in a number with only four digits unless the number occurs in a column containing numbers in which commas are used.)

COMMAS USED ARBITRARILY

14. A comma is used between the last and the first name of a person when the last name appears first, and also after the first name unless the sentence structure calls for some other mark.

Please insert the name Fitzgerald, Duane, in the proper place in the alphabetical list.

15. A comma is sometimes used to indicate the omission of one or more words.

July will be devoted to writing; August, to revision.

16. A comma may be used whenever it is necessary to force a pause for the sake of clearness.

Inside, the building was in better condition.
To begin with, his data are questionable.

The Semicolon

The semicolon is an intermediate mark, less emphatic than a period but more emphatic than a comma. There are many times when the rules would permit either a semicolon or a period, and a writer's choice must depend on judgment about whether it would be better to continue the sentence or end it.

1. A semicolon is used sometimes between main clauses that are not joined by any connective.

Privately endowed schools must not be underestimated; they fill a genuine educational need.

2. A semicolon is used between main clauses connected by a conjunctive adverb rather than by a coordinating conjunction. (Conjunctive adverbs are such words as *also, accordingly, consequently, furthermore, hence, however, indeed, moreover, nevertheless, otherwise, still, then, therefore, thus*.)

He had shown fine managerial ability; consequently his promotion was rapid.

3. A semicolon is used between main clauses when the second clause begins with an explanatory term such as *in fact, for example, that is*.

All the costs have increased; for example, the cost of raw material has increased by 15 per cent.

Note. Rules 1, 2, and 3 might be summarized: A semicolon is normally the proper mark to use between independent clauses occurring in a single sentence except when the clauses are connected by a coordinating conjunction.

4. Even if independent clauses are connected by a coordinating conjunction, a semicolon may be used between them if it is desirable to set them apart more sharply than usual—for example, to set off one clause from two or more others to which it stands in contrast.

Its paint was damaged, its lights were broken, and its fenders were a complete loss; but the motor was in good condition.

5. Coordinate elements of any type, clauses or otherwise, are often separated by semicolons if any of them contains commas.

Some economies, perhaps, may be possible; but however hard we try to hold down expenses, a considerable increase will be unavoidable.

The inspection was made by William Smith, representing the company; Walter Brosser, president of the union; and Boyd Anderson, inspector for the Bureau of Mines.

The Question Mark

1. A question mark is used at the end of a direct question.

Is the price level rising or falling?

2. A question mark may be used to show that any expression is intended as a question, whether the form is interrogatory or not. (The use of a noninterrogative form to ask a question is seldom seen in any writing except reproduction of conversation.)

You claim the records have been altered?

3. A question mark is replaced by a period at the end of a "courtesy question," which is actually a request though it may be interrogatory in form.

Will you please send us this information as soon as possible.

4. A question mark, enclosed in parentheses, is used to express doubt.

Thomas Hooker, a founder of Connecticut, lived from 1586 (?) to 1647.

The Exclamation Point

An exclamation point is used after a word, phrase, or sentence to indicate intense feeling or forceful utterance. There are few occasions, however, to use an exclamation point in technical writing.

Ridiculous! The signature is forged!

The Colon

1. A colon is used before a long direct quotation that is being introduced formally.

John Stuart Mill expressed his doubts as follows:

2. A colon is used to introduce a formal enumeration—especially after *follows* or *the following*.

Bids were offered by the following contractors: Wilson and Taylor Construction Company, Toledo; Central Builders, Incorporated, Akron; Herman L. White and Company, Cleveland.

3. A colon is used between two phrases, clauses, or even sentences when the second is actually the equivalent of the first. In this use it conveys a meaning similar to that of *namely* or *that is*.

The method that they used had one decisive advantage: It could be used by personnel who had received only one week of special training.

The Dash

1. A dash can be used instead of a comma before introductory words such as *namely, in fact,* and *that is,* and before abbreviations such as *i.e.* and *viz.,* to introduce an enumeration. (See also Semicolon, Rule 3.)

It is superior in three respects—namely, economy of operation, safety, and comfort.

2. A dash can be used to set off an informal enumeration or a list of examples that are separated by commas.

Some of the accessories—the heater, the fog lights, and the bumper—are really necessities.

3. A dash is used after a list that is followed by a summarizing expression.

Colds, influenza, sore throats—all the winter ailments were prevalent.

4. A pair of dashes may be used to set off interpolated material. In this use, dashes create sharper separation than commas but less sharp than parentheses.

Most of the additional cost—approximately 90 percent—was passed on to the consumers.

5. A dash may be used to indicate incomplete or interrupted thought. (This use of the dash would be unlikely in technical writing.)

The reasons for our decision—but no, I'll not bore you with them.

Note. The dash is a mark to use when clear-cut rules make it correct. It should not be used haphazardly, merely because one is uncertain what mark would be appropriate. Unless one is writing dialogue or writing very informally, the use of dashes to show interruption of thought indicates lack of smoothness and continuity.

Note. In typewritten material, a dash is indicated by a double hyphen, or by a single hyphen preceded and followed by a single space.

Quotation Marks

1. Quotation marks are used to enclose direct quotations.

"The modern automobile," he pointed out, "sells at about the same price per pound that is charged for beefsteak."

Note. If a quotation is more than one paragraph long, opening quotation marks are used at the beginning of each paragraph but closing quotation marks are not used until the end of the quotation.

Note. Quotation marks are not used when the material quoted is set in smaller type or when it has wider margins than those of the regular text.

Note. Quotation marks are not used to enclose widely known proverbs, such as *Honesty is the best policy*, or other well-known quotations that the reader will recognize as quotations without assistance.

2. Quotation marks are used to enclose titles of short poems, articles, short stories— in general, the titles of writings that are not printed as independent publications.

The statistics come from an article entitled "Science and Public Relations."

3. Quotation marks are sometimes used to enclose the names of ships, trains, airplanes, and the like, and to enclose words used as words. (Italics are used more frequently.)

He had secured reservations on "The Portland Rose."
In the fourth paragraph, the word "unique" is used incorrectly.

4. Quotation marks may be used to indicate that a word or phrase is used to convey the meaning it has acquired in some special field. For example, in writing addressed to readers in general you might use quotation marks with "pickle" as used in the metals industry, "flip flop" as used in connection with computers, or "mark sense" as used in connection with IBM cards.

This use of quotation marks should be held to a minimum, for it is obtrusive if it occurs very often. In writing addressed to readers who are familiar with the special vocabulary of a field, quotation marks are not needed around the terms in question. In writing addressed to readers who lack that familiarity, it is better to use standard English if it will convey the meaning efficiently. If frequent use of such terms is unavoidable, it is better to omit the quotation marks and insert a parenthetical explanation of each term the first time you use it.

5. A quotation within a quotation is indicated by single quotation marks.

The instructions say, "You are to write 'Rejected' on the top of every imperfect copy."

Quotation Marks in Relation to Other Punctuation

1. A period or a comma ordinarily precedes closing quotation marks, even though it might logically belong outside.

One of his poems, "Fuzzy Wuzzy," was especially popular.

2. A colon or a semicolon ordinarily follows closing quotation marks.
3. A question mark or exclamation point is placed inside the closing quotation marks if it applies to the quotation, but outside if it applies to the sentence as a whole.

His exact words were, "Why was the gun concealed?"
Had he ever read "The Third Ingredient"?

The Apostrophe

1. An apostrophe is used to form the possessive case of nouns and indefinite pronouns. (Indefinite pronouns include such words as *one, everybody, everyone, nobody,* etc.)

(a) Singular or plural words that do not end with the sound of *s* form the possessive by adding *apostrophe* plus *s*.

the company's property   everybody's business
one's conscience   the men's wages

(b) Singular words ending with the sound of *s* also form the possessive by adding *apostrophe* plus *s*.

Mr. Jones's desk   the horse's age
the boss's office

Exception. If the form created by following this rule would be difficult to pronounce, it is permissible to add only the apostrophe, especially if the word in question is a person's name.

Moses' people (not Moses's people)   Mr. Jones' desk
Dickens' novels

(c) Plural words ending with the sound of *s* form the possessive by adding only an apostrophe.

the companies' policies   the workers' houses

2. An apostrophe is never used to form the possessive case of a personal or relative pronoun. (The possessive of *it* is *its*, not *it's*. The possessive of *who* is *whose*, not *who's*. *It's* means *it is*; *who's* means *who is*.)

3. The apostrophe to show the possessive case is often omitted in a formal title. In using any title, the form to follow is the form that has official sanction.

The Teachers Retirement Act   The Farmers Cooperative

4. A compound term such as *director of information* or *everybody else* is made possessive by adding the *apostrophe* plus *s* (or apostrophe only, if appropriate) to the last word.

Director of Information's statement   his father-in-law's tool chest

5. Joint possession is indicated by a change in the ending of only the last of two or more nouns.

Fred and Henry's office (Joint possession.)
Fred's and Henry's offices (Each has a separate office.)

6. An apostrophe is used to indicate the omission of one or more letters in a contraction.

can't   o'clock   they'll
isn't   it's (it is)   I'll
you're   who's (who is)

7. An *apostrophe* plus *s* is used to form the plural of words used as words, letters as letters, figures as figures, etc.

The sentence contains too many *and's*.
The *e's* could not be distinguished from the *i's*.
The *8's* were blurred and looked like *3's*.

Note. There is a growing tendency to omit this apostrophe when clearness would not be reduced. It would be clear, for example, to write Bs, Cs, 8s. Note, however, that omitting it before adding *s* to *A*, to *U*, or to *I* would result in *As*, *Us*, and *Is*.

Parentheses

Parentheses are used to set off a word, phrase, or clause that constitutes a definite interruption in continuity.

The store used "loss leaders" (articles priced below cost to attract customers) only on exceptional occasions.

Note. If material in parentheses is inserted into a sentence, no other punctuation precedes the opening of the parentheses, but the closing of the parentheses should be followed by whatever punctuation would have been used if the parenthetical material had not been inserted. If the parenthetical material is itself a sentence and is inserted at the end of a sentence, a period is used in the normal manner at the end of the sentence preceding the parentheses, and a period is placed inside the parentheses at the end of the parenthetical sentence.

Brackets

Brackets are used to mark off material that is inserted into a quotation but is not part of the quoted material.

"This year [1949] the outlook is less favorable."
"It is definitely established that he [Mr. Schenley] signed the contract."

The Hyphen

A hyphen is used at the end of a line when lack of space makes it necessary to break a word and complete it on the line below. The break must come at the end of a syllable. It is best, when possible, to avoid breaking a word where only two letters would either precede or follow the hyphen. For hyphenation of compound words, see Chapter 4.

CAPITALIZATION

1. The first word of a sentence is capitalized, as is the first word of a line of poetry, although contemporary poets sometimes use lowercase letters.

> The air was smoky.
> "A voice by the cedar tree
> In the meadow, under the hill."

2. The first word of a quotation is ordinarily capitalized, but a capital letter is not used when the quotation merges into the structure of the enclosing sentence.

> His exact words were, "The request is refused."
> He objected to "the capitalization of unrealized anticipations."

3. Proper nouns, their derivatives, and common nouns used as proper nouns are capitalized.

(a) Capitalize personal and geographical names, names of races, nations, languages, and the like.

Henry A. Collins	American
English	France
Negro (but white; black)	Hudson River
King County	Lake Erie
The National Biscuit Company	Reed College

Note. Such words as lake and bay are always capitalized when they precede the word that indicates the specific bay or lake. Some magazines and most newspapers, however, do not capitalize them when they follow the word that indicates the specific name (for example, *Payette lake, Hudson bay*). Many other such words—for example *mountain, street,* and *railroad*—are similarly treated.

Note. When the terms discussed above are preceded by two proper nouns they are never capitalized.

> The Mississippi and Missouri rivers.

(b) Capitalize the first word and each important word in the names of organizations, businesses, governmental bodies, etc.

> Veterans of Foreign Wars
> The Builders Supply Company
> The Federal Communications Commission

(c) Capitalize the days of the week, months, holidays.

Monday   June   Memorial Day

(d) Capitalize each important word in the titles of books, magazines, articles, and other written materials. The word *The* is capitalized when it is the first word in the title of a book, but is not capitalized at the beginning of the title of a newspaper or magazine unless it is part of the name, such as *The New Yorker*, but *the Saturday Evening Post*.

Essentials of Microwaves   the Saturday Review
The World Almanac   Science and Public Relations

(e) Capitalize the title of a person if it precedes his name, but do not capitalize a title that follows a name unless it is a title of distinction.

Professor Page
Judge William E. White
President Reagan
Ronald Reagan, President of the United States
Sandra O'Connor, Justice of the United States Supreme Court
Sergeant Keppler
Dr. Springer

but

Norman Page, professor of chemistry
William E. White, police judge
John Moore, president of the Rotary Club

Note. The illustrations above indicate capitalization in ordinary text. In the address of a letter or in most display printing, most titles would be capitalized.

(f) Capitalize words that are derived from the names of persons, places, or other proper names.

The crew lived in a Quonset hut.
The Plimsoll mark was visible.

Note. Eventually, such capitalization is usually discontinued, as in *sandwich* and *boycott*. Since the discontinuation is gradual, there will always, at any given time, be many terms in which usage differs in different publications. A dictionary will give guidance, but dictionaries necessarily lag behind technical books and magazines.

4. Nouns and pronouns referring to the Deity, and numerous other words with sacred significance are capitalized.

"Earth changes, but thy soul and God stand sure."
"So take and use Thy work."

5. The pronoun *I* and the interjection *O* are capitalized, but *oh* is not capitalized except at the beginning of a sentence or quotation.
6. In ordinary text, capitalization merely as a means of emphasis is poor form. Inexperienced writers tend to capitalize too frequently. The following precautions, especially, should be observed:

(a) North, south, east, and west are capitalized only when they are used as proper nouns, that is, the names of specific geographical regions. When these words indicate mere direction, they are not capitalized.

He lived in the South.
He lived south of the tracks.

(b) Such words as chemistry, psychology, and history are not capitalized except when they occur as part of a title—for example, the title of a specific university course.

Being interested in physics, he decided to take The Physics of Sound.

ITALICS

Italics (indicated in typing or longhand by an underline) show that words are used in some special manner and hence should be set off from the rest of the material.

1. Italics are used for the titles of books, magazines, newspapers, bulletins, or other separately issued publications.

The Business Letter in Modern Form
Webster's Seventh New Collegiate Dictionary

Basic Engineering Metallurgy
the *Chicago Tribune*
the *Farm Science Reporter*
The American College Dictionary

Note. If the word *A, An,* or *The* is the first word of the title of a book, it is italicized; preceding the title of a periodical, it is not italicized unless it is part of the title.

2. Italics are used for the names or titles of ships, trains, airplanes, works of art, etc. Some authorities, however, place such material in quotation marks.

the *Lusitania*
El Greco's *Cleansing of the Temple*
the *Twentieth Century Limited*

3. Italics are used for words considered as words, phrases considered as phrases, and letters or numbers considered as letters or numbers. (Quotation Marks are sometimes used for the same purpose. See Rule 3 under *Quotation Marks*).

The writer used *effect* incorrectly.
The phrase *in case of* is overworked.
The *i* looks like an *e*, and the *8* is so dim it looks like *3*.

Note. A letter used to indicate shape, as in "V-shaped" or "I-beam," is not italicized.

4. Foreign words or phrases are italicized, whether written out or abbreviated. (This rule covers Latin scientific names). To determine whether some words are considered foreign or are accepted as English, it may be necessary to consult a dictionary.

coup d'état   *ipso facto*   *de facto*

5. Italics are used to indicate special emphasis, but their use for this purpose should be limited to occasions when the emphasis is justified or when special emphasis is really necessary to make sure that the reader grasps the intended meaning.

Do you mean to tell me that you met *the* Walter Huston?
He definitely gave instructions that the deduction should *not* be made.

ABBREVIATIONS

The following rules for abbreviation apply to ordinary writing only. The additional abbreviations needed for technical style, and the use of abbreviations in footnotes, bibliography, and technical parts of letters, are discussed with those particular topics.

In ordinary writing, abbreviations should be used sparingly. Unless one of the following rules clearly calls for abbreviation, a word should be written out.

1. The abbreviations A.M., P.M., A.D., and B.C. are always used in place of the terms for which they stand.
2. The abbreviations *e.g.* (for example), *i.e.* (that is), and *viz.* (namely) may be used when informality is appropriate. At any other time, their English equivalents are preferable.

3. The abbreviation *etc.* is so extremely informal that it should usually be avoided in regular text that is offered as finished writing. It is acceptable, however, in a series of brief rules—as you see it used, for example, in Rule 2 of *Italics,* above.

4. In recent years many abbreviations such as NATO, TVA, and FBI have come to be used as if they were actual names. Their acceptance results from the length of the names they stand for. When the readers addressed are sure to understand their meanings, they are acceptable except on occasions so formal that nothing except full, official names would be appropriate.

5. Such abbreviations as *Jr., M.A., Ph.D.,* and *C.P.A.,* when following names, are preferred to the full words or phrases. Four abbreviations for titles—*Mr., Ms., Mrs.,* and *Dr.,* are used before names. In ordinary text, however, all other titles are in better form if written out whether they precede or follow a name. Examples of these titles are *Professor, Captain, President, Senator, Governor, Major.* Like these titles, terms of respect such as *reverend* and *the honorable* (these are not titles) should be written out.

6. Such words as *figure* and *number* preceding a number may be abbreviated: *Fig. 3; No. 7.*

7. Abbreviations are not used for the names of persons: *James, William, Robert,* not *Jas., Wm., Robt.*

8. Abbreviations are not used in ordinary text for the names of states or similar terms (except for *D.C.* and sometimes *N.Y.*); nor are they used in text for words such as *street, avenue, company, volume,* and *chapter.*

SPELLING

Among all possible errors in English, few are so damaging as errors in spelling. Mistakes in spelling are noticed and found annoying by readers who pass over all but the most glaring errors. No person who expects to do technical professional work can afford to shrug and say "I never could learn to spell."

The first necessity in overcoming weakness in spelling is the development of an adult attitude, a determination not to go through life handicapped by a juvenile weakness. Once such an attitude is acquired, improvement may not be easy, but it is hardly so formidable a task as might be anticipated. Even weak spellers, if they keep a list of words that they miss, are usually surprised at its shortness. The frequency of errors does not ordinarily result from misspelling a great number of words, but from repeatedly misspelling certain frequently used words. Weak spellers can often eliminate most errors by learning to spell no more than 40 or 50 words.

If you are weak in spelling, you should be able to make steady improvement if you will carry out the following instructions:

1. Check your work carefully for the express purpose of correcting the errors that result from sheer carelessness rather than ignorance.

2. Keep a list of every word you miss, and have that list in front of you when you write. Your individual list will be more valuable than a list taken from a book.

You will soon come to recognize any word on the list when you start to write it, and can check the way you spell it without taking time to consult a dictionary.

3. Learn just a few rules—the ones that are violated most frequently.

4. Give special attention to certain words that are misspelled because of sloppy pronunciation, and to pairs of words that are confused because of their similarity.

The Rules

The following rules do not cover everything; the list is the result of eliminating as much as possible, so that the rules that are most helpful will not be buried in the mass of rules and exceptions that would be necessary for completeness.

1. *ie* and *ei:* When the pronunciation is long *e*, use *i* before *e* except after *c*, and use *e* before *i* after *c*. (When the pronunciation is not long *e*, the spelling is usually *ei;* but few words except those where the pronunciation is long *e* cause trouble.)

believe	receive	weigh
chief	conceit	neighbor
field	ceiling	foreign

Exceptions. either, neither, seize, leisure, weird, financier, species.

2. The effect of prefixes: When a prefix is placed before a word, the spelling of the word itself remains unchanged. (This rule will solve many troublesome questions about double letters.)

dis + appoint = disappoint	un + worried = unworried
dis + appear = disappear	un + noticed = unnoticed
dis + satisfied = dissatisfied	grand + daughter = granddaughter

3. Treatment of final silent *e* when a suffix is added:

(a) When a suffix beginning with a consonant is added to a word ending in silent *e*, the *e* is retained

hope + ful = hopeful	shame + less = shameless
move + ment = movement	like + ly = likely

Exceptions. *duly, truly, argument, awful*. Also, dropping the *e* is preferred to retaining it in the spelling of such words as *judgment, acknowledgment,* and *abridgment*.

(b) When a suffix beginning with a vowel is added to a word ending in silent *e*, the *e* is dropped. (This covers the spelling of hundreds of words to which *ing* may be added.

hope + ing = hoping   quote + able = quotable
move + ing = moving   promote + ion = promotion
locate + ion = location   change + ing = changing

Exception. The letters *c* and *g* always have a hard pronunciation before *o* and *a* (*cat, coat, gave, go*). Hence a final silent *e* must sometimes be retained to keep the pronunciation of *c* or *g* soft.

enforce + able = enforceable   courage + ous = courageous
change + able = changeable

Exceptions. *Mileage, saleable,* and *useable* are exceptions to the rule for dropping the silent *e*. (*Milage* is an error. *Salable* and *saleable* are equally correct. *Useable* is acceptable but *usable* is preferred.) Also, there are a few words such as *singeing* or *dyeing* in which the *e* is retained to prevent confusion with other words (*singing, dying*).

4. Doubling a final consonant: When a suffix beginning with a vowel is added to a word ending in a consonant, the final consonant of the word is doubled under the following condition: (a) The word must end in only one consonant, preceded by only one vowel, and (b) the word must be accented on the last syllable. (This would include all one-syllable words.)

FINAL CONSONANT DOUBLED

occur + ed = occurred   stop + ed = stopped
refer + ed = referred   drag + ed = dragged
begin + ing = beginning   lag + ing = lagging

Exceptions. gas + es = gases   gas + eous = gaseous

FINAL CONSONANT NOT DOUBLED

rent + ing = renting   (Word ends in two consonants.)
read + ing = reading   (Two vowels precede the consonant.)
offer + ed = offered   (Accent not on last syllable.)
refer + ence = reference   (Accent shifted back to an earlier syllable.)

Note. Though the rule just discussed seems somewhat complicated, its application may be greatly simplified, for the misspellings that may be corrected by applying the rule are almost exclusively the result of overlooking the part of the rule referring to the accent. Characteristic errors take place with *occur, begin,* and *refer*. For purely functional purposes, the rule could be simplified to read: *If you are in doubt as to whether to double a final consonant when you add a suffix, notice where the accent lies. If it lies on the last syllable* (this obviously covers all words of one syllable) *double the final consonant. If it is not on the last syllable, do not double the final consonant.*

The cases where this simplification would not apply will not cause trouble. Even the weakest spellers would spell words such as *needed* or *spending* correctly without even considering the question of doubling a final consonant.

5. Adding a suffix to a word ending in *y:*

(a) When a suffix is added to a word that ends in *y* preceded by a *consonant,* the *y* is changed to *i* unless the result would be double *i.*

worry + ed = worried	mercy + ful = merciful
accompany + ment = accompaniment	hurry + ed = hurried
pity + able = pitiable	worry + ing = worrying

Exceptions. There are several exceptions, including words such as *shyness, citylike, secretaryship.* None of the exceptions is likely to be misspelled, or even to occasion doubt. Hence the rule may be applied, in case of doubt, without danger of causing an error in spelling.

(b) When a suffix is added to a word that ends in *y* preceded by a *vowel,* the *y* remains unchanged.

joy + ful = joyful	assay + ed = assayed
annoy + ance = annoyance	employ + ment = employment

Exception. *laid* (from lay + ed); *said* (from say + ed); *daily* (from day + ly).

Note. The most valuable fact to remember about errors resulting from addition of suffixes to words ending in *y* is that practically all these errors result from a single cause: failure to change *y* to *i* when it should be changed. The section of the rule expressed in (a) should receive major attention.

6. Words ending in *ede* or *eed:* Among the troublesome words ending in *ede* or *eed,* only three end in *eed,* namely *exceed, succeed,* and *proceed.* All the others end in *ede*—for example, *concede, recede, precede, supersede.* It is further worth noting, if one is doubtful whether *c* or *s* precedes the *ede,* that only one word, *supersede,* ends in *sede.*

7. For rules that concern the apostrophe, consult the discussion of punctuation.

Pairs of Words That May Be Confused

Many misspellings result from confusion between the word desired and some other word that resembles it in spelling or pronunciation. Some of the words commonly confused are as follows:

accept, except	affect, effect
advice, advise	all ready, already

all together, altogether
born, borne
brake, break
breath, breathe
capital, capitol
choose, chose
cite, sight, site
coarse, course
complement, compliment
dual, duel
dyeing, dying
emigration, immigration
eminent, imminent
formally, formerly
forth, fourth
incidence, incidents
its, it's
lead, led

lightening, lightning
loose, lose
manufacturers, manufactures
passed, past
perform, preform
personal, personnel
plane, plain
precedence, precedents
predominate, predominant
presence, presents
principal, principle
prophecy, prophesy
stationary, stationery
their, there, they're
through, thorough
to, too, two
whose, who's
your, you're

Words Misspelled Because of Pronunciation

Many words give trouble in spelling because of pronunciation. Some of the most troublesome words of this sort are listed below:

accidentally (not accidently)
athletics (not atheletics)
boundary (not boundry)
disastrous (not disasterous)
drought (not droughth)
foundry (not foundary)
height (not heighth)
hindrance (not hinderance)
incidentally (not incidently)
irrelevant (not irrevelant)
laboratory (not labratory)
mathematics (not mathmatics)

miniature (not minature)
mischievous (not mischeevious)
nuclear (not nucular)
performance (not preformance)
perseverance (not perserverance)
prescription (not perscription)
quantity (not quanity)
similar (not similiar)
sophomore (not sophmore)
temperament (not temperment)
temperature (not temperture)

MISCELLANEOUS SPELLING LIST

absence	aggravate	appearance
accommodate	allotted	appropriate
accumulate	all right	argument
acquaint	amateur	auxiliary
across	analyze	beginning
address	annual	believe

beneficial	exercise	occurrence
calendar	exhaust	omitted
ceiling	existence	optimistic
changeable	familiar	parallel
coarse	feasible	particle
committee	finally	partner
comparative	fluorescent	perform
comparison	foreign	permissible
competent	foremost	personal
completely	forty	personnel
conscience	gases	practically
conscious	gauge	precede
continuous	government	preferred
controlled	grievous	prejudice
convenience	guarantee	procedure
course	height	proceed
criticism	hindrance	quiet
criticize	hypocrisy	quite
curriculum	immediately	receipt
cylinder	inconvenience	receive
deferred	independence	recognizable
definite	indispensable	recommend
dependent	indivisible	repetition
description	interest	representative
desperate	leisure	reservoir
develop	liable	schedule
disappear	lightning	seize
disappoint	maintenance	separate
dormitories	mathematics	sincerely
embarrass	mileage	soluble
emphasize	miniature	sponsor
employees	missile	studying
environment	nineteen	therefore
equipped	ninety	transferred
especially	ninth	unnecessary
exaggerate	noticeable	whatever
exceed	occasion	
excellent	occurred	

GLOSSARY OF STANDARD USAGE

The following list includes the most common words and phrases that persistently trouble many writers.

The term *colloquial* is used with the liberal meaning *informal* rather than being restricted to its original meaning of *conversational*.

Accept, Except.   *Accept* means *to receive willingly*. If you use *except* with that meaning, you are in error.

Affect, Effect.   *Affect*, as a verb, means *to influence*. *Effect* as a verb means *to achieve or accomplish*. *Affect* is never correct as a noun (except in one highly technical usage in psychology). *Effect* as a noun means *result* and is essentially the noun form suggested by *affect* as a verb.

The weather *affected* their plans.

The attorneys *effected* a settlement of the suit.

The weather had no *effect* on their plans.

Already, All ready.   *Already* means *previously*. *All ready* means *entirely prepared*.

Among, Between.   *Among* is used in reference to more than two persons or objects. *Between* is used in reference to two only.

The profits were divided *among* the employees.

The profits were divided *between* the two partners.

The profits were divided *between* the company and the employees.

Amount, Number.   *Amount* refers to quantity. *Number* refers to objects that can be counted.

The *amount* of beef in storage was increasing.

The *number* of cattle on the range was increasing.

And/or.   Useful though this expression seems, it is frowned on by most authorities. Never use it in ordinary text if you can avoid it.

UNDESIRABLE: It had been colored by the use of ammonium carbonate, and/or ammonium chloride.

BETTER: It had been colored by the use of ammonium carbonate, ammonium chloride, or both.

As.   Often ambiguous when used to introduce a clause that might be introduced by *because* or *since*.

AMBIGUOUS: *As* the time of departure was drawing near, he became very nervous.

CLEAR: *Since* (or *because*) the time of departure was drawing near, he became very nervous.

CLEAR: *When* the time of departure came he grew very nervous.

As per.   Commercial jargon, undesirable in normal use as a substitute for *in accordance with*.

UNDESIRABLE: The work was done as per instructions.

BETTER: The work was done in accordance with the instructions.

Bad, Badly.   See *Adjectives and Adverbs*.

Balance.   The word *balance*, except in connection with bookkeeping or accounting, should not be used for *remainder*.

UNDESIRABLE: We will do the *balance* of the work next week.
CORRECT: Our *balance* in the bank account had decreased.

Because.   Do not misuse *because* for *for*. *Because* should be used to introduce a clause that tells the cause of some result rather than a clause that merely tells the reason for a belief or knowledge.

WRONG: He has not been here today, *because* his tools are still in his locker.
CORRECT: He has not been here today, *for* his tools are still in his locker.
CORRECT: He stayed away *because* he was ill.

Beside, Besides.   *Beside* means *by the side of*. *Besides* means *in addition to, except, moreover*.

WRONG: *Beside* the reason mentioned, there are other reasons.
CORRECT: *Besides* the reason mentioned, there are other reasons.

But what, But that.   The word *but* in these expressions is meaningless and should be omitted in formal writing.

WRONG: We had no doubt *but what* (or *that*) he would be acquitted.
CORRECT: We had no doubt *that* he would be acquitted.

Can, May.   *Can* indicates ability. *May* indicates permission. In recent years, however, the objections to using *can* for *may* have grown less vigorous.
Can't hardly, Can't scarcely.   These terms are illogical just as a double negative is illogical but are common in some dialects. Taken literally, either of them would cause a statement to mean the exact opposite of what it was intended to mean.

DIALECT: We *can't hardly* see it from here.
CORRECT: We *can hardly* see it from here.

Can't help but.   This expression cannot be defended from the grammatical standpoint and is avoided by careful writers except in extremely informal use. It is a distortion of the extremely formal expression, *cannot but*.

INFORMAL: I *can't help but* agree with him.
CORRECT: I *can't help* agreeing with him.
CORRECT, BUT EXCESSIVELY FORMAL: I *cannot but* agree with him.

Come.   Use of *come* for *came*, as in "We *come* here ten days ago," is dialect usage in the oral language of many people who would not write it.
Considerable.   *Considerable* is an adjective. It should not be used as a noun or adverb. As a noun it is dialect usage; as an adverb, illiterate.

DIALECT: The company lost *considerable* during the first year.
CORRECT: The company lost a *considerable* amount of money during the first year.
WRONG: We were influenced *considerable* by what he said.
CORRECT: We were influenced *considerably* by what he said.

Continual, Continuous.   *Continual* means *constantly recurring* (as in *continual* interruptions). *Continuous* means *without cessation* (as in the *continuous* increase in pressure).

Could of, Would of, Should of, etc.   In all such expressions, *of* is misused for *have* as the result of careless pronunciation in speech.

Data, Phenomena, Strata, Media, Bacteria.   These forms are all plurals. Their singulars are *datum, phenomenon, stratum, medium,* and *bacterium*. There is a strong tendency, at present, to use *data* as a singular when referring to a mass of facts considered as a whole; but most authorities still insist that it should be treated as a plural.

Different from, Different than.   *Different from* is preferred.

Differ from, Differ with.   *Differ from* means *to be dissimilar. Differ with* means *to disagree.*

Apple trees *differ from* pines in appearance.

The manager *differed with* the president as to who should be promoted.

Enable.   *Enable* means *to make able, to give ability to.* It should not be used in such a sentence as, "This appropriation *enables* the road to be paved," for the road is not given any new ability. It would be preferable to say, "This appropriation *enables* the highway department to pave the road."

Farther, Further.   Careful writers prefer *farther* in reference to actual distance and *further* in reference to quantity or degree.

Fewer, Less.   *Fewer* should be used when numbers are referred to; *less,* when quantity is referred to.

CORRECT: *Less* wheat was harvested this year than last; consequently *fewer* (not *less*) freight cars will be needed.

Good, Well.   See *Adjectives and Adverbs.*

Had of.   *Had* is sufficient.

WRONG: I wish I *had of* learned the news earlier.
CORRECT: I wish I *had* learned the news earlier.

Had ought.   This is an undesirable form, mistakenly substituted for *ought* or *should.*

WRONG: *He'd* (he had) *ought* to resign.
CORRECT: He *should* resign.
CORRECT: He *ought to* resign.

Hardly.   See *Can't hardly.*

Have got.   A redundant way of saying *have,* and an undesirable substitute for *must* or *have to* except on occasions of extreme informality.

REDUNDANT: We *have got* many things to be thankful for.
CORRECT: We *have* many things to be thankful for.
UNDESIRABLE: *We've got* to reach the town before morning.
CORRECT: We *have* to reach the town before morning.
CORRECT: We *must* reach the town before morning.

Hopefully.   Using *hopefully*, as in "Hopefully it will stop raining," is widely deplored by most authorities. Such usage has become so widespread, however, that it seems likely to stay. Objections are justified when *hopefully* replaces *I hope* (or *he, she,* or *everyone hopes*). Objections may be somewhat less justified when *hopefully* refers to a general feeling rather than that of some specific person or persons. In this case *hopefully* replaces *it is hoped that,* an expression that may be criticized if used as a hedge by an author who does not wish to make a commitment.

Regardless of whether the use of *hopefully* that now raises objections does or does not gain general acceptance, it is highly advisable at present to avoid that usage on all but the most casual occasions.

If.   When *if* would be ambiguous because it would not imply *or not* with sufficient emphasis, replace it with *whether*.

AMBIGUOUS: Let me know *if* you expect to visit us.

CLEAR: Let me know *whether* you expect to visit us.

Imply, Infer.   A speaker or writer implies. A listener or reader infers. It is inaccurate to say, "I *implied*, as I read your letter, that. . . ." Similarly, it is wrong to say, "Did you mean to *infer*, by your remarks, that. . . ." The same distinction applies to *implication* and *inference*.

Irregardless.   *Irregardless* is an unacceptable alternative for *regardless*. Superficially, it resembles *irresponsible* and *irrelevant*, in which the *ir* means *not*, but the *ir* in *irregardless* has no effect on meaning.

Is when, Is where, Is because.   These phrases are often misused, as seen in the following sentences.

WRONG: Insolvency *is where* one cannot pay one's debts.

CORRECT: Insolvency is a lack of funds to make payments that are due.

WRONG: Bankruptcy *is when* one's property is administered for the benefit of one's creditors.

CORRECT: Bankruptcy is a state *in which* one's property is administered for the benefit of one's creditors.

WRONG: The reason for his absence *is because* he is ill.

CORRECT: The cause of his absence is *that he is ill*. (or *is illness*.)

Less.   See *fewer*.

Lie, Lay.   *Lie* is an intransitive verb meaning *recline*. *Lay* is a transitive verb meaning to *place*, or *cause to lie*.

Principal parts: lie, lay, have lain
      lay, laid, have laid

CORRECT: He lies on the bed. (present)

CORRECT: He lay on the bed. (past)

CORRECT: He *has lain* on the bed all afternoon.

WRONG: He *lays* on the bed.

WRONG: He *laid* (or *has laid*) on the bed all afternoon.

CORRECT: He *lays* the newspaper on the bed.

CORRECT: He *has laid* the newspaper on the bed.

Note. The confusion results primarily because the past tense of *lie* is identical with the present tense of *lay*. Some errors also result from the use of *laid* as the past tense or past participle of *lie*.

Like, As if, As though. Careful writers do not use *like* as a conjunction, preferring *as if* or *as though*. *Like* is correct, however, as a preposition. In conversation, *like* as a conjunction slips into the language even of many literate people.

CORRECT: He looked *like* a man of forty.

CORRECT: He spoke *as if* (or *as though*) he had a cold.

Note. Being a preposition, *like* cannot introduce a clause; and the usual sign that the construction is a clause is the presence of the subject and verb. An easier way to avoid misuse of *like* is to try *as if* in its place, and use it if it sounds right. You may be sure that when *as if* is wrong it will sound wrong.

Most, Almost. These are two distinct words, with different meanings. Sometimes, however, *most* is used colloquially for *almost* as in "He made an error in *most* every copy." The simplest way to avoid this error is to remember that *most* is never correct when *almost* would convey the desired meaning.

CORRECT: He made errors on *most* of the copies.

CORRECT: It was a *most* unfortunate statement.

COLLOQUIAL: Most everyone in the crowd saw the incident.

Of. Do not use *of* for *have* as in "He must *of* gone," for "He must *have* gone." Other examples of this error are *could of, should of,* and *might of.*

Percent. This term should not be loosely used for *portion* or *part*. Ordinarily, it should be used only following a number. It should not be used for *percentage*, though to save space many writers disregard this distinction in tables and figures.

CORRECT: The company earned a 10 *percent* profit.

CORRECT: The *percentage* (not *percent*) of profit increased.

WRONG: A large *percent* of the profit resulted from sales abroad.

WRONG: The *percent* elongation was excessive.

Practicable, Practical. *Practicable* is used to indicate that a method *can* be used or that a result *can* or *could* be accomplished. *Practical* goes further; it indicates that whatever is being considered not only is a possibility but would actually give results that would justify action. For example, building a steam-powered automobile is *practicable* but has not proved to be *practical*. In most sentences either word makes sense, but they convey different meanings to well-informed readers.

Principal, Principle. *Principal* as an adjective means *main* or *chief,* and as a noun means the *main* or *controlling person,* as the *principal* of a school or one of the *principals* in a lawsuit. *Principle,* always a noun, means *fundamental truth, basic law*.

CORRECT: The *principal* reason for its failure was defective workmanship.

CORRECT: The *principles* of democracy must be preserved.

Reason is because, reason why. Both of these expressions are redundant, for neither *is because* or *why* adds to the meaning already conveyed by *reason*. Careful writers avoid them even in informal writing.

Rise, Raise. The distinction between *rise* and *raise* is essentially the same as the distinction between *lie* and *lay*. *Rise* is an intransitive verb, *raise* is transitive.

> *Principal parts:* rise, rose, have risen
> raise, raised, have raised

CORRECT: The prices *rise* when the supply becomes scarce.

CORRECT: The dealers *raise* the prices when the supply becomes scarce.

WRONG: The prices *raised* when the supply became scarce.

CORRECT: The prices *rose* when the supply became scarce.

Seldom ever. Redundant; the full meaning is expressed by *seldom* alone.

Shall, Will. The careful discrimination between *shall* and *will* is not so common as in past times, and even literate writers frequently use *will* on some occasions when *shall* is called for by the rules. The main points to remember in order to make the traditional discriminations are as follows:

(a) To indicate simple future or expectation, use *shall* in the first person and *will* in the second and third persons.

> I *shall* be a senior next year.
> He *will* spend August in Omaha.

(b) *Will* in the first person and *shall* in the second and third person indicate determination, command, or prohibition.

> I *will* do it in spite of your objections.
> You *shall* change your itinerary..
> No substitutions *shall* be made by the contractor.

(c) *Will is used in all persons to express willingness.*

> I *will* go if no one else *will*.

Should, Would. Traditionally, the difference between *should* and *would* is the same as that between *shall* and *will*. Consequently, those who were meticulous would write, "I *should* enjoy reading the book," and "He *would* enjoy reading the book." Each of the two words has special uses, however, as the following comments show.

Should is used in all persons in the sense of *ought to* or in a conditional clause, which is usually a clause beginning with *if*.

> I *should* give him credit for working overtime.
> He *should* give me credit for working overtime.
> If I *should* refuse, no one could blame me.
> If he *should* refuse, no one could blame him.

Similarly, *would* is used in all persons to indicate customary action. It is also used whenever *should* would be misleading in that it might suggest *ought to*.

Customary action: Long before closing time, I would begin preparing to go home. *Avoidance of the meaning of ought to:* Under the circumstances, I would have refused.

Sit, Set.   The distinction between *sit* and *set* is essentially the same as the distinction between *rise* and *raise*, and *lie* and *lay*. *Sit* is an intransitive verb; *set* is transitive. Exceptions, of course, are such uses as "The sun *sets* in the west" and "The hen *sets* on the eggs."

Principal parts: sit, sat, have sat
set, set, have set
CORRECT: He *sits* at my desk, where he *sat* yesterday, and where he *has sat* so many days before.
CORRECT: He *sets* the case on the desk, where yesterday he *set* it down so carelessly. He *has set* it there every time he has come in the office.

Some, Somewhat.   *Some* should not be used as an adverb with the meaning of *a little*. The word for this use is *somewhat*.

DIALECT: I was feeling *some* better.
CORRECT: I was feeling *somewhat* better.

These kind.   This expression is wrong. The plural *these* should not modify the singular *kind*.
Try and.   This expression should not be substituted for *try to*. *Try and* is just as illogical as "*I am able and*" for "*I am able to*."

WRONG: I will *try and* go.
CORRECT: I will *try to* go.

Type.   The common tendency among technical writers to use *type* as an adjective does not conform with the rules of good use. *Type* is a noun or verb, not an adjective. It should not directly modify a noun unless, in a hyphenated term, it becomes part of a compound adjective.

WRONG: We installed the new *type* machinery.
CORRECT: We installed the new *type of* machinery.
CORRECT: The design calls for an *injector-type* condenser.

Very.   It is a poor style to overwork *very* as an intensifier. In a phrase such as *very nauseating*, *very* is superfluous.
While.   See 2.b. under *Coherence*.
Who, Whom.   Traditional grammar calls for the use of *who* in the nominative case and *whom* in the objective (the object of a verb or preposition or the subject or object of an infinitive). Use of *who* for *whom* has been a common problem and has been widely corrected for years. In recent years, however, some liberal authorities

have modified their objections to *who* in place of *whom* at least at the beginning of a sentence or of the element of a sentence where it appears.

The misuse of *who* for *whom* has been corrected so much that the result has been to cause the opposite problem—the misuse of *whom* for *who* as for example, "Please let us know *whom* is coming." This is not only wrong (*whom* is used as the subject of a verb) but sounds affected.

The following sentences illustrate typical errors and the corrections called for by traditional grammar.

WRONG: Let me know *who* you have invited. (*Who* is used as the object of a verb.)
CORRECT: Let me know *whom* you have invited.
WRONG: I don't know *who* he was working with. (*Who* is used as the object of a preposition.)
CORRECT: I don't know *whom* he was working with.
WRONG: I don't know *who* to blame. (*Who* is used as the subject of an infinitive.)
CORRECT: I don't know *whom* to blame.
WRONG: Please let me know *whom* is to be invited. (*Whom* is used as the subject of a verb.)
CORRECT: Please let me know *who* is to be invited.

With.   *With* should not be loosely used to establish some vague, undefined thought relationship.

WRONG: *With* the plans completed, the president of the company went on a vacation.
CORRECT: The plans having been completed, (or, When the plans had been completed,) the president of the company went on a vacation.
WRONG: The production was increasing, *with* the prices holding steady.
CORRECT: The production was increasing, *and* the prices were holding steady (or, *but* the prices were holding steady).

EXPLANATION
OF GRAMMATICAL TERMS

Absolute.   An expression that is not grammatically connected with the sentence where it occurs; sometimes called a "nominative absolute."

The job being completed, the crew was laid off.

Adjective.   One of the parts of speech. A class of words used to modify (describe or limit) nouns or pronouns.

rough road; *this* year; *reasonable* profits

Adjective clause.   A clause used as an adjective to modify a noun or pronoun.

The tire *that blew out* was defective. (The clause *that blew out* modifies the noun *tire*.)

Everyone *who came* enjoyed himself. (The clause *who came* modifies the pronoun *everyone*.)

Adverb. One of the parts of speech. A class of words used to modify verbs, adjectives, or other adverbs.

He drove *cautiously*. (*Cautiously* modifies the verb *drove*.)
The problem is *very* hard. (*Very* modifies the adjective *hard*.)
The frame was *very* strongly built. (*Very* modifies the adverb *strongly*.)

Adverb clause. A clause used as an adverb to modify any part of speech that an adverb might modify, usually a verb. Such a clause often tells how, when, or where.

If the cost is excessive, we shall cancel the project. (The italicized clause modifies the verb *shall cancel*.)
They will burn the slashing *when the weather grows damper*. (The clause modifies the verb *will burn*.)

Agreement. The necessary correspondence between a subject and verb in person and number; between a pronoun and its antecedent in person, number, and gender; and between a demonstrative adjective *(this, these)* and the noun that it modifies, in number.

Antecedent. The noun or the other pronoun to which a pronoun refers.

After reading the *report* he returned *it*. (*Report* is the antecedent of *it*.)

Appositive. A noun or other substantive placed next to some other noun, used in the same way grammatically, and referring to the same thing or person.

The main plant, *a four-story building*, stood at the edge of town.
It was discovered by the watchman, *a reliable employee*.

Article. Any of three words—*a, an,* or *the*.

Auxiliary. A word used as part of a verb to assist in indicating the tense, voice, mood, etc. of the main verb.

are working; *have* answered; *must have* heard; *shall* or *will* arrive

Clause. A grammatical unit containing a subject and a finite verb. At least one independent clause is necessary in any complete sentence. (Sometimes part of the clause may be implied rather than expressed, in which case it is called an elliptical clause.)
A main or independent clause is one that makes an independent assertion.
A subordinate or dependent clause is one that is not self-sufficient and does not of itself make an assertion, but is used as part of a main clause in the sentence where it occurs. It is always used as if it were a single word—a noun, adjective, or adverb.

Independent clause: The sky is blue.
Subordinate clause used as a noun: We heard *that the request was refused*.

Subordinate clause used as an adjective: I know of a hotel *that will be satisfactory*.
Subordinate clause used as an adverb: Please close the door *when you leave*.

Comparison. The change of form in an adjective or adverb, indicative of change of degree, as in *good, better, best; long, longer, longest; easily, more easily, most easily*.

Complement. An element that ordinarily follows a verb and completes the assertion made by the verb about the subject. Types of complements are shown in the following examples:

Direct object: The farmer raised *wheat*.
Indirect object: The manager gave *him* a check.
Predicate adjective as subjective complement: The tractor is *old*.
Predicate adjective as objective complement: The news made him *happy*.
Predicate noun as subjective complement: The tree is an *oak*.
Predicate noun as objective complement: The president made John Jones the *manager*.

Complex sentence. See *Sentence*.
Compound sentence. See *Sentence*.
Compound-complex sentence. See *Sentence*.
Conjunction. One of the parts of speech. A word used to connect words, phrases, or clauses. Coordinating conjunctions connect elements that are equal in grammatical rank. Subordinating conjunctions are used to connect subordinate clauses to main clauses.

Coordinating conjunctions: and, but, or, nor, for; sometimes *yet* and *so*. See also *Correlative Conjunctions*.
Subordinating conjunctions: after, as, because, since, when, where, etc. For example, in the sentence "He did not come because he was ill," *because* makes the clause it introduces subordinate.

Conjunctive adverb. An adverb that functions also as a conjunction to join main clauses (*also, however, moreover, consequently, furthermore,* etc.). Use of a comma between independent clauses joined by a conjunctive adverb rather than a co-ordinating conjunction results in a comma fault. See "Unity of Structure" under " Unity and Coherence in the Sentence."
Correlative conjunctions. Conjunctions used in pairs to join elements of equal rank.

either . . . or; neither . . . nor; not only . . . but also; both . . . and

Ellipsis. The omission of words that are necessary for the grammatical completeness of the sentence, the words omitted being implied by what is expressed.

While (I was) working out of doors, I gained weight.
The new model is sturdier than the old (model is).
(You) Send me the answer promptly.
The subject of an imperative sentence is almost always *you*, implied. Hence an imperative sentence is almost always elliptical.

Expletive.   A word that bears no real meaning but is used to fill out a sentence. An expletive at the beginning of a sentence is sometimes necessary, but sometimes causes postponement of the word that should be the subject.

It is certain that production will increase.
It is snowing.
It is nine o'clock.

Gender.   The status of a noun or pronoun as masculine *(man, waiter, he)*, feminine *(woman, she, waitress)*, neuter *(tree, house, it)*, or common *(mouse, person, you)*.
Gerund.   A verb form used as a noun. A gerund plus its object is a gerund phrase.

Gerund: Walking is good exercise.
Gerund phrase: Pruning a tree demands judgment.

Idiom.   An expression that is accepted as correct even though it may not be in accordance with the normal patterns of the language, and that often conveys a meaning that would not be conveyed by the literal meaning of its component words.

The more the merrier is a typical idiom. It conveys more meaning than can be accounted for by the words and is acceptable as a sentence even though it does not have the elements a sentence calls for. *Make off with* is an idiomatic expression meaning *take away. Of yours* (as in "that car *of yours*") is idiomatic, for it is acceptable even though the object of a preposition is in the possessive case. *Many a person is* is an idiom for it is accepted as correct even though *many*, a plural, modifies a singular word and though a subject with a plural meaning takes a singular verb.

When we refer to the *idiom of a language,* we refer to its particular manner of using words to convey ideas—a manner that cannot be entirely accounted for by rules or by the meaning of individual words. The use and meaning of prepositions, especially, depend upon the idiom of the language. When some expression in your speech or writing is corrected on the grounds that it is not idiomatic of the language, it is usually impossible to justify the correction by citing rules.

Infinitive.   The form of a verb that is used after *to,* though sometimes the *to* is implied rather than expressed. (He hoped *to be* present. He can *go.*) When it is part of the verb phrase, as in the examples above, an infinitive is called a complementary infinitive because it completes the verb.

An infinitive can also be used as a noun, an adjective, or an adverb.

Infinitive as a noun: To err is human. (*To err* is used as the subject of the verb *is.*)
Infinitive as an adjective: I have a confession *to make.* (*To make* modifies the noun *confession.*)
Infinitive as an adverb: They were eager *to accept.* (*To accept* modifies the adjective *eager.*)

Inflection.   Variation in the form of words to indicate change of number, gender, person, tense, or mood. Inflection of nouns is referred to as declension; inflection of verbs as conjugation; inflection of adjectives and adverbs as comparison.
Interjection.   One of the parts of speech. A type of word placed at the beginning of a sentence or inserted into the sentence (whence *interjection—thrown in*) but not

connected with the grammatical structure. Words such as *oh, alas,* and *well* (as an exclamation) are interjections.

Linking verb. A verb that expresses neither action nor condition, but merely establishes a connection between its subject and a noun, pronoun, or adjective in the predicate. A noun or pronoun following a linking verb is in the nominative case. The commonest linking verb is the verb *to be.* Other characteristic linking verbs are *become, appear, seem, smell* (in the sense of "possess an odor"), *taste* (in the sense of "possess a flavor"), and *feel* (in the sense of "experience or provides a sensation").

The seed *is* pure.
The price *seems* right.
The warmth *feels* good.

Modifier. A word is a modifier when it is used to limit or change the meaning of some other word, or of a phrase or clause. A phrase or a clause, as well as a word, can be a modifier.

Word as a modifier: His *slightly* embarrassed manner was amusing. (*Slightly,* an adverb, modifies *embarrassed,* a participial adjective, which in turn modifies the noun *manner.*)
Phrases and clauses as modifiers: Construction *of the bridge* will be postponed *until the weather is warmer.* (*Of the bridge,* an adjective phrase, modifies *construction. Until . . . warmer,* an adverb clause, modifies the verb *will be postponed.*)

Mood. The form, in a verb, that indicates how the action or condition expressed by the verb is conceived by the writer or speaker. In a statement of a fact or in a question, the indicative mood is used. In a command, the imperative mood is used. In a statement expressing doubt, wish, or condition contrary to fact, the subjunctive mood is used.

Indicative: The rock *is* heavy. *Can* you *lift* it?
Imperative: Examine the surface carefully.
Subjunctive: If I *were* you, I *should accept* the offer.
I wish he *were* not so young.

Nonrestrictive modifier. A modifier that merely gives information about the term modified rather than limiting or identifying it.

The roof, *which was very old,* was damaged by the wind.
His oldest son, *who was more experienced,* handled the case efficiently.

Noun. One of the parts of speech. A word used to name a person, place, thing, quality, etc.

John; Aristotle; engineer; Atlanta; wood; courage

Many words can be used, of course, as more than one part of speech. Such words cannot be classified until they are actually used. For example, *green* is a noun in the sentence, "Green is its natural color" but an adjective in "The green grass is attractive."

Noun clause. A subordinate clause used as a noun.

I know *what you mean*. (Noun clause used as object of the verb *know*.)

How he had escaped was a puzzling question. (Noun clause used as subject of the verb *was*.)

Object. The word, phrase, or clause identifying the person or thing that receives the action indicated by a transitive verb, or the substantive referred to by a preposition.

Direct object of a verb: The company installed a *lathe*. (*Lathe* is the direct object of *installed*.)

Indirect object of a verb: He offered *me* a bribe. (*Me* is the indirect object of *offered*, for it indicates the person to whom the offer is made. *Bribe*, like *lathe* in the preceding example, is a direct object.)

Object of a preposition: He worked on a *farm*. (*Farm* is the object of *on*.)

Participle. The form of a verb ending in *ing* (though this form may also be a gerund) or the form used after *have*. A participle may be used as the main element in a verb phrase or as an adjective.

Participles as verbs: I am *following* the route I have *followed* before.

Participles as adjectives: Running water purifies itself. The *rejected* casting was returned. The guard was *tired*.

Parts of speech. The categories under which all words are classified. The eight parts of speech in traditional grammar are nouns, pronouns, verbs, adjectives, adverbs, prepositions, conjunctions, and interjections. Each of these terms is defined in its proper alphabetical place in this Glossary.

A word qualifies as one or another part of speech only on the basis of its use in a sentence or phrase; it is possible for the same word to be used as more than one part of speech. Good usage, however, limits the *permissible* uses of most words. To determine whether it is correct to use a word as a certain part of speech, it is sometimes necessary to consult a dictionary.

Person. Some pronouns and most verbs vary in form to indicate whether they refer to the person speaking or writing (first person), the person addressed (second person), or the person spoken or written about (third person).

FIRST PERSON	SECOND PERSON	THIRD PERSON
I come	you come	he comes
I have	you have	she has
I shall find	you will find	she will find
We shall find	you will find	they will find
I am	you are	he is
I can	you can	he can

Phrase. A group of related words that does not include a subject and predicate and that is usually used as a single part of speech—adjective, adverb, verb, or noun. Phrases may be prepositional, participial, gerund, infinitive, or verb.

Prepositional phrase: He discovered the address *of the owner.* (Adjective phrase, modifying the noun *address.*)
Prepositional phrase: The road will be closed *in December.* (Adverbial phrase, modifying the verb *will be closed.*)
Participial phrases: Entering the room, I noticed that the fan *standing in the corner* was noisy. (Adjective phrases, modifying, respectively, *I* and *fan.*)
Gerund phrase: Pitching hay is strenuous work. (Noun.)
Infinitive phrase: To predict the results is impossible. (Noun.)
Verb phrase: The method *has been found* successful.

Predicate. The portion of a clause consisting of the verb, its complements, and the modifiers of both. Except for interjections and absolute phrases, neither of which appears frequently, the predicate consists of everything in a clause or sentence except the subject.

Preposition. One of the parts of speech. A word used to introduce a noun or pronoun and establish its relationship to the sentence.

He will come *in* the morning *on* the plane *from* Chicago.

Principal parts. The three forms of a verb from which, by help of auxiliary verbs, the various tenses are derived. These forms are the present infinitive, the past tense, and the past participle. (The past participle is the form used after *have.*)

PRESENT INFINITIVE	PAST TENSE	PAST PARTICIPLE
work	worked	worked
sing	sang	sung
begin	began	begun
catch	caught	caught

Pronoun. One of the parts of speech. A word used in place of a noun.

Personal pronouns, I, you, he, she, it, we, they
Interrogative pronouns: who, which, what
Relative pronouns: who, which, what, that
Demonstrative pronouns: this, that, these, those
Indefinite pronouns: each, either, neither, any, anyone, some, someone, one, no one, few, all, none
Reciprocal pronouns: each other, one another
Reflexive pronouns: myself, yourself, himself, herself, itself, themselves
Intensive pronouns: myself, yourself, himself, herself, itself, themselves

Relative clause. A clause, always dependent, introduced by a relative pronoun. The clauses used as restrictive modifiers below are examples.

Restrictive modifier. A modifier that narrows, and thus restricts, the meaning of whatever it modifies.

All persons *who drive recklessly* should be arrested.
The cases *that were damaged* have been returned.

Sentence. A group of words expressing a thought and containing, either actually or by implication, a subject and a finite verb. Grammarians have never defined the term *sentence* to their own complete satisfaction, but this definition will serve our purpose. In grammatical form a sentence may be simple, compound, complex, or compound-complex.

A simple sentence consists of a single independent clause, though as indicated by the second example below, its subject, its verb, or both may be compound.

Simple sentence: Honesty pays.
Simple sentence: Both industry and labor support the new ruling and oppose the old.

A compound sentence consists of two or more independent clauses.

Compound sentence: Sales were falling off, and inventories were increasing.

A complex sentence contains a single independent clause and one or more dependent clauses.

Complex sentence: The police recovered the property that had been lost.

A compound-complex sentence contains at least two independent clauses, hence being compound, and at least one dependent clause, hence being complex.

Compound-complex sentence: The wind was rising and the sky had clouded over when we finally reached the shelter.

Subject. The substantive that a verb or verbal makes an assertion about, asks a question about, or gives an order to.

Subject of a sentence: Birds fly south in the winter.
Subject of an infinitive phrase (a verbal): I know *him* to be honest. (*Him* is the subject of *to be.*)

Substantive. A noun, or any word or group of words used as a noun. Any of the following, in addition to a noun, may be a substantive: pronoun, infinitive phrase, gerund, or noun clause. (See separate listing for each of these terms.)

Tense. The distinctive form in a verb that indicates time.

Present tense: It *runs* well.
Past tense: It *ran* well.
Future tense: It *will run* well.
Present perfect tense: It *has run* well.

Past perfect tense: It *had run* well.
Future perfect tense: It *will have run* well.

Tenses appear in the progressive and emphatic forms of verbs, as well as in the simple forms.

Present progressive: I *am reading*.
Present emphatic: I *do read*.

Verb.   One of the parts of speech. A word or phrase that indicates action, being, or state of being. In a declarative sentence the verb is the word or phrase that actually makes the assertion.
A transitive verb takes an object.

He *bought* the *house*.

An intransitive verb cannot take an object.

The sun *rises* in the east.

Some verbs may be either transitive or intransitive.

Transitive: I *read* the letter.
Intransitive: I *was reading*.

Verbal.   A word or phrase derived from a verb but not making an assertion. Infinitives, gerunds, and participles (participles used either in participial phrases or as adjectives) are the three types of verbals. Each is listed as a separate entry.

Voice.   The form of a verb that shows whether the subject of a verb or verbal performs the action indicated (active voice) or has the action performed upon it (passive voice). Only transitive verbs can be in passive voice.

Active voice: The government *collected* the taxes.
Passive voice: The taxes *were collected* by the government.

APPENDICES

A

ABBREVIATIONS
FOR SCIENTIFIC
AND ENGINEERING TERMS

The number of abbreviations in current use is so great and is expanding so rapidly that in a book like this only a small fraction of them can be listed. For those in positions that make it necessary for them to use long and highly specialized lists, such lists are readily available. Also it is worth noting that both collegiate and unabridged dictionaries provide abbreviations for hundreds of technical terms.

Since minor differences persist in actual use, the abbreviations in different lists are not always the same in form, but those in the list that follows are supported by well-regarded authorities.

As you use the list you will observe that some terms are abbreviated in one way when they appear alone and otherwise when they appear in the abbreviation for a phrase, especially if the phrase contains *per* or a diagonal line (/) standing for *per*. For example, *ft* is the abbreviation for *foot* or *feet* and *sec* for *second;* but *feet per second,* which includes both words, may be shortened to *fps.* Other examples of abbreviations that are used as combining forms are *m* for *miles* and *h* for *hour* in *miles per hour* (*miles* standing alone is not abbreviated and *hour* is abbreviated as *hr* or *h*); *g* for *gallons* and *m* for *minute* in *gallons per minute* (the ordinary abbreviations for *gallons* and *minute* are *gal* and *min*); and *rf* for *radio frequency* (*radio* is not abbreviated standing alone, and *frequency* is abbreviated as *freq*). Even in abbreviating a phrase, however, it is better not to use the combining forms if there is any doubt about their being easily understood by the readers. For example, *ft per sec* is sometimes better than *fps.*

The abbreviations in this book do not include arbitrary signs and symbols. (Such signs and symbols are not actually abbreviations and should rarely be used in ordinary text.) They comply, so far as facts permit, with the sound generalization that capital letters are ordinarily used in abbreviations only when the terms for which the letters stand would be capitalized if written out. Periods are omitted except when omitting them might cause the abbreviation to be mistaken for a regular word, or when the term is abbreviated in general as well as in technical writing. You should not hesitate to use periods, however, if personal observation convinces you that it is customary to use them in the field that includes your subject.

abampere	abamp
absolute	abs
acoustic	acst
acre	(spell out)
acre-foot	acre-ft
air horsepower	ahp
air position indicator	api
air speed indicator	asi
alternating-current (adj.)	a-c
American Standard	Amer Std
American Wire Gage	AWG
ammeter	am.
ampere	amp
ampere-hour	amp hr
amplitude modulation	am.
analysis of variance	anova
angstrom	A
approximate	approx
atmosphere	atm
atomic weight	at wt
audio frequency	a-f
auxiliary	aux
azimuth	az

balanced incomplete block designs	bib designs
barometer	bar
barrel	bbl
Baumé	Bé
best linear unbiased estimator	blue
billion electron volts	bev
biochemical oxygen demand	bod
bits per inch	bpi
board feet	fbm
boiler horsepower	bhp
brake horsepower	bhp

brake horsepower-hour	bhp-hr
Brinell hardness	Bh
Brinell hardness number	Bhn
British thermal unit	Btu

calorie	cal
candle-hour	c-hr
candlepower	cp
carrier wave	cw
cathode ray	cr
Celsius	C
centigrade	C or c
centimeter	cm
central processing unit	cpu
chemically pure	cp
circumference	circ
cologarithm	colog
continuous wave	cw
counts per minute	cpm
cubic	cu
cubic centimeter	cc
cubic feet per minute	cfm
cubic feet per second	cfs
cubic foot	cu ft
cubic inch	cu in.
cubic yard	cu yd
cycles per minute	cpm
cycles per second	cps

data control block	dcb
decibel	db
decimeter	dm
degree	deg*
dew point	dp
direct-access storage device	dasd
direct-connected	dir-conn

*The symbol (°) is sometimes used in text as an abbreviation for degrees of latitude, longitude, or angles, but should not be used in place of the regular abbreviation for degrees of temperature. Since it is permissible, however, to use F, C, and K respectively for *degrees* Fahrenheit, Centigrade, and Kelvin, there is a strong tendency to do so and to omit *deg*.

direct-current (adj.) . . . . dc or d-c
distilled water . . . . dw
dry bulb . . . . db

effective horsepower. . . . . . . ehp
efficiency . . . . eff
electromagnetic units . . . . . . emu
electromotive force . . . . emf
electron volts . . . . ev
electronic data processing . . edp
elevation. . . . . el
equation . . . . eq
extremely high frequency . . ehf

Fahrenheit (or degrees
    Fahrenheit) . . . . F
farad . . . . f
feet board measure (board
    feet). . . . . fbm
feet per minute . . . . fpm
feet per second . . . . fps
fluid. fl
fluid ounce . . . . fl oz
foot . . . . ft
foot candle . . . . fc
foot per minute . . . . fpm
foot pound . . . . fp
fractional horsepower . . . . . . fhp
free on board . . . . fob
frequency . . . . freq.
frequency modulation. . . . . . fm
fusion point . . . . fp

gallon . . . . gal
gallons per day . . . . gpd
gallons per hour . . . . gph
gallons per minute . . . . gpm
gallons per second . . . . gps
gram . . . . gr
gravities (acceleration of
    gravity) . . . . g
ground position indicator . . . gpi

hectogram . . . . hg
henry . . . . h
high explosive . . . . he
high frequency . . . . hf
high frequency current . . . . hfc
high-potential test . . . . hipot

horsepower . . . . hp
horsepower hour . . . . hphr
hour . . . . hr or h
hundredweight . . . . cwt

inch . . . . in.
inch-pound . . . . in-lb
input/output . . . . i/o
inside diameter . . . . id
inside dimension . . . . id
intermediate frequency . . . . i.f.
International Critical Tables ICT
international unit . . . . iu

jet propulsion . . . . jp
job control language . . . . . . jcl
joule . . . . j

Kelvin (or degrees Kelvin). . K
kilo . . . . k
kilovolt . . . . kv
kilovolt-ampere . . . . kva
kilowatt. . . . . kw
kilowatt-hour . . . . kwh
kinetic energy . . . . ke

lambert L
latitude . . . . lat
linear. . . . . lin
linear foot. . . . . lin ft
liquid . . . . lq
liquefied petroleum gas . . . . lpg
liter l
logarithm (common) . . . . . . log
logarithm (natural) . . . . ln
longitude . . . . long.
low frequency . . . . lf
lumen . . . . lm
lumen hour . . . . lhr
lumens per watt . . . . lpw

maximum . . . . max
meter . . . . m
miles per gallon. . . . . mpg
miles per hour. . . . . mph
millimeter . . . . mm
minute . . . . min
modulator. . . . . mod

modulus . . . . mod
molecular weight . . . . mol wt
motor-generator . . . . mg
multiple programming with
    variable number of tasks   mvt

National Aircraft Standards NAS
National Bureau of
    Standards . . . . NBS
National Electrical Safety
    Code . . . . NESC
National Electric Code
    Standards . . . . NEC
negative . . . . neg
net weight . . . . nt wt
nickel-silver . . . . ni-sil
normal temperature and
    pressure . . . . ntp
number . . . . no.

octane . . . . oct
optical . . . . opt.
optimum working frequency owf
oscillate . . . . osc
ounce . . . . oz
outside diameter . . . . od

parts per billion . . . . ppb
parts per million . . . . ppm
permeability . . . . perm
perpendicular . . . . perp
phase . . . . ph
phase modulation . . . . pm
polar . . . . pol
potentiometer . . . . pot.
pound . . . . lb
pound-foot . . . . lb-ft
pound-force (avoirdupois) . . lbf
pound-mass (avoirdupois) . . . lbm
pounds per cubic foot . . . . . . pcf
pounds per square foot . . . . psf
pounds per square inch . . . . psi
power amplifier . . . . pa
primary sampling unit . . . . . psu
probable error . . . . pe
pulse-amplitude modulation pam

qualitative . . . . qual
quality . . . . qual

radio direction finder . . . . . . rdf
radio frequency . . . . rf
Rankine . . . . R
Réaumur . . . . Ré
reference line . . . . ref l
remote job entry . . . . rje
revolutions per minute . . . . . rpm
revolutions per second . . . . . rps
rheostat . . . . rheo
Rockwell hardness . . . . Rh
roentgen . . . . r

scleroscope hardness . . . . . . sh
second . . . . sec
second-foot . . . . sec-ft
sequential access method . . . sam
shaft horsepower . . . . shp
shortwave . . . . sw
signal-to-noise ratio . . . . snr or s/n
socioeconomic status . . . . ses
solenoid . . . . sol
specific gravity . . . . sp gr
spherical candlepower . . . . . scp
square . . . . sq
square meter . . . . sq m
square mile . . . . sq mi
standard metropolitan statis-
    tical area . . . . smsa

tachometer . . . . tach
tangent . . . . tan
temperature . . . . temp
template . . . . temp
tensile strength . . . . ts
tension . . . . tens.
thermal . . . . thrm
thermocouple . . . . tc
thermometer . . . . therm.
thousand . . . . m or M
thousand cubic feet . . . . mcf
time sharing option . . . . tso
tolerance . . . . tol
transceiver . . . . xvr

ultrahigh frequency . . . . uhf
United States Standard . . . . USS

vacuum . . . . vac
vacuum tube . . . . vt
variable . . . . var
variable-frequency oscillator vfo
very high frequency . . . . . . vhf
video frequency . . . . vdf
volt v
volt-ampere . . . . va

watt . . . . w
watt-hour . . . . wh or whr
wavelength . . . . wl
wet bulb . . . . wb

yard . . . . yd
year . . . . yr
yield point . . . . yp
yield strength . . . . ys

B

HELPFUL PUBLICATIONS FOR TECHNICAL WRITERS

Technical Writing and Reports

BALACHANDRAN, SAROJINI, *Technical Writing: A Bibliography*. Champaign, Ill.: American Business Communication Assn., 1977.

BEAKLEY, GEORGE C., JR., and DONALD D. AUTORE, *Graphics for Design and Visualization*. New York: The Macmillan Company, 1973.

FEAR, DAVID E., *Technical Writing*. New York: Random House, 1973.

GARLAND, KEN, *Graphics Handbook*. New York: Van Nos Reinhold, 1966.

GIESECKE, FREDERICK E., *et al.*, *Technical Drawing* (2nd ed.). New York: The Macmillan Company, 1967.

GRAVES, HAROLD F., and LYNE S. S. HOFFMAN, *Report Writing* (4th ed.). Englewood Cliffs, N.J.: Prentice-Hall, Inc., 1965.

HARWELL, GEORGE C., *Technical Communication*. New York: The Macmillan Company, 1960.

HOELSCHER, R. P., *et al.*, *Graphics for Engineers: Visualizations, Communication and Design*. New York: John Wiley & Sons, 1968.

LUZADDER, WARREN J., *Basic Graphics* (2nd ed.). Englewood Cliffs, N.J.: Prentice-Hall, Inc., 1968.

MENZEL, DONALD H., HOWARD MUMFORD JONES, and LYLE G. BOYD, *Writing a Technical Paper*. New York: McGraw-Hill Book Company, 1961.

MILLS, GORDON H., and JOHN A. WALTER, *Technical Writing* (4th ed.), New York: Holt, Rinehart & Winston, Inc., 1978.

RIVERS, WILLIAM L., *Finding Facts*. Englewood Cliffs, N.J.: Prentice-Hall, Inc., 1975.

SOUTHER, JAMES W., *Technical Report Writing* (2nd ed.). New York: John Wiley & Sons, 1976.

TURNER, R. P., *Technical Report Writing* (2nd ed.). New York: Holt, Rinehart & Winston, Inc., 1972.

ULMAN, JOSEPH N., and JAY R. GOULD, *Technical Reporting* (3rd ed.). New York: Holt, Rinehart & Winston, Inc., 1972.

WEISS, HAROLD, and J. B. McGRATH, *Technically Speaking: Oral Communication for Engineers, Scientists, and Technical Personnel*. New York: McGraw-Hill Book Company, 1963.

ZETLER, ROBERT L., and ROBERT GEORGE CROUCH, *Successful Communication in Science and Industry*. New York: McGraw-Hill Book Company, 1961.

General English

COPPERUD, ROY H., *American Usage: The Consensus*. Princeton, N.J.: Van Nostrand Reinhold, 1970.

CORBETT, EDWARD P., *Classical Rhetoric for the Modern Student* (2nd ed.). Fair Lawn, N.J.: Oxford University Press, 1971.

KIERZEK, JOHN M., and WALKER GIBSON, *The Macmillan Handbook of English*. New York: The Macmillan Company, 1977.

LEGGETT, GLENN, C. DAVID MEAD, and WILLIAM CHARVAT, *Prentice-Hall Handbook for Writers* (7th ed.). Englewood Cliffs, N.J.: Prentice-Hall, Inc., 1978.

A Manual of Style (rev. 12th ed.). Chicago: University of Chicago Press, 1969.

PERRIN, PORTER GALE, and WILMA R. EBBITT, *Writer's Guide and Index to English* (5th ed.). Chicago: Scott, Foresman & Company, 1972.

STRUNK, W., JR., and E. B. WHITE, *The Elements of Style* (3rd ed.). New York: The Macmillan Company, 1972.

U.S. Government Printing Office Style Manual (rev. ed.). Washington, D.C.: Government Printing Office, 1967.

Indexes

Applied Science and Technology Index (1958–)
Bibliography of Scientific and Industrial Reports
The Biological and Agricultural Index
Business Periodicals Index
Engineering Index
The New York Times Index
Readers' Guide to Periodical Literature
Social Sciences and Humanities Index

Sources of General or Special Information

The Census Volumes (Department of Commerce, Bureau of Census)
    Agriculture
    Construction Industry
    Drainage of Agricultural Lands
    Irrigation of Agricultural Lands
    Manufacturers
    Retail Distributors
    Unemployment
    Wholesale Distribution
Encyclopedia Americana
Encyclopaedia Britannica
The Foundation Directory
Minerals Yearbook (Department of the Interior, Bureau of Mines)
The Statesman's Yearbook
Technical Book Review Digest
U.S. Government Publications Monthly Catalog
Van Nostrand's Scientific Encyclopedia
The World Almanac

Index